How to Build Small Barns & Outbuildings

How to Build Small Barns & Outbuildings

Monte Burch

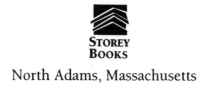

STOREY
BOOKS

North Adams, Massachusetts

*The mission of Storey Publishing is to serve our customers
by publishing practical information that encourages personal independence
in harmony with the environment.*

Cover design by Meredith Maker and Michelle Arabia

Text design and production by Cindy McFarland

Production assistance by Michelle Arabia, Carol Jessop, Nancy Lamb, Susan Moore and Meredith Maker

Cover photographs by Monte Burch
All uncredited text photographs by Monte Burch

Illustrations on pages 2, 4, 26, 30, 34, 45, 55, 56 bottom, 57, 58 bottom, 59 bottom, 60, 62 bottom, 67 top, 79 bottom, 89, 91, 92, 93 top left, 100 top and middle, 102, 103, 109, 110, 111 top, 120, 121 top left, 124, 125, 131, 136 top, 137, 141, 145 middle, 157, 158 top, 160 top, 166, 177, 178, 179, 180, and all following p. 185 by Raymond Wood from sketches by the author

All other illustrations by Bob Vogel

Copyright © 1992 by Monte Burch

The information in this book is true and complete to the best of our knowledge. All recommendations are made without guarantee on the part of the author or Storey Publishing. The author and publisher disclaim any liability incurred with the use of this information. For additional information please contact Storey Books, 210 MASS MoCA Way, North Adams, MA 01247.

Storey books are available for special premium and promotional uses and for customized editions. For further information, please call Storey's Custom Publishing Department at 1-800-793-9396.

Printed in the United States by Vicks Lithograph & Printing Corporation
20 19

LIBRARY OF CONGRESS CATALOGING-IN-PUBLICATION DATA

Burch, Monte.
    How to build small barns & outbuildings / Monte Burch.
        p.   cm.

    Includes bibliographical references and index.
    ISBN 0-88266-774-2 — ISBN 0-88266-773-4 (pbk.)
      1. Barns—Design and construction—Amateurs' manuals.   2. Farm buildings—Design and construction—Amateurs' manuals.   3. Garden structures—Design and construction—Amateurs' manuals.   4. Barns—Designs and plans—Amateurs' manuals.   5. Farm buildings—Designs and plans—Amateurs' manuals.   6. Garden structures—Designs and plans—Amateurs' manuals.   I. Title.
    TH4930.B87   1992
    690'.8922—dc20   91-57948
                                                     CIP

CONTENTS

Introduction

Little did I know when I was writing *Building Small Barns, Sheds & Shelters* back in the late seventies that I was creating a "classic." I guess, though, that any nonfiction book that lasts almost a decade, and sells more than 100,000 copies deserves that title. And I've been pleasantly surprised and pleased at the reception that book has received. In fact, this new book, *How to Build Small Barns & Outbuildings,* was written in response to requests for more farm and small building projects.

Readers of my earlier volume may find some of the general construction information contained in the first few chapters of this book somewhat repetitive. But they'll also notice a lot of new information: things do indeed change in the course of a decade. New materials, tools, and techniques are now available that make construction easier, faster, and more enjoyable than ever before, and I've added information on these here. Plus, we all learn from experience, and in ten more years of barn-building I've learned some new techniques and easier ways of doing things myself.

I consider myself extremely lucky. I grew up on a small farm in central Missouri and can remember many changes, such as the first electricity and running water coming to our community, as well as the "miracle" of television. One of the advantages garnered from that childhood (although I didn't think of it as an advantage at the time!) was the necessity of self-reliance. If you needed a barn, shed, or other outbuilding, you built it yourself. If you needed electricity or plumbing in a building, you ran the wiring or did the plumbing yourself. At times it could be a real hassle, because you often had to make do with the tools or materials you had on hand.

I learned a great many of these basic skills from my father at an early age. Later on, a succession of contracting jobs added to my building skills: a summer job with a concrete contractor, another summer job building motels with my father, then a stint with a general contractor. I constructed a few buildings on my own, landed a job as an associate editor at *Workbench* magazine, and finally turned to freelance work — and to my own farm in southern

Building your own barns and outbuildings offers many benefits: the first, of course, is saving money. These days about half the cost of any building project is the labor, and, by doing that yourself, you can construct the building more economically, or even put up a larger building with the same budget. A greater asset, however, is the satisfaction you get from creating your own surroundings. A well-constructed barn or outbuilding can be pointed to with pride, and will also add to the value of your property.

Missouri, which when we bought it consisted of an old rundown farmhouse and no outbuildings. Nowadays we have lots of outbuildings!

I feel strongly that this kind of hard-earned knowledge should be passed along to others. You'll notice photographs of several youngsters helping with the construction of projects in this book. Building has always been a family affair in our household. Ever since our oldest son, Mark, was able to swing a hammer he has been helping Joan and me with our building projects. Now he helps whenever he happens to be at home. Daughter Jodi, youngest son Michael, and nephew Morgan all put in plenty of time with saw and hammer. In fact the three of them, plus myself, formed the entire carpentry crew that framed the greenhouse/garden shed shown in this book. It took us four days in the heat of midsummer. We pay the youngsters an hourly wage, and they enjoy not only the money they earn, but also the satisfaction of working with their hands and challenging their minds to create something they can look at for a long time and be proud of. And, although the youngsters don't put a lot of stock in it now, they are gaining something even more important—knowledge and valuable skills.

Some of the projects in this book are a bit more "urbanized" than those in my earlier volume. The garden woodshed/greenhouse, for instance, would fit well in almost any suburban backyard. On the other hand, several general barn plans, a horse barn, sheep barn, chicken house, and one chapter on finishing barn interiors provide plenty of projects and ideas for rural landowners and homesteaders.

The general construction information in the first part of this book illustrates the techniques, materials, and tools needed to construct any number of different buildings and building styles. Then the projects offer plans that you can use to construct specific buildings. In many cases, you may wish to adapt buildings to suit your own needs, something that can easily be done.

Just remember that, in addition to their usefulness, even humble barns and outbuildings can be quite distinctive and special. I'm always meeting or getting letters from folks who have built such structures, and I know of several projects from my first book that are now considered classics. I wouldn't be at all surprised if some of the plans in this new book qualify as well. But most of all I hope that the information presented here will help you in planning and constructing your own "classic" barn or outbuilding.

CHAPTER 1

Planning

Careful planning is necessary for a homestead that is convenient, safe, economical to operate, and aesthetically attractive. Once a permanent building is up it's too late to change it, no matter how much you wish it faced another direction, or didn't block the view of your pond, etc. Do some extremely careful planning before the actual construction of the building.

Unfortunately you may often have to contend with existing buildings that are unattractive, inefficient, or haphazardly arranged. Even if you don't have that particular problem, constructing a building without a "master" homestead plan can result in a great deal of headaches. We've lived on our present farm for about 20 years. When we purchased it there was nothing but an old abandoned house and a completely rundown barn. Several piles of rubble and stone indicated the locations of former outbuildings. Unfortunately we made the mistake of starting to rebuild without any plan or direction. We planted fruit trees only to discover they were where we wanted the horse pasture, and I constructed a large barn too close to our house and office. One day, some stored hay caught on fire and the barn burned to the ground, nearly taking all of our other buildings with it.

CREATE A MASTER HOMESTEAD PLAN

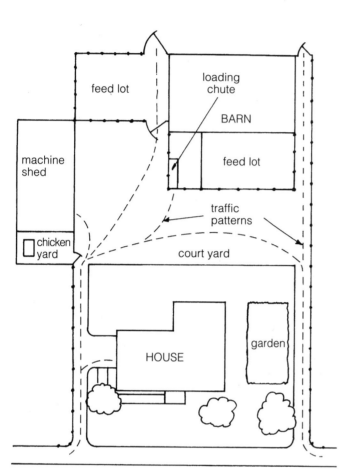

Regardless of whether you're starting with nothing but bare ground or must build around existing buildings, create a master homestead plan. Make a rough sketch showing the relative scale of existing surrounding buildings, and/or those you intend to construct. This will not only show the accessibility, but the most efficient traffic pattern for using the new building or buildings. The sketch or working plan should be done to scale, although it doesn't have to be fancy. A small scale ruler can be used to determine the exact sizes to put on the sketch. Note any existing buildings you wish to keep and then draw in the new buildings, fencing, lots, corrals, drives, etc.

One of the best layouts for a small farm operation is the old-time courtyard. This usually consists of a small fenced lot with the house situated on one side, then the other outbuildings constructed on the other three sides. Incidentally, you don't want to use the main courtyard as a livestock lot, except in an emergency, since you will have to get out and open and close gates each time you drive through or into it.

A circular-drive layout is also extremely effective, as I have learned from experience. After backing over gas cans, small fruit trees, and many other items with a large farm truck and cattle trailer, I finally put in a circular drive. Now I can drive through without having to spend a great deal of time and gasoline turning a long stock trailer around.

1-1. The old-fashioned courtyard provides one of the best farm layouts, if you have the advantage of starting fresh. Note the traffic patterns. One access through the courtyard to fields beyond, and another through a feed lot. There is also ample space for turn-around in the courtyard and for access to the machine shed as well as the livestock loading area.

DISTANCES BETWEEN BUILDINGS

Buildings should be located at least 75 feet apart to provide fire protection and prevent a fire from jumping from one building to another. This will also provide room for the local fire department to get between the buildings if needed. On the other hand buildings shouldn't be located so far apart that it's an inconvenience just to go between them to do chores, or get a tool or feed that you need. Naturally some buildings, such as lawn and garden sheds, will probably be constructed fairly close to the house. Swine and other animal buildings on the other hand should be constructed further away. Swine buildings should also

2

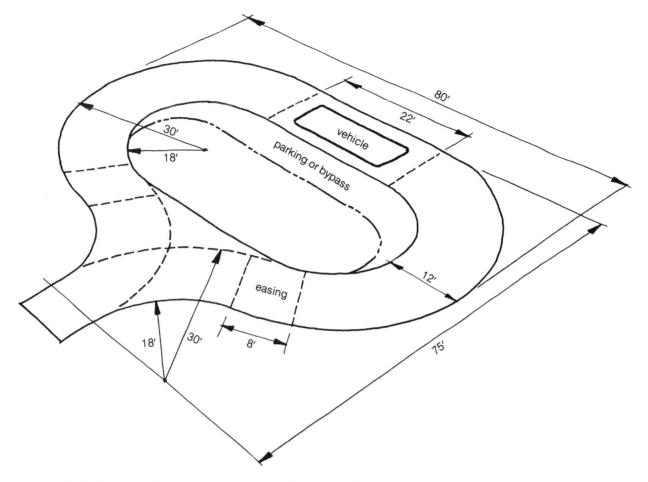

1-2. A circular drive is a convenient way to approach a house and outbuildings. Adjust dimensions to suit your property and vehicles.

be located downwind or you're going to suffer on a hot, muggy summer day. If you're new to the area you can get information on the prevailing wind as well as other general weather information from your local cooperative extension office.

WEATHER PROTECTON

Buildings to house livestock such as swine, poultry, sheep, goats, etc., should be situated to provide winter protection and summer ventilation. In most instances this means that the buildings should face the south or southeast and that the majority of the openings should be in the front to take advantage of the south sun and the south to southwest prevailing winds. The back should be closed off to provide winter protection from the north winds. This, of course, will depend on your geographic location and again a check of general local weather conditions is important.

3

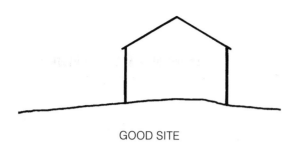

GOOD SITE

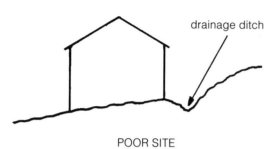

drainage ditch

POOR SITE

1-3. A good site for a barn or outbuilding would provide a slight drainage to all sides. A steep hillside is a poor site, but if it must be used, dig a drainage ditch on the uphill side which can be used to divert water from the building.

This normally faces the long dimensions of the building to the south and north. If possible place your buildings running along these "true" directions rather than angled.

You should also take advantage of any shade trees that are available, but buildings should not be constructed where they will be continuously shaded. This allows for a damp, cold condition.

TOPOGRAPHY AND SOIL TYPES

Another important consideration in building location is the topography and type of soil. Locations that are reasonably flat and well drained are best. Steep, sloping locations can result in problems as well as higher construction costs. Buildings to house animals should not be constructed uphill of your water supply. They also should not be located to allow runoff from lots or feeding areas to contaminate creeks, streams, and ponds. Even a small number of animals can produce an amazing amount of manure and create a serious sanitation problem unless the area is kept properly cleaned or adequate waste disposal systems are provided. In many areas local zoning laws regulate the waste disposal from feedlot runoff. Check with the local Extension Service office for laws pertaining in your area.

The type of soil is also important. Although in many instances you won't be able to change the soil make-up you can utilize different types of construction to take into effect soil types or specific soil problems. If you have a choice, the best type of soil for supporting a building is a coarse sand or gravel and clay mixture beneath the topsoil.

ACCESS

Buildings should be located far enough apart so that you can easily get farm equipment between them. Plan feed storage rooms in buildings so you can get to them easily with purchased feed, hay, or other supplies. They should also be located within easy reach of livestock areas. In fact, if you plan to enlarge your farm operation in the future, you might wish to make the feed preparation area centrally located in relation to other buildings such as a hog confinement building. You should also have easy access to marketable products. For instance if the loading chute is situated in the middle of a lot or back down behind a barn, it may be too muddy or wet for a truck to get to it during bad weather months.

UTILITIES

Of utmost importance to the homestead is an adequate supply of water. Use Table 1–1 to calculate the amount of water needed for your homestead and, if you don't have an adequate supply, drill a new well and run water lines to the building or buildings before you begin construction. This will save a lot of headaches later on.

In most instances you will also need electrical service to the barn or outbuildings. Contact your local Extension Service office and your local electrical utility company for information regarding location of service entrances, costs, and local zoning laws. You will also have to determine from the zoning laws and your own capabilities whether you will be hiring an electrician to run power to various buildings or be doing the job yourself. In many instances electricity is run underground, rather than utilizing above-ground poles and wires. This kind of installation offers a more aesthetic appearance and solves the wind and storm damage problem, but it often requires the services of an expert electrician and one with the proper ditch-digging equipment.

THE BUILDING PLAN

Not only is it important to create a master plan of your buildings, but the proper building should be constructed to suit the specific purpose. Although a chicken house can double as a hog house if needed (and many a substitution like this has been made here in the Ozarks), a building designed for a specific purpose is best.

It is also important to fit the size of the barn or outbuilding to its purpose. For instance, each type of animal requires a minimum amount of space. If this is not provided you can have some serious management problems in the way of disease and, in poultry and swine, even cannibalism. In addition, sheds to store equipment and bins for grain should be sized properly to be efficient and effectively protect their contents. Tables 1–2, 1–3, and 1–4 list the minimum space requirements for most animals, as well as standard dimensions for most farm equipment.

Of course, style and aesthetic appearance of the building is also important, and you'll want to ensure that it is attractive and matches existing building styles. Ease of maintenance is another important consideration.

Once again, the best approach is to make a plan. Regardless of whether you intend to utilize an existing

TABLE 1-1

Approximate Daily Water Requirements.*

Water Use Per Animal	Gal/Day
Milking cow	20–25
Dry cow	10–15
Calves (1–1½ gals./100 lbs. body weight)	6–10
Swine, finishing	3– 5
nursery	1
sow and litter	8
gestating sow	6
Beef animal	8–12
Sheep	2
Horse	12
100 chicken layers	9
100 turkeys	15
Rabbit (10 lbs.)	1 pt.
Rabbit doe (10 lbs.) with litter	2 qts.

* *Midwest Plan Service Structures and Environment Handbook* (Ames, Iowa: Midwest Plan Service, 1980).

TABLE 1-2

Outbuilding Space Requirements for Livestock, Beddings and Feeds*

Space Requirements for Poultry

Type of Bird		Floor Space[1] (square feet)
Chicks	0–10 weeks	.8–1.0
	10–maturity	1.5–2.0
Layers	brown egg	2.0–2.5
Layers	white egg	1.5–2.0
Layers	meat-type breeders	2.5–3.0
Broilers	0–8 weeks	.8–1.0
Roasters	0–8 weeks	.8–1.0
Roasters	8–12 weeks	1.0–2.0
Roasters	12–20 weeks	2.0–3.0
Turkeys	0–8 weeks	1.0–1.5
Turkeys	8–12 weeks	1.5–2.0
Turkeys	12–16 weeks	2.0–2.5
Turkeys	16–20 weeks	2.5–3.0
Turkeys	20–26 weeks	3.0–4.0
Turkeys	breeders (heavy)	6.0–8.0
Turkeys	breeders (light)	5.0–6.0
Ducks	0–7 weeks	.5–1.0
Ducks	7 weeks maturity	2.5
Ducks	breeders (confinement)	6.0
Ducks	breeders (yarded)	3.0
Geese	0–1 week	.5–1.0
Geese	1–2 weeks	1.0–1.5
Geese	2–4 weeks	1.5–2.0
Geese	breeders (yarded)	5.0

[1] Many factors determine the floor space requirements including the type of management system, the type of house, the number and kind of bird, the climate and even the management the birds receive.

Space Requirements for Horses
Dimensions of Stalls Including Manger

	Box Stall Size		Tie Stall Size
Mature animal	10' x 10'	small	
(mare or gelding)	10' x 12'	medium	5' x 9'
	12' x 12'	large	5' x 12'
Brood mare	12' x 12'	or larger	
Foal to 0-year-old	10' x 10'	average	4½' x 9'
	12' x 12'	large	5' x 9'
Stallion[†]	14' x 14'	or larger	
Pony	9' x 9'	average	3' x 6'

[†] Work stallions daily or provide a two to four acre paddock for exercise.

Space Requirements for Beef

Feedlot, sq. ft./head

20 in barn, 30 in lot	Lot surfaced, cattle have free access to shelter
50	Lot unsurfaced except around waterers, along bunks and open-front buildings, and a connecting strip between them
150–800	Lot surfaced, no shelter
20–25	Sunshade

Building with Feedlots, sq. ft./head

20–25	600 lb. to market
15–20	Calves to 600 lb.
½ ton/head	Bedding

Cold Confinement Buildings, sq. ft./head

30	Solid floor, bedded
17–18	Solid floor, flushing flume
17–18	Totally or partly slotted
100	Calving pen
1 pen/12 cows	Calving space

Space Requirements for Swine

Building floor space
Sows and boars: 8 sq. ft. indoors, 18 sq. ft. outdoors
Sow and litter: 26 sq. ft. slotted floors or 32 sq. ft. indoors, 42 sq. ft. outdoors
Pigs to 60 lb.: 3 sq. ft.
    60 to 125 lb.: 6 sq. ft.
    125 and up: 8 sq. ft. or 5 sq. ft. indoors, 13 outdoors

Pasture space
10 gestating sows/acre
7 sows with litters/acre
50 to 100 growing-fininshing pigs/acre
    depending on fertility

Shade space
15 to 20 sq. ft./sow
20 to 30 sq. ft./sow andlitter
4 sq. ft./pig to 100 lb.
6 sq. ft./pig over 100 lb.

*Sources: American Plywood Association; *Midwest Plan Service Structures and Environment Handbook* (Aimes, Iowa: Midwest Plan Service, 1976); Len Mercia, *Raising Poultry the Modern Way* (Garden Way Publishing, 1975).

Space Requirements for Sheep
Shelter space

Open-front building with lot: 10–12 sq. ft./ewe
12–16 sq. ft./ewe and lambs
6– 8 sq. ft./feeder lamb

Lot: 25–40 sq. ft./ewe
25–40 sq. ft./ewe and lambs
15–20 sq. ft./feeder lamb

Solid floor (confinement): 12–16 sq. ft./ewe
15–20 sq. ft./ewe and lamb
8–10 sq. ft./feeder lamb

TABLE 1-3

Hay, Beddings and Feeds

Weights and Storage Space Requirements

Material	Pounds Per Cubic Foot	Cubic Feet Per Ton
Hay—loose in shallow mows	4.0	512
Hay—loose in deep mows	4.5	444
Hay—baled loose	6	333
Hay—baled tight	12	167
Hay—chopped long cut	8	26\50
Hay—chopped short cut	12	167
Straw—loose	2–3	1000–667
Straw—baled	4–6	500–333
Silage—corn	35	57
Silage—Grass	40	50
Barley—48# 1 bu.	28	72
Corn, ear—70# 1 bu.	28	72
Corn, shelled—56# 1 bu.	45	44
Corn, cracked or cornmeal—50# 1 bu.	40	50
Corn-and-cob meal—45# 1 bu.	36	56
Oats — 32# bu.	26	77
Oats, ground — 22# 1 bu.	18	111
Oats, middlings — 48 # 1 bu.	39	51
Rye—56# 1 bu.	45	44
Wheat—60# 1 bu.	48	42
Soybeans—62# 1 bu.	50	40
Any small grain[1]	Use ⅘ of wt. of 1 bu.	
Most concentrates	45	44

[1] To determine space required for any small grain use wheat (60# = 1 bu.) for example. Then: 60 (⅘ = 48# wheat per cubic foot volume. To find number cubic feet wheat per ton, Then:

$$\frac{2000\# \text{ (Wt. of one ton)}}{48\# \text{ wheat per cubic foot volume}} = 42 \text{ cu. ft.}$$

TABLE 1-4

Approximate Space Required for Machines in Storage

Machine	Length	Width	Height (total if above 8')
Tractors			
Row-crop types			
2-plow	11½	7½	
3-plow	12	7½	
General-purpose types	11	6	
Tillage machinery			
Moldboard plow			
2-bottom	12½	6	
3-bottom	15	6½	
4-bottom	16½	7½	
Disk harrow, tandem			
7-ft.	10½	7	
8-ft.	11	8	
9-ft.	11	9	
10-ft.	11½	10	
corn cultivator			
2-row	8	8	
4-row	8½	12	
Field cultivator	10½	15	
Rotary hoe	7	7½	
Planting and seeding machinery			
Grain drill			
12 x 6	7	8	
20 x 6	7	13	
23 x 6	9	16	
18 x 7	9	12½	
Corn planter			
2-row (with hitch)	10	8	
(without hitch)	5	8	
4-row (with hitch)	12	14	
(without hitch)	6	14	
Harvesting machinery			
Combine, self-propelled			
7-ft. cut	18	9	12
9-ft. cut	18	10	13½
12-ft. cut	23½	13½	13
14-ft. cut	21	15½	
Ensilage harvester, 1-row	13	7	
Ensilage cutter and blower	12	5½	
Forage crop blower	10–13	6	
Corn picker, drawn	15	9½	8½

plan, such as those shown in this book, customize such a plan to your needs, or make your own design, a working plan of the building will save time, money and a lot of hassles. As with the master plan, these drawings don't have to be fancy, but they should be made to scale using a scale ruler. List the activities you will be using the building for, then scale the size and shape to the space needed according to Table 1–2 through 1–4. You will actually need at least three drawings and maybe five. The first is a floor plan indicating dividers, floor space, etc. The others are elevation drawings. These show the dimensions of the sides, back, front, with roof, etc. They also show any door or window openings. You may wish to make one elevation drawing for each side. You can also make up construction drawings indicating spacing of studs, rafters, etc., if you desire. This not only enables you to visualize the construction but helps you make up a materials list as well.

The final step is to make a drawing of any special details such as complicated roof angles, trusses, etc. A little time spent with these can save a great deal of time and money later on when you're actually constructing the building and may have a crew of carpenters on hand.

PERMITS

Many towns, municipalities, or counties require a building permit before the construction of any permanent building such as a barn or shed. In many instances these buildings will also have to adhere to local and sometimes regional building and zoning codes. These codes are written to ensure that buildings are safe, inoffensive, and don't infringe on the rights of others. The codes usually specify the type of construction allowed, the minimum distance of new buildings from property lines, the types of uses that may be allowed, and even who may or may not construct the building, or whether the site is suitable. For instance a hog confinement building may not be allowed near a heavily populated residential area. These codes can also be a help in determining special needs for specific areas, for instance the depth of foundations needed.

Your detailed homestead plan and building plans can be invaluable in helping you wade through the often burdensome task of getting the various permits (see also the checklist in Table 1–5). Distances from property lines, roads and bodies of water should be included in your plan. This allows the building inspector and zoning administrator to see exactly what you intend to construct and where.

Contact your local county or municipal officials for advice on what is needed.

Basically there are three building code associations in existence in the United States and one in Canada. Each of them has developed a code that specifies how buildings should be constructed. Most municipalities, counties, and provinces have adopted one or the other of these codes. If you're interested in reference information the code books are available, although they're fairly expensive, averaging about $50.00 each.

For more information contact:

Southern Building Code Congress, International, 900 Montclair Road, Birmingham, AL 35213 (205) 591-1853.

Building Officials and Code Administrators, International, 4051 Flossmoor Road, Country Club Hills, IL 60478 (708) 799-2300.

International Conference of Building Officials, 5360 South Workman Mill Road, Whittier, CA 90601 (213) 699-0541.

Canadian Building Code Agency, Institute of Research in Construction, National Research Council, Publications Sales, Administrative Services Branch, Building M20, Ottawa, ON K1A OR6 (613) 993-2054.

FINANCING

The first step in financing is determining your building cost. Take your drawings or plans to your local building supply dealer. He will often provide a free estimate of the materials cost, and in many instances can also offer good advice on local construction, availability of materials, where you can get help with special problems you can't handle, etc. You will also need to talk to a concrete supplier for a cost estimate on the amount of concrete needed for footings, foundations, and slabs.

Not only should you get the cost of the various materials but also of any other jobs required, such as excavation, running in electricity, water, etc. Make sure you have a firm estimate for any jobs you're considering hiring out, and have the contractor submit it in writing. Then make up a construction budget as shown in Table 1–6.

The best way of financing is through your personal savings. Even a fairly large project can be constructed in several stages, saving the cost of interest on a loan.

If you must obtain a loan, take your drawings, plans, and estimated construction budget to the bank or loan officer of the institution where you apply for the loan. If

TABLE 1-5

Planning Checklist

Siting
- ❑ locate building on site to complement other structures
- ❑ consider soil composition, topography and drainage
- ❑ consider sun, shade and prevailing winds
- ❑ plan for access and utilities
- ❑ consider relationship of building to outdoor functions

Building Planning
- ❑ organize activity list for building's functions
- ❑ determine style: shed, lean-to, gable, gambrel roof, etc.
- ❑ determine position of doors and windows considering lighting
- ❑ determine width of aisles, interior doors, stalls, etc.
- ❑ draw scale foundation and floor plans and side elevations
- ❑ choose siding, roofing and finish materials
- ❑ draw plans for electrical and other services as needed
- ❑ draw details of special or hard-to-build items
- ❑ prepare materials list

Permits and Financing
- ❑ apply for building permit or zoning variance as necessary
- ❑ get estimate on materials from building supplier
- ❑ prepare itemized budget for entire project

your farm is financed at that institution you may be able to get a loan on the outbuildings. It is, however, sometimes fairly difficult to get a loan on an outbuilding alone unless you are able to put up tangible assets in addition to the outbuilding to assist in securing the loan.

TABLE 1-6

Construction Budget

Item	Estimated Cost	Actual Cost
Excavation		
Footings		
Foundation		
Concrete Floor		
Framing		
Roofing		
Siding and Trim		
Windows and Doors		
Fastenings		
Plumbing		
Wiring		
Heating/Air Conditioning		
Paint		
Utilities		
Septic System		
Building Permits		
Landscape & Grading		
Equipment Rental		
Additional Work Hired		
Allowance for Overruns		
TOTAL		

CHAPTER 2

MATERIALS

American settlers and homesteaders have traditionally been quite adaptable when it comes to finding materials for constructing buildings. Sod, stone, logs, and sawn wood along with whatever tools were available have been used. Salvaged and recycled materials are also popular for outbuildings. In fact, the garden tool shed in this book was constructed using some framing materials salvaged from an old chicken house that was torn down on the spot where the building was constructed.

These days, however, most barns, sheds, and other outbuildings are constructed using more modern materials, tools, and techniques. Many of these materials provide longer lasting buildings that are easier to construct. The materials you select to construct your outbuilding will depend on the availability of some materials, the amount you wish to spend on the building, construction techniques you wish to tackle, and the aesthetic appearance desired.

NATIVE LUMBER

Native lumber, sometimes called green lumber, is dimensional wood that has been rough-sawn at a local sawmill. It is neither planed nor dried and is therefore quite wet, heavy, and sometimes hard to work. Its primary advantage is that it costs about one-third to one-half less than kiln-

2-1. Barns, sheds and outbuildings can and have been constructed of almost anything you can think of. Traditional materials in the past have been native materials, but most of today's buildings are constructed of dimension lumber.

11

2-2. Native, or green lumber, has been a choice of homesteaders for years. It's easy to work when fresh, hardens like iron and provides a substantial building for years, although it does have a tendency to shift and "sag."

dried wood purchased from a building supplier, and even less if you have the timber and can do the milling yourself.

Native lumber is perfect for barns and outbuildings, where its slightly uneven dimensions and rough texture do not matter. It is also ideal for traditional board and batten siding. Because green lumber shrinks when it dries, cracks will develop between siding boards. But battens cover these cracks and seal out the weather. Always use cement coated or galvanized ring-shank nails when building with green lumber because regular nails will pull out when the lumber shrinks.

Most sawmills cut all types of wood, from hardwoods such as oak and maple to softwoods such as pine, spruce, and hemlock, depending on local availability. In the Northeast and Northwest, spruce and hemlock are the most common building materials. In the South, pine and even oak are popular choices. For a more complete description of various woods and their principal characteristics, see Table 2–1.

Native lumber is normally cut at the mill into 1- and 2-inch-thick stock that is usually available in widths up to 10 to 12 inches. Large beams such as 6x6s or even 12x12s are also readily cut at most mills, while building suppliers seldom carry such items. If you're thinking of post and beam construction you may wish to consider native materials.

Store green lumber carefully if it is not used as soon as it comes from the mill. Lay boards in "stickered" piles so they dry properly without warping, twisting, and cracking. A sticker is a small piece of 1 inch wood that separates each layer of boards so air can circulate through the pile.

If you're lucky enough to own some land with marketable timber, you can have this custom sawn for your own buildings. Our farm has about 60 acres of timber with a good number of marketable saw logs. In the past we have cut and hauled these logs to a local mill where they were sawn to the exact dimensions we needed.

Then I purchased a one-man sawmill from Foley/ Belsaw and we sawed our own dimension lumber. If you try this, use caution: for the unskilled, this is a hard and

TABLE 2-1

Common Wood Characteristics

Species	Work-ability	Shrink-age	Strength[a] (Bending Stress At Prop. Limit)	Weight[b] (Lbs. Per Cu. Ft.)	Decay Resistance	Insulation[c] (R-Factor Per Inch)	Uses
Balsam Poplar	easy	low	very weak (5000–6000 psi)	26	low	1.33	walls
Northern White Cedar	easy	very low	very weak	22	high	1.41	walls, posts
Hemlock	mod.	low	weak	28	low	1.16	walls
Black Spruce	mod.	low	weak	28	low	1.16	walls
Basswood	easy	high	weak	26	low	1.24	walls
Red Cedar (east)	easy	very low	weak	33	high	4.03	walls, shingles
Red Cedar (west)	easy	very low	weak	23	high	1.09	walls, shingles
Redwood	easy	very low	weak	28	high	1.00	walls, shingles, trim
Cypress	mod.	low	weak	32	high	1.04	walls, posts
Aspen	mod.	low	weak (6000–7000 psi)	26	low	1.22	walls
Cottonwoods	mod.	med.	weak	24–28	low	1.23	walls
BalsamFir	mod.	med.	weak	25	low	1.27	walls
White Pine	easy	very low	fair (7000–9000 psi)	25	mod.	1.32	general, trim
Ponderosa Pine	easy	low	fair	28	mod.	1.16	walls, trim
Jack Pine	easy	low	fair	27	mod.	1.20	walls
Red Pine	easy	low	fair	34	low	1.04	walls, joists
Tamarack	fair	med.	fair	36	mod.	0.93	general
Yellow Poplar	easy	med.	fair	28	low	1.13	general
Elm, soft	hard	high	fair	37	low	.097	fuel, floors
Maple, soft	hard	med. high	fair	38	low	0.94	fuel, floors
White Birch	hard	high	fair	34	low	0.90	fuel, floors
Black Ash	hard	high	fair	44	low	0.98	fuel, floors, furn.
Douglas Fir	mod.	med.	strong (9000–11,000 psi)	34	mod.	0.99	general
Yellow Pines	hard	med. low	strong	36–41	mod.	0.91	floors, joists
White Ashes	hard	med.	strong	38–41	low	0.83	fuel, furn.
Beech	hard	very high	strong	45	low	0.79	fuel, furn.
Rock Elm	hard	high	strong	44	low	0.80	fuel
White Oaks	hard	high	strong	47	high	0.75	fuel, floors
Red Oaks	hard	very high	strong	44	low	0.79	fuel, floors
Sugar Maple	hard	high	strong	44	low	0.80	fuel, floors
Black Locust	hard	low	very strong (11,000–13,000 psi)	48	high	0.74	fuel, posts
Yellow Birch	hard	high	very strong	44	low	0.81	fuel, floors, furn.
White Ash (2nd growth)	hard	high	very strong	41	low	0.83	fuel, floors, furn.
Hickory, Shag	hard	very high	very strong	51	low	0.71	fuel, floors, furn.

[a] Stress at which timber will recover without any injury or permanent deformation.

[b] At 12 percent moisture content.

[c] Calculation for 12 percent moisture. Value *varies* greatly with moisture: variation is 43 percent for softwoods, and 53 percent for hardwoods (see USDA FPL handbook No. 72) for moisture ranging from 0-30 percent. Values given *per inch* of thickness in direction of heat flow; normal to grain.

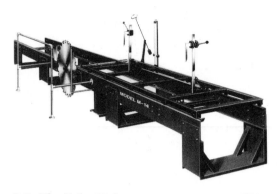

2-3. The Foley/Belsaw, one-man sawmill is an excellent choice for sawing logs to dimension lumber for any number of purposes.

2-4. Bandsaw mills make the chore of sawmilling much safer and easier. Bandsaw mills are available from Foley/Belsaw and Wood-Mizer, among others.

2-5. Most buildings, these days, are framed with dimension lumber — softwood 2x4s, 2x6s, etc.

sometimes dangerous job. A safer and easier option is the new bandsaw-type sawmills also available from Foley/Belsaw. The Wood Mizer, another bandsaw mill from Wood-Mizer Products comes on its own trailer and can be hauled to the woods where the trees are felled. Instead of using a large circular saw blade, which is quite dangerous and hard to work properly without some skill, the bandsaw mill uses a wide bandsaw blade to saw the logs. If you have plenty of native trees on hand, this is an excellent and less expensive alternative to buying materials. Investing in a planer will enable you to easily plane or surface your native lumber.

KILN-DRIED LUMBER

The choice for the framing of most construction projects is standard softwood, kiln-dried dimensional lumber. This is usually readily available and easy to work with. It's kiln-dried and milled to exact dimensions (features adding to its cost), and not as heavy as green native lumber. The better grades won't crack, twist, or warp as badly as native green lumber.

Softwood dimensional lumber comes in several different grades, so before selecting the wood, think about how you will use it. For instance, you might not be particular about materials for a portable chicken house, but if you build a hay barn or horse barn, you would want attractive, durable materials so the building will last a lifetime and be nearly maintenance-free. Understanding these grades is important, but difficult. There are more than a dozen associations or trade organizations, all with different grading systems. I'll briefly describe a grading system I'm familiar with. The one used by your supplier may differ.

Yard grade lumber, most often sold by local building suppliers, includes 1x12s, 2x4s, 2x6s, 2x8s, 2x10s, 2x12s, and sometimes 4x4s and 6x6s. Common lengths are from 8 to 20 feet. Also available are pre-cut studs that are 91½ inches long, the exact length needed for studs between a bottom and the two upper plates to create the standard 8 foot high wall.

Most building supply dealers purchase the grades most commonly used, sorting the material into No. 1, 2, 3 and 4 common, depending on the knots, blemishes, and defects found in the board. No. 4 common is the cheapest grade and usually has a lot of open knots and other structural weak spots.

Structural grade lumber is harder and much more dense than the yard lumber and is used primarily for heavier framing members. It may not even be available in some local yards, although you can probably order it. Structural lumber is graded according to its density. There are five general classifications:

1. Dense Select...........Southern pine and Douglas fir

2. Select.....................Douglas fir

3. Select.....................Other softwood species, except Southern pine.

4. Dense Common.....Douglas fir and Southern pine

5. Common................ All softwoods

In almost all light framing, I use lumber 1 to 1½ inches thick and No. 1 common yard materials. For wall framing, spruce, pine, or fir No. 2 common are most often used. For heavier beams, posts, and girders I use common structural or dense common.

All kiln-dried dimension lumber is cut to the full dimension at the mill, but after drying and planing dimensions are smaller. For example, a 2x4 usually measures 1½" x 3½", 2x6s measure 1½" x 5½"; 2x8s are actually 1½" x 7¼", and a 2x10 is really only 1½" x 9¼". In short, everything is half an inch or more smaller than its nominal measurements. The shrinkage dimension may range from ½ to ⅜ to ¼, depending on the mill and supplier. It's a good idea to measure the stock before beginning your project just to make sure.

Treated Lumber

Use lumber that has been pressure-treated for all framing members that will come in contact with water, the earth, or a concrete foundation. This includes the bottom plate or sill plate if you wish to add longevity to your building. All poles for pole barns should be made of pressure-treated lumber. Stall linings and skirts should also be made of treated lumber. Pressure-treated lumber is available in all dimensional sizes as regular kiln-dried lumber as well as in plywood. Pressure-treating protects the wood from rot, decay, and insects.

Not all pressure-treated woods are alike. It depends on the chemicals used and the amount of penetration. It's also important to purchase brand name woods that have a warranty or guarantee. For instance Wolmanized Pressure-treated wood from the Hickson Corporation has a lifetime limited warranty. Their warranty is good from the date of purchase for as long as you own the property on

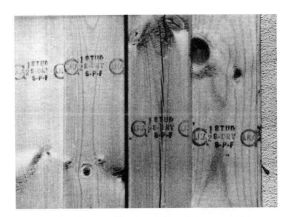

2-6. Dimension lumber is graded, and it's important to fit the grade of lumber to the specific job.

What's a Board Foot?

Lumber is usually measured by the board foot (bdf), which is a volume measurement of 144 cubic inches.

Thus a piece of 1-inch board that is 12 inches long and 12 inches wide is exactly 1 board foot.

Similarly, an 8-foot board that is 2x6 measures 2 inches times 6 inches times 8 feet divided by 12, or 8 board feet.

Kiln-Dried Lumber

Nominal Size	Actual Size
1x2	¾ x 1½
1x3	¾ x 2½
1x4	¾ x 3½
1x6	¾ x 5½
1x8	¾ x 7¼
1x10	¾ x 9¼
1x12	¾ x 11¼
2x4	1½ x 3½
2x6	1½ x 5½
2x8	1½ x 7¼
2x10	1½ x 9¼
2x12	1½ x 11¼
4x4	3½ x 3½
6x6	5½ x 5½
8x8	7¼ x 7¼

2-7. Pressure-treated lumber is one of the joys of today's materials. It can be used for a number of purposes, won't rot, is insect resistant and can be used for in-ground purposes, such as the bottom framing of the greenhouse on this garden shed.

2-8. Make sure you purchase the right treated wood for your project. Brand name woods have a stamp on the end indicating their use.

which your new Wolmanized or outdoor wood structures are built. Look for an identifying label and/or stamp on the wood. Some pressure-treated woods are for use only above ground while others can be used for ground or fresh-water contact. The label or stamp should indicate the use by the American Wood Preservative Bureau.

Because the wood is treated with a pesticide, some limitations are involved, depending on the type of chemi-

SAFETY PRECAUTIONS

Pressure-treated wood does require more care in handling than untreated lumber.

1. Dispose of treated wood by ordinary trash collection or burial. Treated wood should not be burned in open fires or in stoves, fireplaces, or residential boilers because toxic chemicals may be produced as part of the smoke or ashes.

2. Avoid frequent or prolonged inhalation of sawdust from treated wood. When sawing and machining treated wood, wear a dust mask.

3. Whenever possible, sawing and machining should be performed outdoors to avoid indoor accumulation of airborne sawdust from treated wood.

4. When power-sawing and machining, wear goggles to protect eyes from flying particles.

5. After working with treated wood and before eating, drinking, and the use of tobacco products, wash exposed areas thoroughly.

6. If preservatives or sawdust accumulate on clothing, launder before reuse. Wash work clothes separately from other household clothing.

cal used. Most do not allow the use of treated materials under circumstances where the preservative may become a component of food or animal feed. Examples of such sites would be structures or containers storing silage or food. It should also not be used for cutting boards or countertops. Treated wood should also not be used in those portions of beehives that may come into contact with the honey, nor in any place where it may come into direct or indirect contact with public drinking water, except for uses involving incidental contact, such as docks and bridges.

HOME TREATING

You can treat wood yourself and, in some cases, add 10 years to the life of a pole or framing member that comes in contact with the ground. However, home-treated wood is not as durable as commercial pressure treated lumber and the job is hard, messy, and somewhat dangerous. There are many ways of treating wooden posts to extend their usefulness, but only two of them are practical for the small farm: cold soaking on seasoned posts and end soaking on green posts, which is not as satisfactory.

TABLE 2-2

Preservative Retention Requirements*

Application	CCA As Required By Specifications, Lb./Cu. Ft.
Lumber and Plywood	
Above Ground	0.25
Ground Contact	0.40
Fresh Water Immersion	0.40
Salt Water Splash	0.60
Permanent Wood Foundation	0.60
Salt Water Immersion	2.50
Piling and columns	
Foundations or Fresh Water Immersion	0.80
Structural Poles	0.60
Salt Water Immersion	2.50

*American Wood-Preservers' Association

TABLE 2-3

Lumber Grading

Select Lumber

Select lumber is stock which can be finished with good results.

> **Grade B & BTR** — Almost entirely free from defects. May contain a few blemishes or small defects. May be given a natural finish.
> **Grade C** — Limited number of blemishes or small defects which can be hidden by painting. Not suitable for a natural finish.
> **Grade D** — Unlimited number of blemishes or defects which can be hidden by painting. Not suitable for natural finish.
> **Grade E** — Includes falldown from higher grades of finish with certain characteristics and limitations. Finish grade, not Select. Stock used for cut-up.

Common Lumber

Common lumber is stock which includes enough blemishes or defects to detract from finished appearance. Suitable for utility and construction use where finish is of secondary importance.

> **No. 1 Common** — Tight-knotted, sound stock considered watertight. Defects limited in size.
> **No. 2 Common** — Considered tight-grain stock. Large defects.
> **No. 3 Common** — Occasional knot holes plus larger and coarser defects than those found in No. 2 stock.
> **No. 4 Common** — Low-quality lumber including defects such as decay and holes.
> **No. 5 Common** — Only requirement is that it holds together under conditions of ordinary handling.

End Soaking

If you need a post right now and don't have time to cut your posts ahead and season them before treatment, end soaking is the method to use. Cut round posts and leave the bark on the post. Then use a 15 to 20 percent solution of zinc chloride, or chromatized (chromated) zinc chloride in water. Allow about 5 pounds (or about a half-gallon) of the solution for each cubic foot of post to be treated. The chromated zinc chloride is sold in a granular form that is easy to use, and is less subject to leaching from the posts than the plain zinc chloride. Often it is difficult to find a source for this chemical. If you can't find it, you usually can buy its two ingredients from a chemical company and mix them yourself. Use 80 percent zinc chloride and 20 percent sodium bichromate. For the solution, mix a 20 percent chemical, 80 percent water solution (each by weight). Thus 100 gallons of water weighs about 830 pounds and would require 166 pounds of zinc mixture.

Stand the posts bottom down in a tub or drum of this solution until they absorb about three-fourths of it, which takes from three to ten days. Then stand them on their tops, and let them absorb the rest. They should be seasoned for about a month before using, to allow the treated wood to dry.

Either green or seasoned posts may be soaked, covered completely, and steeped in a 5 percent solution of this same zinc chloride for one to two weeks, but this is not as effective as a pentachlorophenol solution treatment.

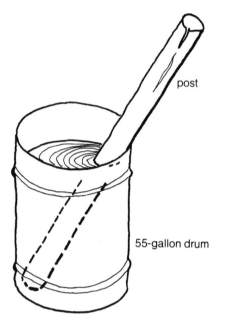

post

55-gallon drum

2-9. With care and attention to details, you can treat lumber yourself, but in most cases, these days, it's not cost (or safety) effective.

TABLE 2-4

Estimated Life of Untreated Wood Posts[1]
(diameter, 5 to 6 inches in size)

Over 15 Years	7 to 15 Years	3 to 7 Years	
Black locust	Cedar	Ash	Honey locust
Osaga-orange	Red cedar	Aspen	Maple
	Red mulberry	Balsam fir	Pine
	Redwood	Beech	Red oak
	Sassafras	Box elder	Spruce
	White oak	Butternut	Sycamore
		Douglas fir	Tamarack
		Hemlock	Willow
		Hickory	Yellow poplar

[1] Split posts, which have more "heartwood," will last longer than the time listed, and larger sized posts also last longer.

Cold Soaking

The best absorption and penetration are obtained by first seasoning the posts. This lets the sap dry out to make room for the preservatives. Peeled posts should be open-piled, so that the air can circulate around each one, and the bottom of the pile should be at least a foot above the ground. The best place for piling would be an exposed area on well-drained ground.

While posts cut in the spring will peel more easily, posts cut in the fall will have a chance to dry more slowly, which prevents some cracking and checking. This is more important with oak posts than with wood from cone-bearing trees.

The seasoning of posts adds little to their life unless they are also treated with preservatives.

Cold soaking of seasoned posts consists of soaking in a solution of pentachlorophenol (called *penta* for short), or in copper naphthenate, diluted with either fuel oil or diesel oil. Always wear protective clothing and rubber gloves, for these solutions irritate the skin. Penta can be purchased in a concentrated solution, in a ready-to-use solution (more expensive), or in flakes. If you use penta in dry flakes, wear goggles and a dust mask when mixing, to avoid irritating the eyes and throat. Penta is made in several strengths, calling for dilution by mixing with two to twelve or more parts of oil, to make a 5 percent solution. The label should specify strength and amount of dilution. Fifty gallons of a mixture of oil and penta will treat 50 posts 6 inches in diameter and 6 feet long. For convenience, they can be soaked in upright drums, soaking the bottom portion longer, for the tops are less subject to decay.

After removal from the solution, posts should be stacked so that the excess solution dripping from each post will be absorbed by the post beneath it. Wear rubber gloves when handling them until they are dry. Penta is highly toxic and can be absorbed through the skin and by way of the lungs. Animals should be kept away from newly treated wood. Do not treat wood for feed troughs. Store penta carefully, away from the reach of children, for there is no antidote for penta poisoning.

PLYWOOD

Plywood, which comes in 4x8-foot sheets and thicknesses ranging from ¼ to 1 inch, is an excellent material for roof and wall sheathing and flooring. It is cheaper per square foot than boarding (solid 1x12 stock) in some parts of the

2-10. Plywood is one of today's most effective building materials. It can be a cost and labor saver in more ways than one.

country; in others, it is more costly. But its strength and ease of installation make it a preferred material by many builders. Plywood is made by shaving thin layers of wood off a log and then gluing these layers together with special glues. Because the grains of the different laminations go in different directions, plywood is much more stable and stronger than solid wood and won't split or crack. It can warp, however, if stored wrong.

Plywood is available with softwood or hardwood faces. Hardwood-faced plywood is most commonly used for furniture and cabinets, while softwood-faced plywood is used for structural purposes. Plywood is available in several different grades and with either interior or exterior glue. Table 2–5 shows the different grades of plywood and their uses. By using the table you can pick the best yet most economi-

---

TABLE 2-5

Plywood Grades for Exterior Uses*

Grade (Exterior)	Face	Back	Inner Plies	Uses
A-A	A	A	C	Outdoor, where appearance of both sides is important.
A-B	A	B	C	Alternate for A-A, where appearance of one side is less important. Face is finish grade.
A-C	A	C	C	Soffits, fences, base for coatings.
B-C	B	C	C	For utility uses such as farm buildings, some kinds of fences, etc., base for coatings.
303® Siding	C (or better)	C	C	Panels with variety of surface texture and grooving patterns. For siding, fences, paneling, screens, etc.
T 1-11®	C	C	C	Special 303 panel with grooves ¼" deep, ⅜" wide. Available unsanded, textured or MDO surface.
C-C (Plugged)	C Plugged	C	C	Excellent base for tile and linoleum, backing for wall coverings, high-performance coatings.
C-C	C	C	C	Unsanded, for backing and rough construction exposed to weather.
B-B Plyform	B	B	C	Concrete forms. Re-use until wood literally wears out.
MDO	B	B or C	C	Medium Density Overlay. Ideal base for paint; for siding, built-ins, signs, displays.
HDO	A or B	A or B	C Plugged or C	High Density Overlay. Hard surface; no paint needed. For concrete forms, cabinets, counter tops, tanks.

* American Plywood Association

cal plywood for a particular job. One fairly standard plywood form used for a great deal of building is "sheathing." This is an economical ⅜ to 1-inch thick plywood used to cover roofs, as a wall covering under siding, and as a floor underlayment.

For all barn and outbuilding construction where plywood will be exposed, use exterior or, better yet, marine-grade plywood. It is bonded with a waterproof glue that is not affected by moisture, urine, or silage. Plywood can be cut with standard circular saw blades, though special fine-toothed plywood blades are made that minimize splintering of the face surface.

GLUED WOOD PRODUCTS

In addition to plywood, there are several other even more economical wood products that can be used for such things as sheathing, floor underlayment, etc. These are all made of wood by-products from sawmills, gluing the wood pieces together with special glues to create a "panel." These include chipboard, particle board, and flakeboard, depending on the brand name, size of chips, etc. Although they are much more economical, they do not have the strength of plywood and should not be used without proper support.

Hardboard is another glued wood product that has a great number of uses in barn and outbuilding construction. It is available in a wide variety of products including prefinished siding, soffit board, and lap siding.

SIDING

Many different types of wood, metal, and synthetic sidings are available. They differ greatly in cost, durability, and appearance.

Board and batten is a traditional barn siding, using 1-inch thick cedar, spruce, pine, or hemlock boards nailed in a vertical position. Western white cedar is the best choice because of its resistance to water and decay. Unfortunately, it is also the most expensive and may not be available from local sawmills. Spruce and pine, the next best choices for siding, are fairly resistant to weathering. Hemlock is a strong, attractive wood, but it is susceptible to grain separation, causing large layers to split and separate from a board.

Two other types of vertical siding include shiplap and tongue-and-groove, sometimes called "V-Groove" or "car

2-11. Flakeboard, particle, or chipboard, utilizing the waste from modern wood milling, is extremely cost effective and can be used for a great number of outbuilding projects. It does need more support in most areas than does plywood or solid wood.

2-12. Hardboard, another type of glued wood product, is also extremely economical and useful in many ways for outbuilding construction. Shown here are Masonite clapboards.

MASONITE CORP

2-13. Solid board and batten siding is a popular, easy-to-install material. Best choices are long lasting woods such as this white cedar.

2-14. Plywood siding is one of the most popular choices in today's building materials. Shown here it is combined with 1×2 white cedar battens.

2-15 Prefinished hardboard siding offers the ultimate. Once installed it's done!

siding." Shiplapped boards have ½- to ¾-inch half-laps on their edges, allowing the boards to overlap and form a seal. Tongue-and-groove boards have a tongue on one side and a groove on the other, allowing them to lock together.

Other wood siding materials include clapboards, drop siding and shingles. Shingles are fairly expensive and labor intensive to install and should only be used on outbuildings when you need to match an existing building. Shingles, as well as clapboards, must be installed over full sheathing, while the vertical board sidings are nailed to nailers. Sheathing applied under the vertical board siding will provide the tightest, most energy-efficient system but will also add to the cost. Clapboards are a popular siding material and available in several different widths, as well as in wood, plastic, and metal. The plastic and metal types are prefinished and require little upkeep. Drop siding is similar to tongue-and-groove vertical siding, but is applied horizontally, imitating the look of clapboards.

Several plywood sidings are available in 4×8 and longer sheets. One of the most popular, T 1–11 or Texture 1–11, a rough-sawn plywood with grooves spaced 8 inches on center is the most common plywood siding. Plywood siding, however, comes in a wide variety of textures and

22

without grooves as well. Plywood can be used with or without wall sheathing, depending on the use of the building. Used without sheathing it's an economical method of enclosing a building, provides a tight structure with an attractive appearance, and cuts labor time dramatically. Installing plywood siding is fast and easy. Exterior plywood should be painted or otherwise sealed to avoid delamination when exposed to the weather.

Another alternative that provides a similar result is the use of prefinished hardboard siding. This is an even faster method of enclosing a building because you don't need to paint or stain the siding. Hardboard siding is available in a wide variety of styles and colors. It can be applied over sheathing for a building that is to be heated, or alone for an economical and attractive building.

Metal siding is also a popular choice that we will cover later in this chapter.

ROOFING MATERIALS

Metal roofing is an excellent roofing material and is quickly becoming one of the most popular, because it is long lasting, economical, and easy to apply. Again, we will cover this later in the chapter.

The old style asphalt shingles (now also available in fiberglass) are still popular roof coverings for some outbuildings and barns, particularly if matching an existing building. They're not as economical as metal roofing and there is much more labor involved. This is because metal roofing can be applied directly to nailing strips set on the rafters while asphalt or fiberglass shingles require a solid deck onto which the shingles are nailed. This is usually plywood sheathing. These shingles come in a wide range of colors and are sold by the bundle, enough to cover 33 square feet or one-third of a square 10x10 foot area. When properly applied, a shingle roof is more waterproof and less likely to produce "sweating" conditions than a metal roof. (This is one major factor to consider in installing an uninsulated metal roof on a woodworking shop.) It is also quieter in rain and hailstorms. Of course, one prime reasons for using asphalt or fiberglass shingles is to match the appearance of a nearby building such as a main house.

Asphalt roofing is also available in roll roofing, sometimes called half-lap or double-coverage. It is one of the most economical of roofings but doesn't last as long as other roofings or provide quite the same protection. It has a composition quite similar to asphalt shingles, but comes

2-16. Asphalt, or more often these days fiberglass shingles are quite often the choice in roofing materials, usually to match an existing building.

2-17. Another type of roofing is Masonite Woodruf. Provides the look of wood shingles without the hassles.

2-18. Concrete is also one of the most common building materials. It can be mixed by hand as nephew Morgan and son Michael are doing here for post embedment.

in 3-foot-wide continuous rolls. It is much easier to apply than shingles, and is available in colors ranging from white to black.

CONCRETE AND CEMENT

Concrete is commonly used for building foundations in the form of solid walls or piers. Concrete or concrete blocks are used for aboveground walls as well. While poured concrete is fairly expensive and concrete blocks require a good deal of labor, they are the sturdiest and most durable of building materials. Milk barns are often built of concrete blocks because they are easy to clean and stand up to regular washings with water and disinfectant.

TABLE 2-6

Asphalt Roofing Materials

Product	Approx. shipping weight per square	Packages per square	Length	Width	Units per square	Side or end lap	Top lap	Head lap	Exposure
3 Tab square butt strip shingle	235 lb.	3 or	36"	12"	80		7"	2"	5"
	300 lb.	4	36"	12"	80		7"	2"	5"
Saturated felt	15 lb.	¼	144'	36"		4" to 6"	2"		34"
	30 lb.	½	72'	36"		4" to 6"	2"		34"
Mineral surfaced roll	90 lb.	1.0	36'	36"	1.0	6"	2"		34"
	90 lb.				1.075	6"	3"		33"
	90 lb.				1.15	6"	4"		32"
19-inch selvage double coverage	110 lb. to 120 lb.	2	36'	36"			19"	2"	17"

Concrete is made by mixing cement, sand, and crushed stone along with water in the proper proportions. Cement is the binding agent that holds together either concrete or mortar. Portland cement is most commonly used today and is manufactured from limestone mixed with other pulverized rock. A 1:2½:3½ concrete mix is the standard mix for footings and foundation walls (1 part cement, 2½ parts sand, and 3½ parts crushed stone or gravel). Concrete comes ready-mixed in 80-pound bags that yield ⅔ of a cubic foot of concrete. You can use bags for small jobs, but when you need more than ½ a cubic yard (27 cubic feet equals 1 cubic yard), either mix your own with a power mixer or hire a concrete truck to deliver it ready to pour.

When mixing mortar for laying a block or brick wall,

2-19. Concrete blocks can be used for foundations, chimney flues, and any number of other outbuilding and barn projects.

TABLE 2-7

Estimating Concrete

When you order concrete by the truck, it is measured in yards. A yard is 27 cubic feet or a volume measuring 3 x 3 x 3 feet. When figuring the amount of concrete needed to fill any square or rectangular area, the following formula can be used if all the measurements are in feet:

Cubic yards =

$$\frac{\text{width} \times \text{length} \times \text{thickness}}{27}$$

For example, the concrete needed to pour a slab 9 x 18 feet and 4 inches thick would be:

$$\text{Cubic yards} = \frac{9 \times 18 \times \frac{1}{3}}{27}$$

$$= \frac{9 \times 18}{27 \times 3}$$

$$= \frac{6}{3} = 2 \text{ cubic yards}$$

If you are trying to figure the volume of a Sonotube foundation, you need to substitute the cross-sectional area of the Sonotube for the length x width measurements. The area of a circle is pi x radius^2. Thus the concrete needed for one 6-inch Sonotube that is 4 feet deep would be:

$$\text{Cubic yards} = \frac{3.14 \times 0.25^2 \times 4}{27}$$

$$= \frac{0.785}{27}$$

$$= 0.029 \text{ cubic yards}$$

You would need 34 Sonotubes just to use up one yard of concrete!

TABLE 2-8

Nattinger Materials Company

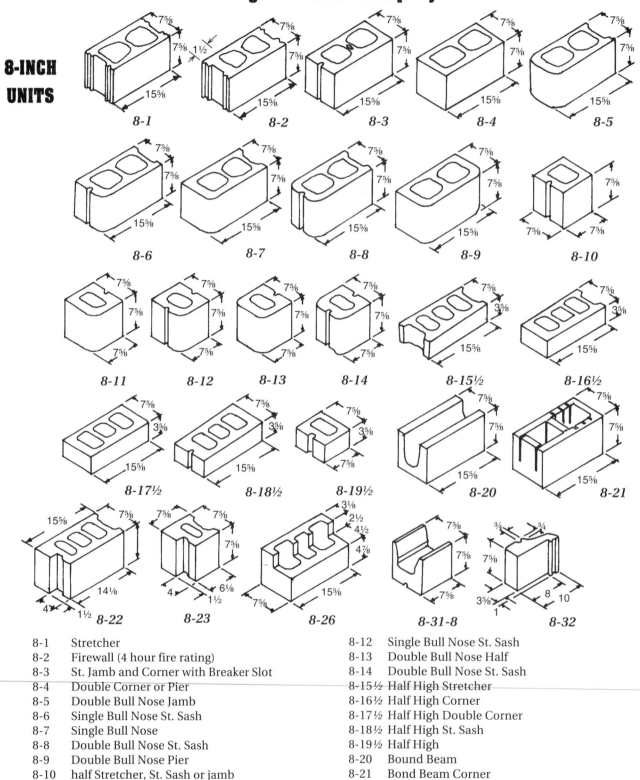

8-INCH UNITS

8-1	Stretcher	
8-2	Firewall (4 hour fire rating)	
8-3	St. Jamb and Corner with Breaker Slot	
8-4	Double Corner or Pier	
8-5	Double Bull Nose Jamb	
8-6	Single Bull Nose St. Sash	
8-7	Single Bull Nose	
8-8	Double Bull Nose St. Sash	
8-9	Double Bull Nose Pier	
8-10	half Stretcher, St. Sash or jamb	
8-11	Single Bull Nose Half	

8-12	Single Bull Nose St. Sash	
8-13	Double Bull Nose Half	
8-14	Double Bull Nose St. Sash	
8-15½	Half High Stretcher	
8-16½	Half High Corner	
8-17½	Half High Double Corner	
8-18½	Half High St. Sash	
8-19½	Half High	
8-20	Bound Beam	
8-21	Bond Beam Corner	
8-22	Wood Jamb	

the proper proportions are 1 part Portland cement, ½ part hydrated lime and 4½ parts clean, screened sand. Mortar can also be bought ready-mixed in 80-pound bags to which you add sand.

Concrete blocks are available in a variety of sizes and shapes. However, the standard blocks for exterior wall construction are 8 inches wide, 7⅝ inches high and 15⅝ inches long. This allows room for a ⅜-inch mortar joint, giving a finished 8x8x16 block space. Blocks used for interior walls are normally 4 inches thick instead of 8.

An unusual pier product is the Dek-Block, a preformed concrete pier block designed with a groove formed into it that allows 2- or 4-inch-thick lumber to be directly supported by the block. A cavity is also cut for a post. Dek-Blocks are available from Shell Rock Industries.

INSULATION

Livestock barns and outbuildings such as tool shops can be insulated to provide comfortable winter working conditions. Three of the most basic insulations are fiberglass batt, loose fill, and rigid foam.

Fiberglass is the most common and least-expensive insulating material. It comes in batts, 16 or 24 inches wide, to fit between studs and rafters and in thicknesses from 3½ inches to 12 inches. Its insulating quality is measured as an R-value, such as R-11 for 3½-inch fiberglass or R-38 for 12-inch. It is now customary in cold northern climates to put at least 6 inches of fiberglass in the walls and 12 inches in the roof of a well-insulated building. For exact insulation specifications in your area, consult an Extension Service office or university agricultural engineer in your area.

Cellulose and vermiculite are loose-fill insulators for block walls or other spaces into which they can be poured or blown. Cellulose may be blown through holes drilled into wall cavities.

Rigid foams are the best insulators per inch of thickness. For example, polyisocyanurate foam has an R-value of 7.4 per inch. Such foams are quite expensive, however, and should only be used where space is a limiting factor such as in roofs or around foundations where moisture might damage other insulating materials. Rigid foam panels can also be

2-20. Insulation available for small barns and outbuildings includes fiberglass batts, loose fill and rigid foam.

used to increase the R-value of existing walls with insufficient insulation. This is done by applying the panels to the interior surface of the wall, and covering them with sheetrock or plywood for protection against wear and tear. Most foams are available in 4x8 rigid sheets. For more on insulations, see Table 2–9, below.

TABLE 2-9

Insulation Types*

Form and Type		R-Value Per Inch	Cost	Characteristics[3]
BLANKET AND BATT				
Fiberglass (spun glass fibers)		3.2	low	non-combustible except for facing difficult with irregular framing
Rock wool (expanded slag)		3.4	low	non-combustible except for facing difficult with irregular framing
LOOSE-FILL[1]				
Fiberglass	attic	2.2	low	non-combustible
	wall	3.3		good in irregular spaces
Rock wool	attic	2.9	medium	non-combustible
	wall	2.9		good in irregular spaces
Cellulose (paper fiber)	attic	3.7	low	combustible — specify "Class I, non-corrosive"
		3.3		can be damaged by water
Perlit (Glass beads)		2.5	high	non-combustible
		3.7		expensive
Vermiculite (expanded mica)		2.4	high	non-combustible
		3.0		expensive
RIGID FOAM BOARDS[2]				
Moded polystyreme ("bead board")		4.0	medium	combustible permeable — do not use below grade
Extruded polystyreme ("Styrofoam")		5.0	high	combustible impermeable — best below grade
Polyurethane/ Polyisocyanurate		6.0	high	combustible
		7.2		used outside framing, protected by siding or sheathing

[1] All loose-fill insulation must be installed at manufacturer's recommended densities as shown on bag to insure proper performance.
[2] All rigid foams are combustible and must be covered with ½-inch drywall or equivalent 15-minute fire-rated material when used on interior.
[3] Data taken from *An Assessment of Thermal Insulation Materials and Systems for Building Applications*, Brookhaven National Laboratory, June 1978, GPO Stock No. 061-000-00094-1.
 Source: Cornerstone Energy Audit.

* Adapted from *Rodale's New Shelter* (Emmaus, Pennsylvania: Rodale Press, Inc., September, 1980).

- Common
- Box
- Casing
- Finish
- Brad
- Nail for general use
- Nail for general use
- Trussed rafter nail
- Pole-construction nail
- Flooring nail
- Underlay floor nail
- Drywall nail
- Roofing nail with neoprene washer
- Roofing nail with neoprene washer
- Asphalt shingle nail
- Asphalt shingle nail
- Wood shingle face nail
- Enameled face nail for insulated siding, shakes
- Nail for applying siding to plywood
- Nail for applying roofing to plywood
- Duplex-head nail

2-21. Sometimes it's hard to know what a particular nail looks like or what it's for. This illustration should help.

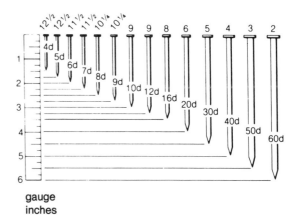

gauge
inches

2-22. A nail is sized according to its length and guage of its wire. The nails you'll use most often will be 6 ds, 8 ds, and 16 ds.

FASTENERS

Using the correct fasteners when building is just as important as choosing the correct materials. Nails are the primary fastener, although screws, bolts, and connecting plates are used in certain instances.

Nails are available in many different size and styles, are measured by the "penny" (abbreviated "d"), and are sold by the pound. An eight penny (8d) nail for example is 2½ inches long, whereas a 10d nail is 3 inches long.

Nails with a special zinc coating are called galvanized nails. Use these for exterior nailing of siding and trim where rust and corrosion will cause staining or weaken the nail. Galvanized "ring shank" or stainless steel nails are also effective. Spiral-grooved nails, called pole-barn nails should be used for pole framing. They are harder to drive and easier to bend than standard nails. *Galvanized roofing nails* are short, flat-headed nails used for fastening shingles or roll roofing in place. *Rubber-grommeted nails* are used for fastening metal sheeting in place.

Concrete screws are used for fastening wood to concrete. In most instances a ³⁄₁₆-inch bit is used to bore the hole through the wood and into the concrete. A power screwdriver or ratchet wrench is then used to drive the screws in place. Wood can also be fastened to concrete

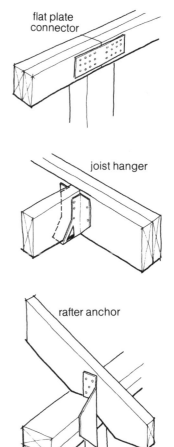

flat plate connector

joist hanger

rafter anchor

2-23. Common steel fastening plates include flat-plate connectors, joist hangers and rafter anchors.

using powder actuated nail guns if the wood member will be subject to minimal load or stress.

Steel fastening plates are available in a wide variety of shapes and sizes. They are used for securing posts and beams, joists, and any number of other fastening chores. The most common are called *joist hangers* and are used to secure floor joists. Not only do these plates make a secure joint, but they cut labor time considerably. Figure 2–23 shows a variety of fastening plates from Kant-Sag, made by the United Steel Products Company, and their most common uses.

Items that have a number of fix it and new project uses
Keep a few around; you never know what needs fixing

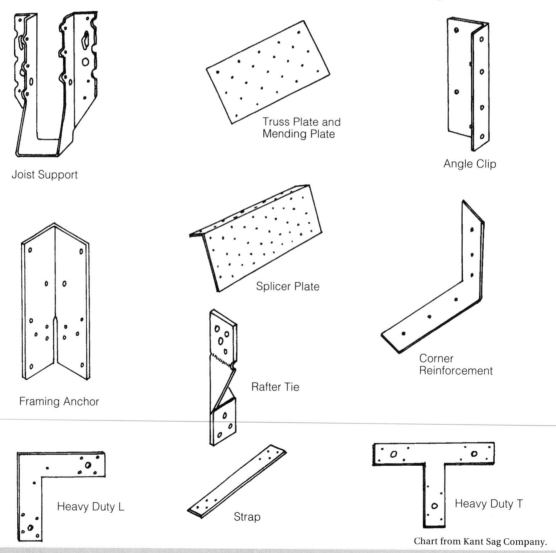

Joist Support

Truss Plate and
Mending Plate

Angle Clip

Framing Anchor

Splicer Plate

Rafter Tie

Corner
Reinforcement

Heavy Duty L

Strap

Heavy Duty T

Chart from Kant Sag Company.

PAINTS AND STAINS

Applying a finish to a barn or outbuilding makes sense from both an aesthetic and an economic point of view. Always cover siding with either paint, stain, or clear oil to protect it from the weather.

Paint, either latex or oil-based, protects wood very well. However, it is fairly expensive, it requires a lot of work to apply, and, in severe northern climates, it must be applied frequently. For those who demand some color on their barn, a more economical choice is an oil stain that is easier to apply and lasts longer. While the initial expense of paint or oil stain might be the same, oil stains cost less in the long run. Oil stains are available in a wide range of colors.

For those who like natural wood colors, clear linseed oil is an excellent preservative and protection. It is probably the least-expensive finish available. In addition, because linseed oil is non-toxic, it can be used freely on wood that livestock come in contact with. You can brush, roll, or spray it on.

2-24. Paint is a very important part of any outbuilding. It offers protection, as well as aesthetic appeal.

CAUTION

Do not use paint containing lead on livestock equipment or on parts of buildings accessible to livestock. Poisoning may result when animals constantly lick or chew objects covered with paint containing lead.

METAL COVERINGS

Metal roofing and siding has fast become a popular covering for barns, outbuildings, and even homes and offices for several reasons. In the past, the material was only available in a grey-galvanized finish, but an almost infinite number of colors for the panels, as well as matching or contrasting trims are available these days, and the new metal coverings are as likely to be found on fashionable shopping malls and downtown offices as on barns and sheds. The first reason for their popularity is the good looks of the finished building. Covered with metal and finished with trim the building looks neat and "finished." The second reason is that metal roofing is long-lasting and requires almost no maintenance. Today's high-tech sheet

TABLE 2-10

Safe Uniform Load Tables
(Lbs. Per Sq. Ft.)

Spans	24"	30"	36"	42"
29 Gage	134	84	54	34

Notes:
1. Table lists safe uniform total load capacity based on unrestrained continuous spans (4 or more supports).
2. Design stress will not exceed 48,000 psi and deflection is limited to $1/100$ of the span.
3. Full hard steel: 90,000 psi minimum strength.

Channeldrain roofing and siding is manufactured in accordance with ASTM specifications noted in "steel Roofing Sheets" latest revision.

Wheeling Corrugating Company reserves the right to change the design and/or specifications of its products without notice.

metal panels are made of durable strong steel, covered with a special zinc coating to prevent oxidation, then quite often another protective corrosion-resistant coating followed by a primer and a baked-on color coating. Most major brand names offer up to a 20 year limited warranty on the panels. These panels are fashioned with corrugations or ridges running their lengths and a wide variety of corrugation designs are also available. Of course the old-fashioned plain galvanized panels are also still available. The third reason for metal roofing's popularity is its ease of application. Metal coverings go on fast and easy, quickly covering a barn or building, and they don't require plywood sheathing.

2-25. Metal coverings, both siding and roofing, have become quite popular because they offer easy applications, low maintenance, and in some cases, aesthetic beauty.

Sheet metal panels are available in widths from 24 to 38 inches. The latter will cover a 36-inch-wide span with the two inches overlapping. Lengths are available from 6 to 40 feet, although you can often have the supplier cut sheets to a specific length, which makes the on-site cutting job easier. The most commonly used thickness for barns and outbuildings is 29 gauge, although more economical, lighter gauges of 28 and 26 are also available. Plain galva-

TABLE 2-11

30-inch Cover Strongpanel® Load Tables

Purlin Or Nailer Spacing
(Center to Center in Inches)

		18	21	24	27	30	33	36	39	42	45	48	54	60
DOUBLE	LOAD Lbs./Sq. Ft.	254	186	143	113	91	76	63	54	47	41	36	28	23
SPAN	DEFLECTION Inches	.052	.071	.093	.117	.144	.176	.207	.244	.286	.328	.373	.465	.582
TRIPLE	LOAD Lbs./Sq. Ft.	317	233	178	141	114	94	79	67	58	51	45	35	28
SPAN	DEFLECTION Inches	.083	.113	.147	.187	.230	.278	.331	.386	.450	.521	.595	.742	.904

NOTE: Limiting steel design stress is 48,000 psi, based upon a safety factor of 1.65 and determined in conformance to standard specifications as published in American Iron and Steel Institute "Light Gage Cold-Formed Steel Design Manual."

nized steel is available in 29, 28, 26, 24, 22, 20, and even economical lightweight 18 gauge.

It is important to have the correct support for the metal, so make sure you have supports spaced according to suggested safe uniform load tables for the material you use. (This information is available from manufacturers.) Table 2–10 shows a span chart for Channeldrain roofing and siding from the Wheeling Corrugating Company. Table 2–11 is a load table and gives purlin or nailer spacing for Strongpanel from National Steel.

Steel roofing and siding is generally calculated and ordered in "squares." This term represents 100 square feet of roofing needs. Table 2–12, from Wheeling Corrugating Company, will help in estimating the amount of metal roofing you'll need. The course calculator from National Steel (Table 2–13) also shows how you can figure the number of panels needed for a specific building length.

Sheet metal panels can be cut with a portable circular saw and metal cutting blade, a sabre saw, snips, or shears. They are fastened in place with special color-coated gal-

SAFETY PRECAUTIONS

Gloves should be worn to prevent injury while handling steel panels. Safety glasses should be worn to prevent eye injury when cutting or drilling steel panels with power tools. Ear plugs or some other form of hearing protection should also be used when cutting metal panels with power saws.

Use care when walking, sitting, or kneeling on a steel roof to avoid a fall. Steel panels may become slippery when wet. Do not work on the steel panels when they are wet or when climatic conditions are not suitable for safe installation.

TABLE 2-12

Quantity Estimation Chart*

38" Wide Sheet

Length of Sheet	Sq. Ft. Per Sheet	Sheets Per Square
6 ft.	19.000	5.2632
7 ft.	22.167	4.5112
8 ft.	25.333	3.9474
9 ft.	28.500	3.5088
10 ft.	31.667	3.1579
11 ft.	34.833	2.8708
12 ft.	38.000	2.6416
13 ft.	41.167	2.4291
14 ft.	44.333	2.2557
15 ft.	47.500	2.1053
16 ft.	50.667	1.9737
17 ft.	53.833	1.8576
18 ft.	57.000	1.7544
19 ft.	60.167	1.6620
20 ft.	63.333	1.5790
21 ft.	66.500	1.5038
22 ft.	69.667	1.4354
23 ft.	72.833	1.3730
24 ft.	76.000	1.3158
25 ft.	79.167	1.2632
26 ft.	82.333	1.2146
27 ft.	85.500	1.1696
28 ft.	88.667	1.1278
29 ft.	91.833	1.0889
30 ft.	95.000	1.0526
31 ft.	98.167	1.0187
32 ft.	101.333	.9868
36 ft.	114.000	.8772
37 ft.	117.167	.8541
38 ft.	120.334	.8310
39 ft.	123.500	.8103
40 ft.	126.666	.7895

*Steel roofing and siding is generally calculated and ordered in "squares." This term represents 100 square feet of roofing sheets. This chart will help you plan for your needs.

TABLE 2-13

Course Calculator

1. Read courses of 30-inch cover width panels under desired building length. Courses allow for 2 inches side lap panel.
2. For building lengths longer than 100 feet, use 100 feet in multiples and/or in combination with other figures.

Building Length (feet)

30" Cover
Strongpanel Courses (panels)

TABLE 2-14

Estimating Nails

Nails per Square

	30-inch cover Strongpanel	36-inch cover Strongpanel
24-inch purlin/girt spacing	78	86
30-inch purlin/girt spacing	65	72
36-inch girt spacing	56	62

vanized steel ring-shank nails with flat rubber washers, or plated steel-drill screws with sealing washers. An estimating chart from National Steel (Table 2–14) gives the nails needed according to purlin/girt spacing.

In addition to the panels a wide variety of accessories are available that can be used to quickly finish and trim out a barn or building. These include ridge roll for finishing off the roof, end wall flashing, rake and corner pieces that can be used to finish the roof rake or building corners, door jamb trim and angles for trimming doors and windows as well as door track dover.

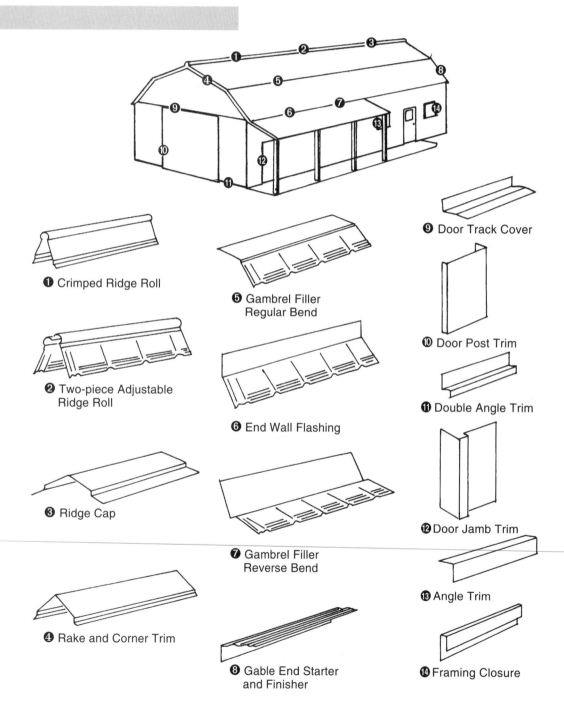

❶ Crimped Ridge Roll

❷ Two-piece Adjustable Ridge Roll

❸ Ridge Cap

❹ Rake and Corner Trim

❺ Gambrel Filler Regular Bend

❻ End Wall Flashing

❼ Gambrel Filler Reverse Bend

❽ Gable End Starter and Finisher

❾ Door Track Cover

❿ Door Post Trim

⓫ Double Angle Trim

⓬ Door Jamb Trim

⓭ Angle Trim

⓮ Framing Closure

CHAPTER 3

Tools

3-1. With the right tools you can build anything. With the wrong tools nothing!

With good tools you can build almost anything and that includes houses, barns, sheds, and other outbuildings. Without the proper tools, or with poor quality tools you can build almost nothing and can endanger your safety. If you do succeed in constructing something it will take a great deal longer to build and may not be as well built.

You will need several specific carpenter's tools as well as some specialized tools for most construction projects. Some larger, more specialized tools can often be rented. Let's start with the hand tools, some of which you may already have, and work through the more expensive power tools and finally the specialized tools.

HAND TOOLS

Hand Saws

You can get by with just one good hand saw for most barn and outbuilding construction. This would be a standard length, coarse-cut saw with 8 points to the inch. This should be hollow or taper ground to provide clearance when cutting green native materials or treated lumber.

Additional saws can, of course, do some jobs easier and better. For better craftsmanship and finer detailed work in trimming out you may also want a 10-point fine-toothed saw. A hand rip saw can also be extremely helpful for cutting trim pieces and other ripping jobs. A timber saw, with 3½ points to the inch can be used for cutting heavy timbers and poles. Or a one man "bucking" saw can also be used for these heavy-duty chores.

3-2. Handsaws are extremely important. In fact, the framing for the garden tool shed, except for the rafters, was cut almost exclusively with this saw.

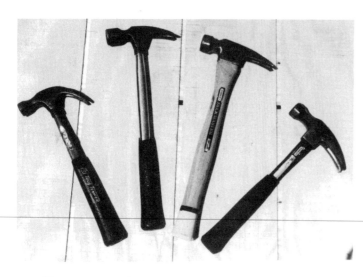

3-3. Hammers are just as important. On the left is a typical 12 ounce general purpose hammer. The second from the left, the author's favorite, a 16 ounce framing hammer. Then a decking hammer and finally a straight claw 12 ounce hammer.

Hammers

Hammers come in a variety of sizes and shapes. Again, you can probably get by with just one hammer. The most common hammer is a 16-ounce carpenter or claw hammer. This can be used for heavy-duty framing or even finish work. If you intend to do a lot of framing, however, a heavier 20 to 22 ounce framing hammer is a better choice. The heavier weight makes driving spikes for framing much easier and is less tiring over a longer period. The claws on the hammer also are straight rather than curved, which provides more leverage for prying, pulling nails, etc. A heavy, short-handled 4 lb. sledge, often called an engineer's hammer is also handy for driving stakes for form boards and other heavy pounding chores.

Hammers come with a variety of handle materials available including wood, steel, and graphite. The wooden handles are less expensive and feel more "traditional." In the larger size framing hammers, however, the steel handles provide a bit more balance and are less tiring after day-long hammering chores. Graphite or fiberglass hammers do a better job of absorbing shock, thus reducing the chances of repetitive motion injuries, particularly to the elbow.

Measuring Tools

Again, although you can get by with just a couple of measuring tools — a 25-foot steel tape and a framing square — a variety of tools are available that will make specific chores easier and more exact. The best choice in a 25-foot tape will have, as well as feet and inches, stud markings every 16 inches. You may also wish a smaller, 12- or 16-foot tape for inside trim use. A 50-foot or, better yet, 100-foot steel

3-4. Measuring tools are just as important. 25 foot tape is a good general purpose choice. Make sure it has feet, as well as 16 inch stud space marks.

3-5. A framing square can be used for any number of purposes, from cutting rafters to squaring boards for cut off.

tape is invaluable for locating and laying out buildings.

You will also need a framing square. Two types are available, a carpenter's square and a framing square. The latter has rafter tables imprinted on it. Both, however, can be used for all framing chores from laying out rafters to marking stud locations, squaring boards for cutting, etc. A combination square is smaller and handier to use for squaring off boards for sawing. It also has a 45-degree angle face and can be used as a level, square, marking gauge, depth gauge, etc. You will also need a T-bevel for making unusual angle cuts.

The Stanley Quick Square is an unusual tool that is designed to provide a quick, accurate, and repeatable means of laying out and marking the various cuts on common, hip, valley, and jack rafters. It comes with a handy, easy-to-read booklet that explains not only how to use the tool but the details of cutting rafters for all different types of roofs. The Quick Square can also be used as a handy adjustable angle or protractor. Used in conjunction with a level, it can be used to determine angles of construction members, for instance the rise of existing roofs and rafters, etc. It also makes a handy saw guide for a portable electric circular saw or electric sabre saw.

3-6. Smaller combination square is ideal for marking cut-offs and intricate cuts.

3-7. Stanley Quick Square makes rafter cutting quick and easy.

Levels

A level is also essential for barn and outbuilding construction, and, again, more than one is usually better. A 2-foot level can be used for a great deal of the work, including laying and leveling concrete blocks and installing doors and windows. A 4-foot level, however, is often best for these jobs and for plumbing long poles, barn corners, leveling sill plates, etc.

A string level can be used for rough lay-out of foundations, concrete block walls, concrete floors, etc. A water level, consisting of a hose with clear plastic tubes on each

3-8. Levels are essential for almost all facets of barn and outbuilding construction. 4-foot model makes precise plumbing of walls easy.

3-9. Brace and bits are used for heavy-duty boring jobs, such as for gates, holes for lag bolts, etc.

end is more versatile and accurate. A plumb bob and mason's string are used to lay out the building and mason's string itself is used for maintaining straight lines on concrete block foundations or buildings.

Boring Tools

A brace and set of bits is invaluable for roughing-in mortises, boring holes for heavy-duty gate and barn door hangers, cutting wiring run holes in studs, etc. Make sure the brace is double-acting (works when rotated in either direction). You may also wish an expansion bit for boring large holes. A hand drill can be used for boring smaller holes for hinge screws, etc. If you intend to work with heavy posts and beams a special heavy duty auger fitted with a T-handle provides more leverage than a brace and is better suited to that particular job.

Planes

Planes are used to smooth down wood surfaces and a No. 5 jack plane is about all that is needed for rough-in carpentry work. It can be used to plane down door edges for a tight fit, smooth and finish boards, etc. A drawknife can be a great help in post-and-beam construction in smoothing tenons for mortises.

Chisels

Cutting door-hinge mortises, squaring mortises in mortise and tenon beam joints, cutting notches, and general wood smoothing are all jobs requiring chisels. Make sure you purchase a set of good quality wood chisels; they'll last a lifetime and provide you with the ability to create finely crafted buildings. A set of four with sizes ¼, ½, ¾, and 1 inch will do most jobs. Again, if you plan to do

3-10. Quite often you'll have to smooth wood surfaces to create a tight fit, and a No. 5 jack plane is the best choice for all around work.

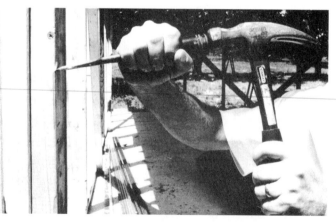

3-11. A sharp, 1½-inch chisel can be invaluable for tight fitting of timbers, cutting mortises for door hinges, etc.

post and beam construction requiring cutting mortises in heavy beams, a large socket-firmer chisel will be required. These range in size up to 19 inches long and are used for cutting deep mortises.

Miscellaneous

A utility knife for sharpening pencils and cutting wooden pins, a pry bar equipped with a nail puller, a crowbar for those heavy prying jobs, and a couple of sets of sawhorse brackets are also needed. I like two sets of sawhorses: one low, or just about 18 inches high for hand sawing, cutting off studs and rafters, etc.; and one pair about 3 feet high to saw paneling, plywood sheathing, and siding sheets. The 3-foot height makes it easier to operate a portable circular saw. I merely set the saw blade depth to about ⅛ inch below the thickness of the plywood being sawn and saw right through the plywood and tops of the sawhorses. This prevents a great deal of the binding problems involved in sawing large sheets of plywood. A pair of 2x4s resting on and perpendicular to the sawhorses will minimize binding when cross-cutting large panels.

A variety of ladders are also needed and lightweight aluminum is the best. Two sizes, a 16-foot extension and an 8-foot folding step, will handle most building chores.

3-12. I like two pair of saw horses, one low for framing cut-off with a hand-saw, and one high for creating rafters, cutting siding and plywood to shape. Note the two wood strips under the siding piece. The saw blade is made to cut just through the siding — an easy tactic for cutting large sheets of plywood or paneling without the usual binding problems because the sheets are always supported.

PORTABLE POWER TOOLS

Power tools can of course make the job of building barns and other buildings much faster and easier. These days new, heavy-duty cordless tools are also available and they're great labor and time savers. If you're using regular cord-powered tools make absolutely sure they're double-insulated. Working in wet conditions with an old-style, single-insulated or antiquated portable power tool can be deadly.

If purchasing portable power tools for barn and outbuilding construction they should also be heavy-duty. Many lightweight portable power tools are fine for repair chores around the house and small woodworking jobs, but they won't hold up to the task of heavy rough-in carpentry.

Drills

A portable electric drill is one power tool you don't want to be without. Many models are available, but you'll need at least a ⅜-inch variable speed, reversing model.

A cordless model is even better, although it lacks the

3-13. Rechargeable portable electrical drill is a necessity. Skil Top Gun model shown has two rechargeable batteries so you can keep one charging while using the other. Hi-Low switch provides plenty of speed for drilling, lots of power for turning in concrete screws.

3-14. Stanley cordless model is also an excellent choice in portable power drilling tools.

power of a good quality cord-powered drill. The Top Gun from Skil is an excellent choice. It is a ⅜-inch variable speed, reversing model with two mechanical gear ranges. Turning a switch provides the speed needed for boring jobs, or the torque needed for large screw fastening chores. It also comes in a handy pack with an extra battery pack. You can keep one pack charging while the other is in use.

Another excellent choice is the Stanley Cordless Drill Kit. It is a rugged ⅜-inch, variable speed, reversing drill.

Screwdrivers

Just as handy are the cordless screwdrivers. You'll find them invaluable for installing hinges and other hardware all around the farmstead. The Skil Super Twist Single Speed Reversing is a good choice, as is the Stanley Cordless Screwdriver Kit that has a ¼-inch hex high-torque driver.

Circular Saws

The portable electric circular saw is the single most important power tool you can have for constructing barns and other buildings. They are available in many sizes; however, the minimum size for most framing work would be a 6½ blade, which cuts 2 inches at 90 degrees and approximately 1¾ inches at 45 degrees. On the upper end of the scale is the 10¼-inch saw such as the Skilsaw Drop Foot Model. This makes the task of cutting poles and heavy timbers easy and fast. Maximum cut on this saw is $3^{15}/_{16}$ inches at 90 degrees. You will need a good carbide-tipped framing blade and a smooth cutting blade for plywood, or use a combination blade that can be used for almost all chores. If you're working with metal you'll also need an abrasive cut-off wheel specifically designed for cutting metal. In addition, make sure you have an adjust-

3-15. A portable electric circular saw makes the chore of barn and shed building easy. Big 10½-inch model shown can easily cut through 4×4s, making pole barn building easy and fast.

40

able rip fence to match your saw. This allows you to rip stock to specific widths.

Sabre Saw

A portable electric sabre saw is handy when cutting door openings, electrical outlet openings in paneling, etc.

SPECIALIZED TOOLS

In addition to these standard hand and portable power tools you'll also need some specialized tools for specific chores, including masonry tools, electrical tools, plumbing tools, and digging tools.

Masonry Tools

Concrete footings, foundations, slabs, and concrete block foundations or walls all require special tools. For concrete work you'll need an old shovel and an old rake for moving concrete around. You'll also need a *screed* or *dragging board*, which can be made from a 2x4; a float for general smoothing; a trowel for finishing; an edger; and a groover. One of the easiest and best methods for finishing off a large slab floor for a barn, garage, etc., is to use a *bull float*. This is a wide wooden or metal float on a long handle and the resulting finish, although not as smooth as a troweled finish, is often slightly rough and perfect for these types of floors. You can usually rent these tools, especially the bull float. If you're interested in an extremely smooth finish, however, a power trowel is the answer. These can also be rented from tool rental dealers.

For concrete block and other masonry work you'll need a pointed trowel, a joint tool, a handmade mason's board or pan, an old hoe with holes cut in the blade for mixing mortar, and a wheelbarrow to mix and hold mortar. A mason's hammer and a chisel are needed for cutting blocks. A heavy-duty sledge is needed for driving stakes for forms.

A powered mixer can be used for mixing large mortar mixes or small concrete pours and is available from tool rental dealers.

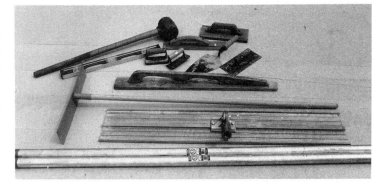

3-16. Concrete tools include a bull float, concrete puller, float, level, trowels, edger, groover and maul. One excellent mail-order source for concrete tools is the Goldblatt Co.

Electrical Tools

Installing wiring requires several tools, including a good pair of linesman's pliers. These should have insulated handles and

41

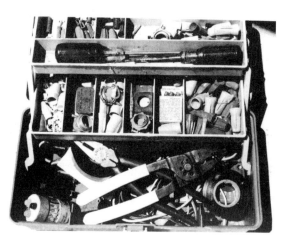

3-17. The right electrical tools can also make the job quicker and easier.

3-18. These days, due to the amount of copper and plastic materials used, the number of plumbing tools required is much less.

are used for cutting plastic-sheathed cable as well as for other heavy-duty cutting and wire chores. A pair of needle-nose pliers makes the job of bending loops for wire screws easy, particularly on the often hard-to-get-to grounding screws of receptacles. A pair of all-purpose electrical pliers that can be used for stripping, cutting, and bending wires is also needed. A wide variety of screwdrivers are necessary, and of course they should have insulated handles. Make sure you have at least one heavy-duty, long-handled driver for installing boxes, making connections with main lines, etc.

You'll also need a small receptacle test light for determining when a receptacle or line is hot, and for trouble-shooting problems.

Plumbing Tools

Not too many years ago a great number of plumbing tools were required, but today, due to a change in materials across most of the country, fewer tools are needed. Most residential plumbing these days utilizes copper or plastic pipe. To work with flexible copper tubing you'll need a tubing cutter and flaring tool. For copper pipe, a propane torch kit and a tubing cutter. Plastic piping requires only a hacksaw. You will also need a pair of adjustable wrenches, a pair of pipe wrenches, and a pair of pump pliers.

Digging Tools

Shovels, including a pointed round-nose shovel, a spade, and a flat shovel are required for digging trenches, moving dirt, etc. You'll probably also need a rock pick in some areas like our Ozarks hills. A steel pry bar, often

3-19. Unfortunately, digging tools, such as a pry-bar, post hole digger, and shovels of all types are a necessity.

3-20. *A small, lightweight chain saw can also be used for many barn and shed building chores, such as cutting the tops off of posts in a pole barn.*

called a "breaker bar" can also be used to dig and pry rocks, lift heavy members in place, etc.

For post holes a post-hole auger or clam-type post-hole digger is necessary. A tractor-powered post-hole digger makes the chore much easier and faster. Or you can often rent gasoline-powered post hole diggers at tool rental dealers.

Chain Saw

Although not a necessity, at times a chain saw can be a blessing, especially when working with large poles or heavy beams. One excellent use is to erect poles for pole barns, fasten the upper framing in place, then cut the tops of the poles level with the upper framing.

Power Wrench

You will of course need wrenches of various sizes for tool maintenance, hanging doors, fastening sill plates to anchor bolts, etc. The Skil power wrench is an excellent portable electric power wrench that provides plenty of power for all these chores.

Regardless of what tools you purchase, keep them in good shape and sharp for ease in cutting and safe use.

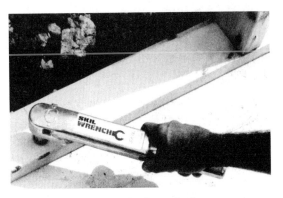

3-21. *Wrenches, including the battery powered portable wrenches are needed for such chores as fastening anchor bolts in place.*

CHAPTER 4

Laying out a Site

4-1. Properly laying out your building is one of the most important jobs. Take your time and make sure the building is situated properly according to plan, and that it is square.

All aspects of building construction are important, but none more so than the site preparation and layout. Once you have established the location, size, and shape of your building it's time to lay it out, or mark the corners and outline of the building. This is regardless of whether the building will be placed on a foundation, on a concrete slab, or will be a pole structure.

PREPARING THE SITE

Usually you must do some site preparation before you can start construction. This can be as simple as clearing away brush and weeds, or may require excavation. If topsoil is removed, be sure to pile it away from the construction site. It can then be replaced after the building is finished to complete and grade. Regardless of all else, make sure the ground allows water to drain away from the building. For clearing and excavation you can roughly estimate the size and location simply by using a long tape and placing stakes. If on the other hand the building is to be supported on a concrete footing and/or foundation, the land clearing and excavation for the footing and foundation is done simultaneously, after the building is precisely laid-out.

44

LAYING OUT THE BUILDING

Again careful preparation will prevent a lot of headaches later on. This is the one chore that many first-time builders hurry through to get to the good stuff of putting up walls and so forth. Laying out a permanent building, whether it is a chicken shed, small workshop, or a large hay or horse barn is relatively simple, but it does take attention to details. If the building isn't constructed square, you'll fight it from start to finish. And, by the same token, if it isn't level you'll also run into a lot of problems.

Using four wooden stakes and a 50- or 100-foot steel tape, locate the approximate corners of the building. The easiest method of locating these corners is to first measure and mark the corners or ends of one long wall of the building and drive stakes at these locations. Drive a nail about halfway into the stakes at each end or corner of the building. Tie a mason's string to the two nails and pull it taut between the two stakes. If the building is to be precisely located according to north, south, etc., use a compass to sight along the two stakes, moving the stakes until the direction of the building is aligned as desired.

One of two methods can be used to layout an adjacent side and make sure it is square with the first. Either one uses the triangle rule: Any triangle that has sides measuring 3 feet, 4 feet, and 5 feet, or multiples of these figures, will make up a right-angle triangle with the square corner where the 3 and 4 foot sides meet. A large wooden triangle (3x4x5 feet) made up of 1x4s can be used as a giant "square." Or you can use two steel tapes and stakes driven along the string lines to establish the triangle, although this method requires more "fiddling" to get an accurate triangle.

To use the latter method, measure 3 feet down the string line just established for one side of the building and drive a stake with a nail in the top at this exact position. Hook a steel tape over the nail on the end of the stake and one over the latter stake and pull them out to where 4 feet and 5 feet meet on the two tapes. Drive a stake at this point. This establishes a squared line for the two adjacent sides of the building. Measure the distance of this wall, extend a string line from the two stakes and establish the side of the building and corner location.

When using this method, it's impor-

4-2. A compass can be used to align the building in the desired direction.

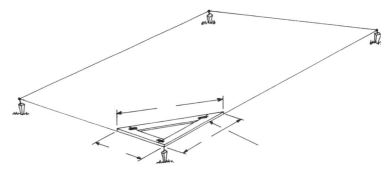

4-3. Four wooden stakes and a string line are used to create a rough outline of the building. A wooden triangle can be used to make sure the string lines of two adjacent sides are square.

45

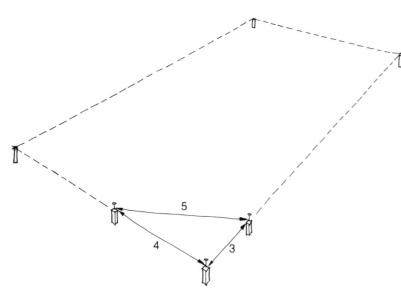

4-4. *Another method is to use a tape measure and the 3-4-5 triangle method. For larger foundations, use greater multiples (9-12-15 for example) to enhance accuracy.*

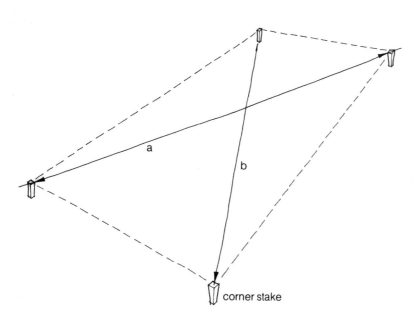

corner stake

4-5. *The best way to check the squareness of a foundation layout is to assure that the diagonals (a and b) are equal.*

tant to use the appropriate maximum expansion of the 3-4-5 rule to fit the building. For a 30- x 20-foot building, for example, use a 9-12-15-foot triangle; otherwise errors can be large. To establish a 3-4-5 triangle and thereby ensure that a second wall is perpendicular to the first, follow the steps outlined in the accompanying box.

With two adjacent building sides and corners established with string lines, measure and mark the opposite two sides by placing a stake in the corner diagonally from the first established corner. With all four corners established with nails driven in stakes and mason's string outlining the building, the next step is to check to make sure the building is square. Measure diagonally from corner to corner, then across the other two corners. If the building is square, the measurements will be the same. If not, shift the stakes in or out to get the same diagonal measurement and remeasure to make sure all sides are the correct length as well. Make sure that the layout stakes are driven well in place and that the nails establish the exact corners.

Now that you have established an outline of your building the next step is to place batter boards in position. These hold a mason's string to permanently mark the corner locations. The mason's string can be removed for excavation, digging holes for poles, etc., then replaced exactly back in position as needed, each time reoutlining the building.

Remember that the string marks the *outside edge* of the foundation wall or piers, *not* their center or the inner edges. This is important so that a building wall designed to be exactly 24 feet (and therefore take exactly six sheets of 4-foot-wide plywood siding) does not turn out to be 24 feet, 8 inches and require a full new sheet of plywood that is mostly wasted. When digging for concrete piers or posts for pole barns, again the string marks the *outside* of the support posts.

THE 3-4-5 TRIANGLE

1. Go to one corner, measure down the string 3 feet (or multiples thereof), drive a stake into the ground, and, at exactly 3 feet, pound a nail into the top of the stake.

2. Ask a friend to take another string (or measuring tape if you have one), attach it to the corner stake, and measure 4 feet. Mark at that length. Extend the string or tape in the direction of the adjoining wall.

3. Hook a measuring tape over the nail on the stake set at 3 feet. Extend the tape 5 feet toward the adjoining wall.

4. When the 5-foot and 4-foot points intersect, a right-angle triangle is formed and the corner is square. At the point of intersection, drive another stake. This establishes a line for the building's adjoining short wall and makes it square with the long wall.

5. Once the right-angle triangle is formed, extend the string or a tape beyond the 4-foot point, as necessary, to reach the full dimension of the wall. If you're planning a side wall of 20 feet, for example, extend the string to that distance. Again, drive a stake into the ground at the new corner, and pound a nail into the stake at the exact wall dimension.

6. Repeat steps 1 through 5 until foundation outline is complete. If you anticipate laying out several foundations, you might make a permanent 3-4-5 triangle from scrap wood to facilitate

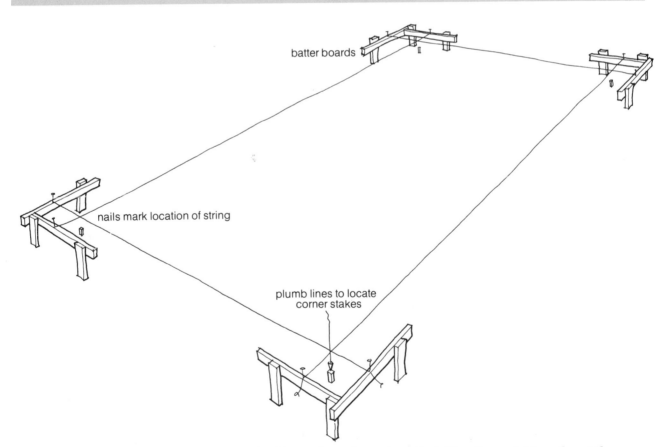

batter boards

nails mark location of string

plumb lines to locate corner stakes

4-6. Once the building outline is established batter boards are installed. They form right angles at the foundation or building corners. To provide room for excavation equipment, they should be set back at least 4 feet from the foundation.

Batter boards are created by first driving pointed 2x4 stakes, four for each corner, 5 to 6 feet away from the building line. If the soil is soft, make sure the stakes are driven well into the ground and braced if necessary. You don't want the batter boards to shift or your building lines will shift. 1x4 ledger boards are then nailed to the stakes at a convenient working height. The top edges of these should be above the top of the proposed foundation or concrete slab. Make sure the ledger boards are level, and that they are approximately level with each other. Pull mason's string between the batter boards, using a plumb bob suspended on the string to locate the position of the nails on the corner stakes. Fasten the string in place to the batter boards by making a shallow saw kerf in the tops of the ledger boards. Then pull the strings tight, place them down into the saw kerf and tie them off to a nail driven into the back of the ledger board. Once all four strings are in position, it's a good idea to recheck the wall, especially the diagonal measurements between the string corners to make sure the stakes or lines haven't shifted and the building outline is still square and of the correct size.

The next step depends on the type of foundation used. If a footing and/or foundation is used, stakes can be driven following the string lines and the area for the footings excavated following these marks. Excavating can be done either by hand or by hiring a backhoe or dirt mover. Small jobs, such as digging a shallow trench, leveling for a slab, and digging holes for concrete piers or poles for pole barns can be done by hand, or with a shovel and post-hole digger.

When digging for piers or poles, locate the outside edge of the pole by the string line, then locate the center of the pier or pole and drive a stake at that location. Do this for all the piers, then remove the building lines from the batter boards. Dig the holes starting at the center mark. The holes should be as wide as a footing, if a footing is to be placed under the piers. For instance an 8-inch pier with a 16-inch footing would require a hole 16 inches in diameter.

Trenches for deep footings are usually dug 4 feet across to allow for

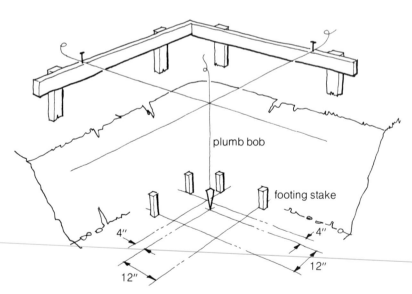

4-7. With the batter boards installed and the string line placed, use a plumb bob to locate the string crossings over the exact position of building corners. To locate the foundation outline, drop a plumb bob, then measure 4 inches to locate one side and 12 inches to locate the other. This establishes the outline for a 16-inch wide footing.

moving around in the area and placing the forms. If the building is to have a basement, the area is excavated including 2½ to 3 feet extra around the perimeter for installing the basement forms.

If the building is to be a pole structure, place stakes at each corner and at each pole location along the lines and you're ready to bore the holes for the poles.

If the structure is to sit on a concrete slab you will probably still need to excavate for a footing, then form the footing or slab/footing combined to the correct height following the string outlines.

Excavation for larger projects, such as footings, foundations and basements will of course require the use of a backhoe. Backhoe operators usually charge around $25 to $40 an hour, but they can do a tremendous amount of work in a short time compared to hand labor. Naturally rocky areas or those requiring removal of large boulders will require heavy-duty equipment.

Always make sure you're present when the excavation work is being done to prevent surprises, such as digging into an old water line or, worse, accidentally cutting into a buried telephone, electrical, or gas line. Check with these utilities prior to excavation to verify location of buried lines. If the excavation is large, such as for a basement or large foundation, have the operator set the topsoil aside. It can then be placed over the finish grade and will make growing grass easier than it would be over rock and gravel from the subsoil.

To find the depth of your excavation, measure to the frost line from the *highest* corner of the building. After the foundation is in, it can be backfilled and graded to create an even soil depth around the perimeter. If you're working with a sloped site that can't be evened out with a little backfill, seek professional advice on the installation of a stepped foundation. With this foundation, footings may be laid below the frost line without extensive excavation.

CHAPTER 5

Footings, Piers, Foundations, Slabs, and Floors

5-1. The support your building rests on is the single most important factor in the life of your building. A good building on a poor foundation will collapse quickly, even a poor building on a good foundation will still be standing — perhaps longer than you wish!

Unless a building is portable it must be anchored to the ground and sit on a solid support of some sort. This is done in a variety of ways depending on the use and type of building, the soil upon which it sits, and the desired cost of the building.

The first consideration is the type of building and its use. For instance a simple shelter for animals with an open side is often constructed using pole barn methods because they're economical, fast, and easy. On the other hand a garage or small workshop is often constructed on a poured concrete slab because this simplifies construction and provides better protection against insects, rodents, etc. Many houses are constructed using footings/foundations with a wooden floor on top of this and a crawlspace beneath. Or they may be constructed with a partial or full basement beneath. Of the above options, pole framing is the least expensive, with full basement the most expensive method of construction. In addition, it's important to make the type of building support follow local building and zoning codes.

FOOTINGS

Size

Footings are the supports below ground level that support the building. They may be a continuous line, or they may consist of concrete piers set in holes in the ground. Footings are normally made of poured concrete.

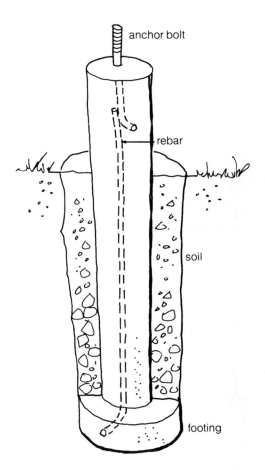

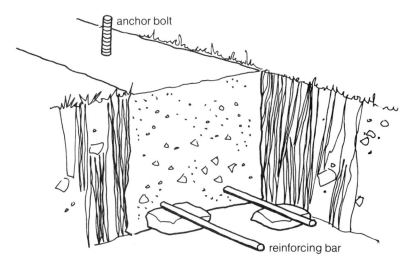

5-2. *Trench footings are strengthened with reinforcing bar (rebar). Anchor bolts are used to secure the sills to the foundation.*

5-3. *Piers, with anchor bolts or brackets, are often used with small barns.*

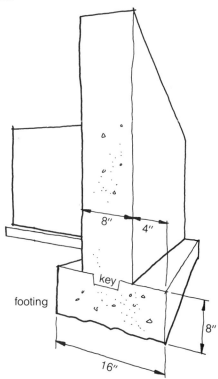

5-4. *Typically, concrete-wall foundations have footings twice as wide as the wall.*

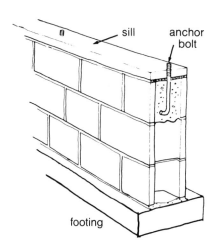

5-5. *A concrete block foundation. Rebar is placed inside the block cores.*

TABLE 5-1

Bearing Capactities of Soils*

Hard Rock	Up to 40 Tons Per Square Foot		
Soft rock	"	8	"
Coarse sand	"	4	"
Hard, dry clay	"	3	"
Fine clay sand	"	2	"
Soft clay	"	1	"

* Eccli, Eugene *Low-Cost, Energy-Efficient Shelter* (Emmaus, Pennsylvania: Rodale Press, 1976).

A foundation is usually added on top of the footings, and this may be poured concrete or concrete block. The size of the footing depends on the location, soil type, frost depth, and other site considerations. Usually for farm and residential construction, footings are twice the width of the foundation wall. Footings are normally as deep as the wall is wide. For instance an 8-inch foundation wall would require a footing 16 inches wide and 8 inches thick. Footings for large barns and on some types of soil need to be made larger.

If building a large barn, check the bearing strength of the soils and footings to make sure they can support the entire load. Tables 5-1 and 5-2 give the customary loads of a typical building and the bearing strength of different types of soils. These tables can be used to get a rough estimate of the total building weight and therefore its

TABLE 5-2

Calculating Building Loads and Footing Sizes

Here is a simple example of how to approximate building loads to determine the necessary size of foundation footings. In the example, we will use a 20 x 24 barn with a dirt floor and hayloft above. The foundation supports are concrete piers set 6 feet on center with footings 16 inches in diameter.

The accompanying illustration shows standard building loads. "Live" loads are usually defined as those not associated with the building or its framing; for example, hay, wind, snow and people. "Dead" loads include the building framing and inherent material weight. Using standard figures for building loads, as shown in the illustration, we can calculate the following loads for a 20- x 24-foot barn:

$$\underset{\text{live load}}{}\qquad \underset{\text{dead load}}{}\qquad \underset{\text{area}}{}$$

1. Roof load = (30 lbs./sq. ft + 5 lbs./sq. ft.) × 670 sq. ft. = 23,450 lbs.

$$\underset{\text{live load}}{}\qquad \underset{\text{dead load}}{}\qquad \underset{\text{area}}{}$$

2. Hayloft floor load = (40 lbs./sq. ft. + 20 lbs./sq. ft.) × 480 sq. ft. = 28,800 lbs.

$$\underset{\text{live load}}{}\qquad \underset{\text{dead load}}{}\qquad \underset{\text{area}}{}$$

3. Exterior wall load (side bearing walls only) = (0 lbs./sq. ft. + 5 lbs./sq. ft.) × (8' x 24' x 2) = 1,920 lbs.

4. Total foundation load on both bearing walls = 54,170 lbs. ÷ 2 = 27,085 lbs. (weight on one wall) ÷ 5 = 5,417 lbs. (weight on one pier)

5. Load per sq. ft. of footing soil = 5,417 lbs. ÷ 1.4 sq. ft. (footing area for 16-inch diameter pad) = 3,869 lbs. per sq. ft.

Since sand/clay or hard clay soils can support 2 to 3 tons per square foot (4,000–6,000 pounds) as shown in Table 5-1, the 16-inch diameter footings here would be adequate for these soils. If you were building in soft clay soils, however, it would be wise to increase the footing size until you were under 2,000 pounds per square foot bearing weight. Footing 24 inches in diameter would effectively do this, having a bearing surface of 3.14 square feet.

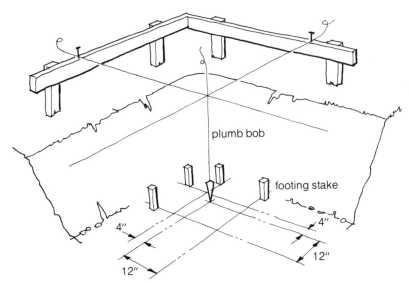

5-6. *Drop a plumb bob to relocate the foundation, then measure 4 inches to one side and 12 to the other. This establishes the outlines for a 16-inch wide footing.*

5-7. *Plumb bob can also be moved to the exact corners of building for slabs, piers, etc.*

bearing weight per square foot of foundation area. Check this weight against the bearing strength of your particular soil to make sure the foundation footings are adequate. The proposed size of the footing can also sometimes be determined by talking to your local county extension agent, building supply dealer, or local building inspector.

Forming Footings

Once the footing and/or foundation trenches have been excavated to the proper depth, reattach the layout strings to the batter boards to mark the outline of the building. Drop a plumb bob from the point where two strings cross to establish the outside corner of the foundation. Measure the distance the footing will extend outside the foundation wall. For instance, on an 8-inch wall with a 16-inch footing, the distance would be 4 inches. Measure from the plumb bob point and mark the correct distance for the inside edges of the footings. For instance, on an 8-inch wall with a 16-inch footing, this would be 12 inches. Drive four temporary stakes at each corner of the building as shown in Fig. 5-7 and tie strings to them. These outline the outside edge of the footing.

In some cases where the ground is absolutely flat you can get by with merely marking the footing before excavation, then digging a footing trench, pouring it full of concrete and leveling it off. But this is very seldom the case.

Usually you must build a form in the trench to construct a proper footing. These form boards should consist of 2-inch-thick material, and they can be used later for

5-8. *Footings and foundations must be absolutely level. Water level is an economical method of achieving this.*

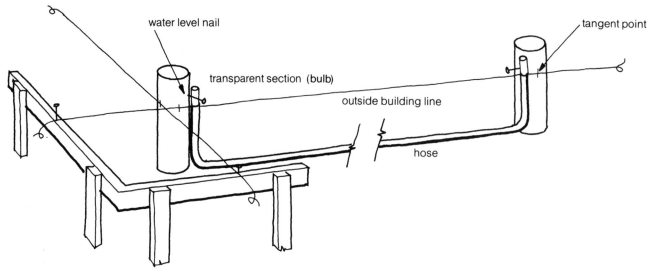

5-9. *Water hose is used to establish level points on all portions of the footing, foundation, etc. Mark each level location with a nail on Sonotubes . Cut off the tubes a few inches above nail markers before pouring concrete.*

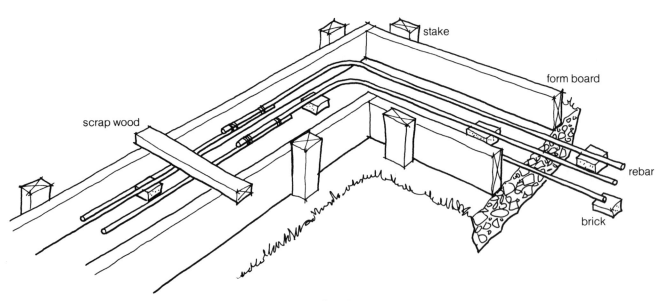

5-10. Either 2x8s or 2x10s held with stakes and scrap wood are suitable for forming concrete footings.

other framing parts of the building if they're cleaned well after the concrete has been poured and they're removed. Form boards can also be sprayed or otherwise coated with a releasing agent to prevent the concrete from bonding to them.

Footing form boards usually consist of 2x8s or 2x10s nailed together at the corners with form nails and held in place with 2x4 stakes driven in place and nailed to the form boards. Make sure the stakes are outside the form boards and the inside faces of the form boards are 16 inches apart or the desired width of the footing. The top edges of all form boards should also be level to insure that the footing will be level. If you need to raise the boards off the ground to level them, you can simply backfill with a little dirt to seal cracks on the bottom where concrete might seep out.

Sometimes, however, footings are poured at different levels. This can only be done if the footing is to have a foundation on top of it. When this is the case, the foundation will normally be of concrete blocks and the drop in elevation must be enough so that a standard concrete block will make up the difference, although occasionally a poured foundation may be created on top of a stepped footing.

It is extremely important to make sure all footings are poured on solid, undisturbed ground that has been scraped free of loose dirt. Footings that have been poured on loose dirt will eventually settle, causing the foundation and the building to settle and sometimes crack. If the soil does become disturbed in the area of the footing it should be tamped solidly before making the pour.

Footings should be reinforced with steel to prevent cracks due to the freezing action of the ground or large point loads. (In cold climates, most codes require footing depths to exceed the frost depth.) No. 4 reinforcing bar (rebar) is the most common reinforcing material for footings. It is placed on rocks or bricks above the bottom of the footing trench and tied together at points of intersection with metal wire.

After the forms are properly assembled, they are filled with concrete, and there are several methods of doing this. A small footing can be poured by mixing the concrete by hand. However, this method usually requires several helpers to keep the concrete mixed and poured in place before the first batches start to set up. Because of the amount of concrete required for most footing jobs, ready-mixed concrete, which is delivered by truck is the best bet. In some cases, though the building site may be inaccessible by truck.

The footing and the foundation wall must be *keyed* or tied together for strength. The most common method is to mold a "key" in the footing by depressing a beveled 2x4 into the wet concrete of the footing after screeding. Push the 2x4 down about 2 inches so part of the board remains above the surface. This makes removal of the board fairly easy after the concrete has set. When the foundation is poured on top of the footing it is "locked" in place by the indentation, which also helps prevent water from flowing between the wall and the footing joint.

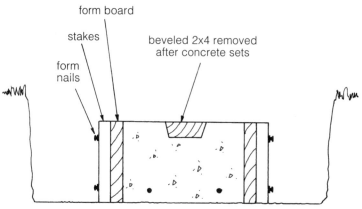

5-11. Footing must be locked to foundations. One method is to press a beveled 2x4 into the footing while concrete is still wet. Remove when cured.

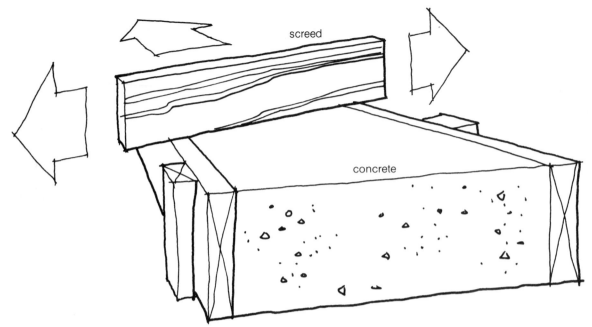

screed

concrete

5-12. *Once footings have been poured, smooth off concrete level with form tops with a screed board. As the board is pushed along the tops of the form boards, it is also moved laterally in a sawing motion.*

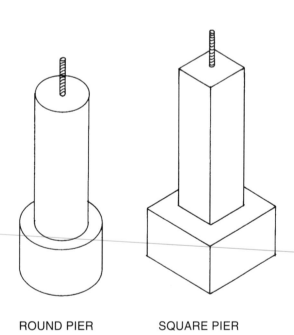

ROUND PIER   SQUARE PIER

5-13. *Piers may be formed in the round with cardboard tubes or square with wooden forms. Note the footings match.*

PIERS

Sometimes you may prefer to use piers instead of a full foundation. In this case the holes are dug in the locations of the piers, the footings formed, then the piers formed and poured. Using piers to support the plates of a small building such as a chicken house is an excellent way of substituting for a full foundation. It doesn't require nearly the expense or preparation and works well for small structures. Careful attention must be paid to insure that the tops of the piers are all level with each other or the plates may sag and cause problems with the building. A transit or water level is invaluable in ensuring all piers are built to the same elevation.

Piers are an alternate method of constructing a pole barn as well. This is an especially effective method if you don't have access to treated poles, since by using piers the poles themselves won't be in contact with the ground. The poles are usually set down on a pin inserted in the concrete pier, or you can use angle irons or special anchor brackets to anchor the poles down to the piers.

When pouring footings for concrete piers or wooden posts you can simply use the circumference of the holes as the forms for the concrete. Before you pour the concrete, however, make sure the hole is where you want it, and that you can get a pier or post plumb and centered on the footing so that its outside edge lines up with the outside edge of the building. Make sure no loose dirt has fallen into the footing holes, which can happen easily with smaller footing holes.

Pier foundations must also be locked to the footings. This is done by inserting a short piece of ⅜-inch reinforcing bar into the wet footing and extending it up to tie into the pier foundation.

Piers extending above ground level are created in several ways. Wooden forms can be constructed to build square piers or round cardboard tubes called Sonotubes can be used to cast round piers in place. Using the layout lines of your building as a guide, place the cardboard tubes on top of the footings and carefully backfill around them with earth until they are well braced and plumb with the outside lines of the building. Using a line level or water level, mark the height the concrete is to be poured in each tube. Then simply cut the tubes off at that height with a handsaw. As you can see, piers are much easier to create on a smaller scale, since you can form the pier and mix small batches of concrete to pour the footings and piers one at a time.

Wooden pier forms are constructed, placed in position

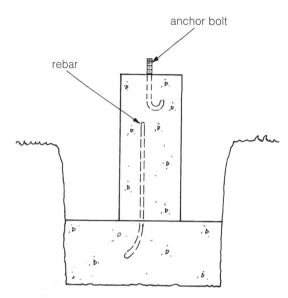

5-14. Piers must also be locked to footings. Rebar is bent and placed in footings, then pier formed over the rebar.

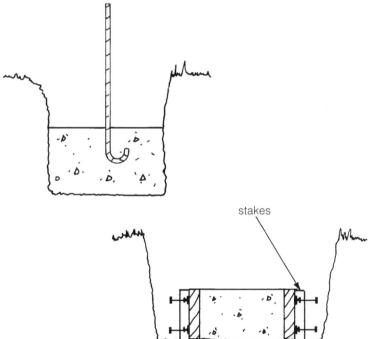

5-15A. Piers must be supported by footings. The simplest method is to dig a hole and pour a flat footing with a rebar tie. 2x8s or 2x10s can also be nailed in a square and placed in a hole to create a footing. Stakes can be driven in by the sides to hold them in place.

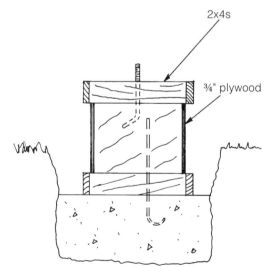

5-15B. Then the piers are formed. Shown is the form for a square pier.

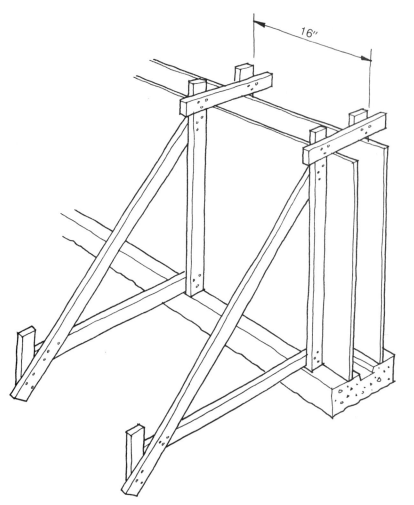

16"

5-16. *Foundation wall forms must be well braced. Shown is a form for a wall up to four feet high. Metal snap ties may also be used to space the form sides consistently and keep them together.*

over the footings, and braced with 2x4 bracing attached to stakes driven into the ground. After the pour, the braces and wooden forming is removed, and the hole is backfilled.

FOUNDATIONS

Concrete foundations, which sit on top of the footings and support the sill plate and walls, are poured on top of the footings after the footings have cured completely and the keys have been removed.

Foundations may be fairly short, just enough to get the structure above the ground level, or may be high in the case of full basement walls. Both are formed in the same manner. The first step is to replace the strings on the batter boards after the footings have cured. Then reestablish the outside lines of the foundation, and build the proper forms.

Forms for concrete foundations up to 4 feet high can be made of 1-

5-17. *Walls more than four foot high require more sophisticated forming. You can often rent forms, or construct them as shown.*

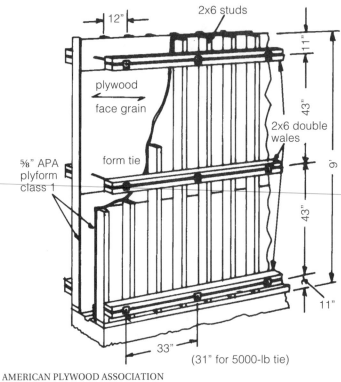

12"

2x6 studs

11"

43'

plywood
face grain

2x6 double wales

⅝" APA plyform class 1

form tie

9'

43"

11"

33"

(31" for 5000-lb tie)

AMERICAN PLYWOOD ASSOCIATION

inch plywood or 1-inch solid stock supported by 2x4 vertical braces every 15 inches and braced at each upright with 2x4s fastened to stakes outside the footing perimeter and across the tops of the form boards.

For forms more than 4 feet high, you'll need double 2x6 horizontal bracing tied with steel rods as well. Because these forms are complicated and expensive to build, it is often wise to rent them from a tool rental company or concrete supplier. Standard panels are available in 4- and 8-foot heights that go together quickly with steel pins.

SLABS

One extremely effective method of supporting a small building such as a shop and providing an easy-to-clean floor simultaneously is to pour a slab, then construct the building on top of the slab. Slabs can be poured on top of footings, or they can be monolithic, which means the slab is poured with a deepened trench around the outside edge to create the footing and slab all in one pour. Figure 5-18 and 5-19 illustrate the two different types.

Most concrete slabs are poured to a thickness of 4 inches, although a barn that will contain heavy equipment may require a 6-inch slab. The first step is to determine how high you wish the slab's top surface to be above ground level or finished grade. This height is normally around 3½ inches, which puts the bot-

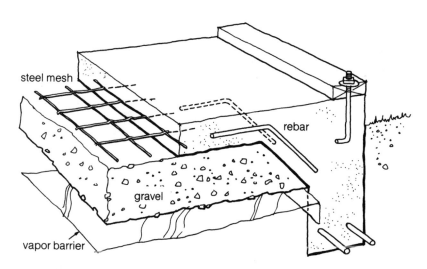

5-18. *A slab rests on a bed of gravel; the thick "turndowns" at its edges keep the slab from shifting.*

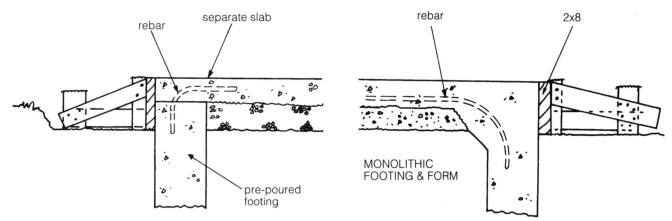

5-19. *Slabs can also be poured on a footing. Shown are the forming methods for combined footing/slab (also called floating slab) and a monolithic.*

tom edge of the building siding approximately 1½ inches away from the finished grade, although in some cases of thick slabs you may prefer to have the slab somewhat higher above grade. Slabs for small, lightweight use are sometimes poured directly on the ground if the soil is hard and level. Usually, however, the slab is poured on a 2- to 4-inch gravel bed.

Regardless of which method is chosen, a footing or edge stiffener is usually formed in place around the perimeter of the slab. This is normally 1 foot wide and 24 inches deep for most buildings. This requires first laying out the boundaries of the slab using the same techniques as for a foundation. Batter boards are installed, string lines positioned, and the slab outlined with stakes or lime. If you're building directly on the ground the footings or edge stiffener excavation is done and you're ready to form up the slab. If pouring on a gravel bed, excavate the topsoil down about 2 inches over the entire slab area, then excavate for the footings.

Any service pipes, conduits for electrical supply, etc., must be installed at this time. Wrapping them with insulation material such as foam or tar paper will help protect them from the shrinkage and expansion movements of the concrete. When pouring the slab, maintain a proper slope so that water can drain away from the slab; the pitch of this slope is usually ⅛ inch per foot.

If pouring directly on the ground, 2x4s are used to form the outside edges of the slab, using stakes and braces to hold the forms in place. If pouring on top of a 2-inch gravel bed with the topsoil excavated, 2x6s are used. For a 4-inch-thick gravel bed, 2x8s are used. All this of course allows for a 4-inch (3½-inch) thick slab. If the slab is large, over 10 or 12 feet wide, it is easier to pour in three separate sections. Divide the slab into thirds, and pour the two outside sections first followed by the inside section, using the appropriate width form boards to create the inside forms.

If the pour is to be on wet ground areas, a 6-mil polyethylene vapor or moisture barrier is placed between the gravel bed and the ground, or between the ground and concrete if no gravel bed is used. Although some construction diagrams illustrate the vapor barrier between the gravel and concrete, I prefer to put it under the gravel. When

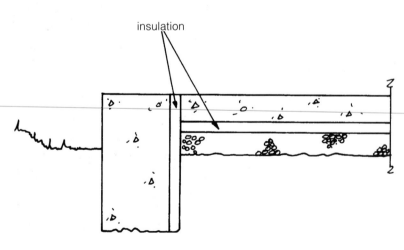

insulation

5-20. In some cold climates slabs must be insulated. Solid foam insulation is used as shown.

the vapor barrier is on top, it is often punctured by the gravel when walking on it to install the reinforcing, pouring the concrete, etc. These holes can allow air to get under the barrier, forming air pockets or huge "bubbles" during the concrete pour. These must be punctured to get the air out, again destroying the purpose of the vapor barrier.

Sometimes a double pour is made, creating a *floating slab*. This is done by first forming and pouring the "footing," extending the height of the footing to the slab height. Then the slab is poured to the same height using the footing as a "form."

This type of construction is used in cold areas when an insulated slab is desired. In this instance the footing is poured first to the height desired. Insulation is installed around the inside perimeter of the footing and over the gravel bed. Then the slab is poured on top of the gravel and insulation and to the top edge of the "footing."

Slabs must also be well reinforced. In most instances No. 10 wire mesh is placed on top of the ground or gravel bed and held suspended in place with stones or bricks. For areas with heavy traffic No. 4 reinforcing bar can be used.

ANCHORS

Anchor bolts are needed on top of concrete slabs, walls, foundations, and piers to attach the building's sill plates. These bolts are usually 8 inches or longer. They have a threaded top for attaching a washer and nut and a J-hook at the bottom that is set in the concrete. In pouring slabs you can often position the anchor bolts in place on special wooden spacers fastened to the forms. This makes it hard to screed off the concrete, or use a bull float, so it's easier to insert the bolts in the concrete while it is still workable. The bolts are pushed down into the wet concrete until they protrude at least 2 inches.

For standard stud framing on a concrete wall or slab, set the anchor bolts at 4- to 6-foot intervals. Lay out your wall framing first, however, and make sure none of the anchor bolts fall where a stud will be located, or a door opening, or at the corner where corner studs will be. These bolts must be placed as soon as the concrete has been poured, so make sure you have them on hand before you start pouring.

Anchor bolts are also required for pier foun-

5-21. *Anchor bolts placed around the perimeter of the slab fasten the building to the slab. The anchor bolts must be positioned to the center of the plate from the outside edge of the slab.*

61

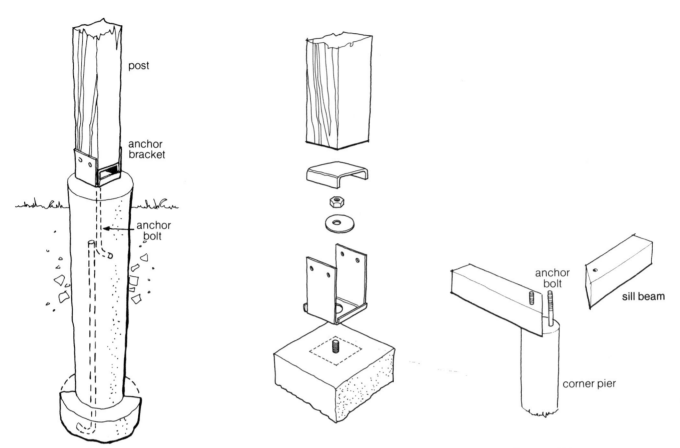

5-22. Anchor brackets like these are available for securing posts, used with pole buildings. The illustration at right shows an exploded view

5-23. For large sill beams, embed two anchor bolts in corner piers.

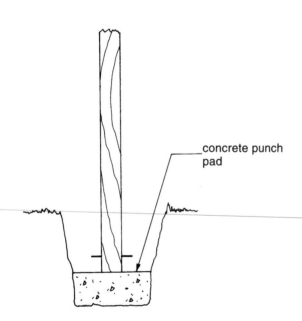

5-24. Posts are often supported on concrete punch pads, then additional concrete poured around the posts once they are plumbed and braced.

dations. These must be long enough to stick into the concrete 4 inches, go through the sill plate and leave plenty of space for washers and bolts. If you're using large beams, such as 8x8 beams for the sills, you'll need 16-inch bolts. If you can't find bolts that long, make them from ½-inch threaded steel rod, bending one end into a J shape. Place an anchor bolt in each pier, except the corner piers where the sill beams come together at a 45-degree angle. This will require two bolts to hold the sills. In this case the corner anchor bolts will unavoidably interfere with the corner beams or studs. One method of remedying this is to counterbore a shallow hole in the top of the sill beam to accommodate the nut and washer. Any excess bolt protruding above the sill beam can be cut off flush with top of the beam.

Special brackets are also available to anchor poles to piers for pole buildings. In most instances, however, pressure-treated poles are set on concrete "punch pads" placed in position and then earth backfilled around them to hold them in place.

Positioning anchor bolts accurately is critical. Even

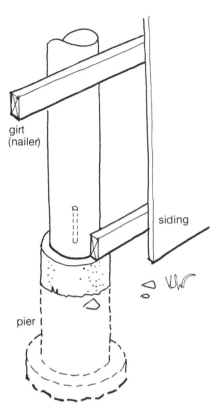

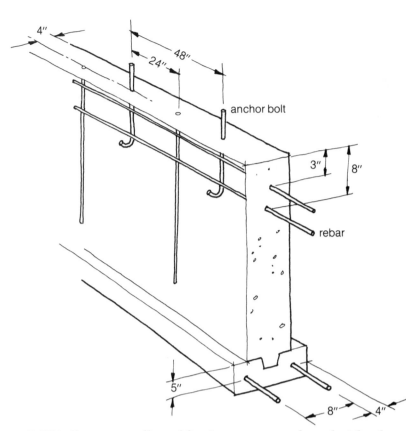

5-25. *When positioning an anchor, compensate for girt width so siding can drop down past the top edge of the pier.*

5-26A. *Concrete walls and footings are strengthened with rebar placed as shown here. Note the spacing of the rebar and anchor bolts.*

when using them on slabs or foundations, make sure the bolts are spaced the same distance away from the outside edge for ease in locating and boring the holes in the sill plates. Proper location is even more important when placing anchor bolts in piers to which poles are attached. If the bolts and/or brackets are not placed correctly, the outside

CONCRETE REINFORCEMENT

Concrete footings, piers, walls, and slabs should be reinforced with steel. Reinforcing helps prevent cracks due to the freezing action of the ground or large point loads such as a tractor. Two standard reinforcing materials are No. 4 reinforcing bar (rebar), and steel wire mesh, usually 6×6 inch No. 10. Rebar, used in footings, walls, and piers, is placed in the proper positions in the concrete when it is poured.

Use No. 10 wire mesh with concrete slabs. To prepare for pouring a slab, first place a 6-mil polyethylene vapor barrier on well-tamped or undisturbed soil. Spread several inches of crushed stone carefully and evenly on top of the barrier, then lay wire mesh on top of the gravel. When pouring the concrete, lift the wire mesh up and position it in the middle of the slab supported by rocks, about 2 inches off the gravel. When you pour a floating slab, use extra reinforcing ties to hold the slab and footings together.

5-26B. *Floors and slabs must be reinforced. Rebar, shown here, is tied together with wire.*

of the pole building framing will not line up with the outside of the foundation piers. Always compensate for the thickness of the girts that attach horizontally to the poles and hold the siding in place. For example, if 2x4 girts are used, the poles must be set 1½ inches in from the outside of the piers so that the siding can drop down past the top of the pier and keep rain from collecting on its top.

MOISTURE PROTECTION

Moisture and especially standing water can wreak havoc on foundations. Water expands clay soils, putting tremendous pressure on foundation walls, and when it freezes the stress is even greater. To avoid heaving and cracking, always locate foundation walls in well drained soil or, if that is not possible, install perimeter drains.

Regardless of the type of soil, the first step is to tar the exterior of the foundation or walls with asphalt coating to seal and protect the walls. This prevents water from seeping through and wetting the interior or basement space. The tar coating, however, should extend just to finish grade or it will look unsightly.

Then lay plastic perforated drainage pipe next to the outside face of the footings on at least 6 inches of clean crushed stone or pea gravel. Connect all sections of the pipe with couplings and elbows to make a continuous loop around the building. The perforated holes should face down and the loop should drain toward one corner at a minimum slope of 1 inch in 20 feet. From this corner, a discharge pipe should lead the water away from the foundation to a dry well or a grade discharge at a lower elevation. Cover the pipes with another 6 to 8 inches of crushed stone to prevent them from clogging with mud and silt, and backfill to grade with dirt.

INSULATION

In cold climates it is best to insulate the foundation of a heated building. Even if the building only has a dirt floor, foundation insulation reduces heat loss through it and keeps the building much more comfortable. Rigid foam, known

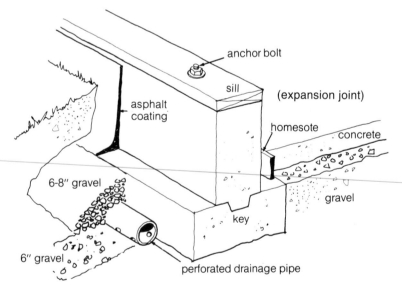

5-27. *Place perforated drainage pipe (holes facing down) next to footings.*

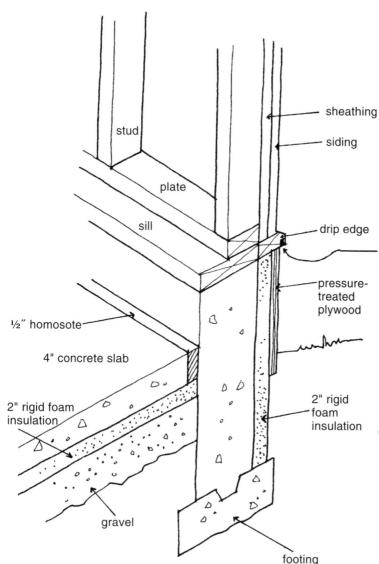

stud

sheathing

siding

plate

sill

drip edge

NOTE: saw kerf underneath drip edge to keep water from running back to plywood. Caulk joint between drip edge and plywood.

pressure-treated plywood

½" homosote

4" concrete slab

2" rigid foam insulation

2" rigid foam insulation

gravel

footing

5-28A. Here's a typical insulated wall foundation with insulated slab. Note that the rigid foam for the wall is protected by a drip edge and pressure-treated plywood.

as *blueboard,* is the standard foundation insulating material. Usually it is 2 inches thick and available in 2x8 foot sheets with a tongue and groove for a tight fit. It is applied with concrete adhesive.

Blueboard is applied after the exterior foundation or wall surfaces have been tarred and the perimeter drain pipe installed. Adhesive is used to fasten the material to the wall surface starting at the top of the footing and installing right up to the building's sill plate or where the wall insulation begins. Where the foam insulation projects above ground, cover it with plywood to protect it from sun and weather.

In northern climates, concrete floors are also insulated. If a warm slab is desired for livestock, insulate the slab completely from the ground. Coupled with foundation insulation it will provide a warmer building and will

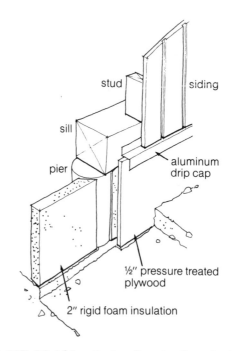

stud

siding

sill

pier

aluminum drip cap

½" pressure treated plywood

2" rigid foam insulation

5-28B. Rigid insulation for pier foundations can be installed between piers, then protected with a ridge cap and pressure-treated plywood.

more than pay for itself in a few years in reduced heating bills.

Insulating pole or stud-framed barns and buildings on piers present a problem: how do you insulate and seal the area between the bottom of the siding and the ground? To prevent rot, the sill plate and siding are at least 6 inches off the ground, leaving a large airspace for wind and snow to enter the barn.

The solution is to nail 2x8s of pressure-treated wood onto the sides of the sill plates so the pieces are buried 2 inches in the ground. Use aluminum drip cap where the treated wood meets the siding to keep water from flowing into the joint. This skirt provides a surface on which to attach the interior insulation. Rigid foam panels can be glued onto the inside of the wood between the concrete piers. Unlike fiberglass, this rigid foam won't absorb moisture and will not degrade. The pressure-treated wood can be painted with an oil-based paint for further protection. Most manufacturers recommend waiting at least six months after construction before applying paint to allow the pressure-treated wood to dry properly.

FLOORS

Slabs are often poured as floors inside a building with other types of supports, say as a floor for a feed room inside a pole building. In other instances, however, floors are poured inside a foundation wall, and this is always the case on a basement floor. The foundation walls are poured first, then the floor is poured inside the walls. Although this eliminates some forming problems, it does make it a bit more difficult to screed and strike off the concrete.

Again a good subsurface is required, using 2 to 4 inches of crushed rock or gravel on top of a polyethylene vapor barrier. In addition, the joints at the walls must be protected with isolation strips to prevent problems with the floor contracting and expanding. Use ½-inch-thick asphalt expansion joints around all sides of the floor. The floor must also be graded and formed to slope toward a door or drain if one is to be installed.

To provide a method of striking off the floor, drive wooden stakes in the ground in parallel rows along the building wall. Then tack 2x4s in place as a guide, making sure the top surfaces of the 2x4 guides are at the

5-29. In many instances floor are poured within buildings.

66

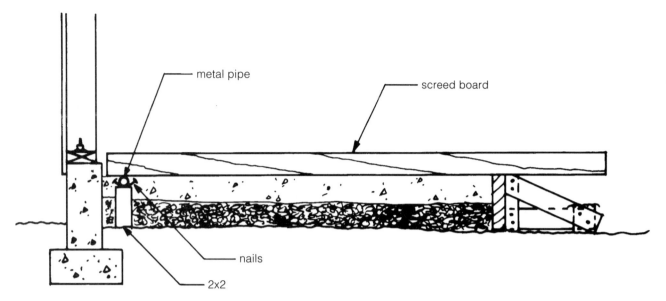

metal pipe

screed board

nails

2x2

5-30. To create a method of striking the concrete off level, place wooden strips on metal stakes or steel piping on wooden stakes against the wall. Pour the floor, strike off, then remove the stakes and strips and fill in with concrete.

correct height for the floor surface and that they are level. If you have rigid steel piping on hand it can be placed on the wooden stakes between nails. Starting at a back corner pour against the wall and screed this off. Remove the first guide-form, fill the opening with concrete, and then make the second pour, screeding and removing the height guides as you go. This is a slow, hard process, and a couple of helpers can be a great deal of assistance on even a small pour.

Like large slabs, large floors are usually divided into thirds. Pour the two outside sections first, allow to cure,

5-31. When pouring a large slab, it helps to divide the job into three sections with form boards. Pour the two outside sections first, then the middle.

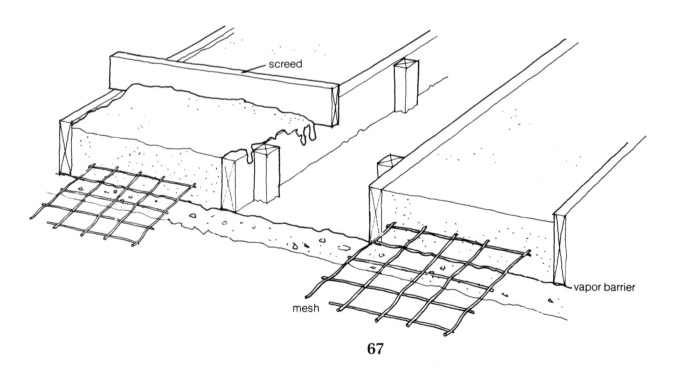

screed

mesh

vapor barrier

then pour the center section.

As you screed the concrete off, use an old rake to tamp it securely or use a professional jitterbug outfit that can be rented. Floors such as horse stalls are often better as hard-packed clay rather than concrete. For chicken houses you may prefer wooden floors or even a dirt floor.

Some floors, such as confinement for hog housing may require slotted flooring. These can be made of either wood or concrete. You can easily pour your own concrete slats if desired. With slotted floors, a concrete pit is placed below the floor level and manure is removed from the pit with a liquid manure pump and a wagon.

CONCRETE

The first step in using concrete is determining whether to mix the concrete by hand or have ready-mix concrete delivered by truck. The decision depends primarily on the size of the job, as well as the cost of ready-mix and whether you can get a concrete truck to the building site.

Excluding labor, ready-mix concrete is quite a bit more expensive than concrete that you mix by hand in a wheelbarrow or machine mix in a power mixer. In most instances, however, when mixing concrete for jobs requiring more than 1 cubic yard, buying ready-mix concrete is the best choice. A concrete truck with chutes can back right up to your foundation, deliver the concrete exactly where you want it and pour your walls in a short amount of time. This will also eliminate the problem of "cold joints," which occurs when small mixes of fresh concrete are poured onto previous batches which have already begun to set up.

Usually, ½ to 1 cubic yard is the minimum load a concrete supplier will deliver, and you should establish beforehand what truck charges or minimum load charges you will be billed for. (The maximum one truck can carry is usually 5 yards.) Everything should be in place and set to go before the truck arrives. Normally, you have one hour to unload the truck and after that a substantial hourly truck fee is charged. Also, the concrete may set up quickly, so make sure that the forms are securely in place, that the truck can reach all sections of the foundation with a 10-foot chute, and that you have shovels and a wheelbarrow handy in case you have to transport concrete by hand to a far corner.

5-32. You can mix concrete for small jobs, but larger ones require ready-mix, delivered by truck to the site.

Mixing Your Own

Concrete for individual piers, small foundations, or even small slabs can be mixed by hand in a wheelbarrow or power mixer. Never attempt to use a wheelbarrow to mix more than 1 cubic yard. A good-sized wheelbarrow will hold only about 1½ cubic feet of concrete, meaning you'll need to mix about 20 wheelbarrows before you'd get 1 cubic yard.

If mixing more than 1 yard of concrete, get either an electric power mixer or one that can be driven by a tractor power take-off (PTO) and buy the concrete ingredients in bulk.

For many small jobs you can purchase pre-mixed concrete, available in 80-pound bags that include the cement, sand, and gravel. Each bag makes about ⅔ cubic foot of concrete. One bag of concrete mix is about right for a 16-inch-diameter pier footing, 6 inches thick.

The concrete used for footings and most other farm construction projects, such as walks, etc., is made from Portland cement. This is available in 94-pound bags from building supply dealers and lumber yards. Each bag contains 1 cubic foot, dry measure of material. Make sure the cement you buy is dry. If it has any lumps in it that you can't break up with your hands, it has been allowed to become damp and is useless. When you purchase the cement make sure you have a dry place to put it until you need it. Don't lay sacks on the ground, as they can pick up enough moisture in this manner to ruin them. Instead, lay boards or pallets down first, then lay the sacks of cement on them.

Opened bags will eventually pick up enough moisture to become lumpy, so plan to buy only enough for the job, but make sure you have enough. It's very frustrating to have to make a trip into town for one bag of cement to finish a job.

The cement is mixed with various aggregates plus water to make up a paste called concrete. The proportions of aggregates to cement and water determine the strength and workability of the paste. In most instances of farm construction the proportions will be one bag of cement to 2 cubic feet of sand and 3 cubic feet of gravel, or 1 part cement, 2 parts sand, and 3 parts gravel. Any method of measuring can be used, though for me the easiest method is to use a shovel: One shovelful of cement, two of sand, and three of gravel. If the sand is damp (and bulked up), use a 1:2:4 mixture.

The sand and gravel should be well cleaned and not

full of trash, leaves, or other debris. Seashore sand should not be used. Most contractors specify the size of sand and gravel they need; in most rural areas, however, you'll probably have to purchase whatever the building supply dealers have on hand. If you have access to a creek with a gravel bar, you can sometimes use that for your gravel aggregate.

Because the size of gravel or crushed stone varies in different locations, it may be necessary to change the amount of cement in your mix. Generally speaking, when gravel is smaller than the normal 1½-inch size, it is good practice to use more cement. When gravel size is a maximum of 1 inch, add ¼ bag of cement to a five-bag mix; when gravel is a maximum of ¾ inch, add ½ bag.

Concrete can be mixed in two ways. You can mix it in a wheelbarrow or concrete trough using a hoe with holes cut in it, or you can mix it in a power mixer. Naturally, the power mixer, regardless of whether it is electric or power take-off driven, is the easiest method and usually results in better mixed concrete.

The first step is to place the cement, sand, and gravel in the mixer in the correct proportions and mix them thoroughly. An old hoe with a couple of holes cut in it and the handle shortened makes an excellent tool for mixing those small jobs in a wheelbarrow.

When the aggregates and cement are thoroughly mixed without any dark or light streaks, add the water. The amount of water is also very important. Too much water makes the finished concrete too weak and will allow it to flake away. Too little water and the solution may not mix thoroughly, or it may set up too quickly. In most instances the proper solution proportion of water is 5 gallons of water to each bag of cement. However, if the sand is wet you will have to use less water.

Don't add the water all at once. Add just a little water and allow the concrete to mix thoroughly, then add more as needed. To test the mix to see if it has the right amount of water, pull the hoe or mixing tool through it in a series of jabbing motions. If the mix is correct the little ridges pulled up will stay. If the mix is too wet, the concrete will slump back quickly. If the mix is too dry, the ridge won't be smooth and even.

When pouring a large amount of concrete such as a footing or foundation, you must keep pouring in a continuous pour, so you won't have cracks or weak spots in the footing or foundation. This is one of the worst problems encountered using home-made concrete for large areas such as foundations.

70

Working with Concrete

When working with concrete, always wear rubber gloves and rubber boots. Rubber gloves protect your hands from the abrasive and caustic action of cement, which can easily wear away skin after a few hours of contact. Rubber boots will allow you to step in the concrete, water, and mud without getting wet, and your good leather work boots won't be ruined.

Once the forms have been filled with concrete use a shovel around the outside edge to poke down the concrete and knock out all the air bubbles to settle the concrete thoroughly in place. A hammer can also be rapped against the outside surfaces of the form boards to help further settle the concrete. If you wish a perfectly smooth outside surface after the form boards are removed, use an old hand saw and jab it up and down in a sawing motion between the concrete and form boards before the concrete sets up. Concrete vibrators can also be rented for large jobs, but must be used sparingly or the aggregate will tend to settle to the bottom of the mix.

Use a 2x4 *striker* or *screed board* to drag the excess concrete off the top of the foundation and smooth it level with the tops of the form boards. Little else is needed for footings, other than inserting the wooden key mentioned earlier. For foundations, use a wooden or steel float to smooth up the surface of the concrete, but don't attempt to make it glass smooth. Set the anchor bolts in place on the foundation, twisting them to make sure they're seated well without holes around them. Use a pointed trowel to smooth up the area around the bolts.

Slabs

A concrete slab (or floor) is finished in much the same way as a foundation, but it requires a good deal more skill and patience to get a good finish. After pouring the slab, use a rake to roughly level out the surface. Then tamp the surface using a home-made, purchased, or rented jitterbug. This drives the aggregates into the mix and brings water to the surface. Again, with the help of another person, use a straight 2x4 to drag the

5-33. Working with concrete require a lot of preparation. Forms must be correct, tools on hand. Use rubber boots for working in the wet concrete. Note the use of a Goldblatt Tool Co. drag to pull concrete into place.

5-34. Once the concrete has been poured a wooden screed board is used to drag excess concrete off. In this case a slab.

71

5-35. Then the concrete must be "settled" in place using a jitterbug, or old rake.

5-36A. In most instances of a barn floor or slab the single best and easiest method of finishing is using a Goldblatt bull float. Float is positioned on the edge of the slab near you, then pushed to the opposite side.

5-36B. Then pick the handle up and pull the float toward you to finish smoothing a 3-foot section. Move to the next section and repeat.

excess concrete off and level the surface with the tops of the form boards. Push the 2x4 back and forth with a sawing motion and slowly move across the slab about 1 inch per stroke. It's a good idea to screed in both directions several times to make sure the concrete is smoothed down level. Fill any holes below the screed board with excess concrete placed in a wheelbarrow.

The next step is to float the surface using the long-handled, large float called a bull float. Again, this can be a home-made, purchased, or rented tool. Place the float on the concrete surface near you, lower the handle, and push the float to the far edge. Then lift the handle slightly so the curved surface of the float rides on the wet concrete and pull the float back toward you, reaching hand over hand on the long handle to make a smooth, even pull. Then pick up the float, move it to the next position down the slab, and do the same thing. For many farm building slabs or floors this is all that is required. If the concrete is extremely wet you may need to do this job several times, waiting a bit between jobs until the concrete just begins to set up.

If the slab surface is to be a barn floor that will handle animals, you may wish to broom-finish the surface to prevent a wet, slippery, and dangerous floor. After all the surface water has disappeared and the concrete has begun to harden, drag a push broom with stiff bristles lightly over the surface to give the floor a lightly textured finish. If pouring a slab, make sure to install anchor bolts before the concrete sets up.

For floors with more finish, such as a garage, troweling is often desired. This can be done by hand, but a rented power trowel will do the job much faster and better. This chore should be done as soon as you can easily put your weight on the concrete without sinking. Use wooden supports to spread out and help support your weight. A piece of plywood two-foot square will do. You will, of course, have to hand-trowel around the anchor bolts and any other protrusions.

5-37. *The bull float will provide an adequate finish for most outbuildings. Smoother finishes can be acquired with a trowel, or better yet, with a power trowel.*

5-38. *Edges should be smoothed and rounded with a hand edge.*

Curing Concrete

Concrete must "cure" or harden before the forms are removed or weight is put on it. Most builders consider 24 hours the absolute minimum before form boards can be removed and the waterproofing and backfilling can begin. It is strongly recommended that you keep the form boards on longer than this, however, because it helps the concrete stay wet, thus slowing the curing time and increasing its strength. After 24 hours concrete is still "green" and susceptible to chipping and cracking.

If possible, leave the form boards on for three or four days and cover any exposed concrete with burlap or old bags that can be wetted down periodically. Plastic sheeting can also be used to cover the concrete. Concrete must cure and dry slowly, otherwise it will be weak and crumbly. The biggest problem with concrete is having it cure too fast. Try to avoid pouring concrete slabs in the hot, noonday sun, and always have a water hose with a spray nozzle on hand so you can dampen the surface if it starts drying too fast.

Concrete must also not be allowed to freeze. If you are pouring in the fall or winter and there is a chance of frost, insulate the concrete with a covering of old tarps and hay. In cold weather, use heated water and aggregate to make sure the concrete is warm enough to set up. Add calcium chloride to the concrete mixture if you want to lower its freezing point. Winter concrete work is best left to experienced contractors, since it is easy to ruin an entire foundation if it does not cure properly.

5-39. *Final finishing can be done with a hand trowel.*

Concrete Block Foundations

Many owner-builders prefer concrete block foundations to poured walls. They're easier to build, in some ways, and they are usually cheaper. This depends on the current price of concrete relative to concrete blocks, and, if you seek help from a professional mason, the cost of his services. A concrete block foundation, however, is an excellent option for the homesteader who has to do the job without a lot of help. You can work on the foundation as you get the time since it doesn't have to be done all at once, as does a poured foundation.

Working with concrete blocks, regardless of whether you are building a foundation or an entire barn, is fairly easy, even for the first-timer. This requires doing the job in a certain way, and attention must be paid to several things, including getting the proper mortar mix and keeping the walls straight and plumb.

If you use a block foundation, make the wall lengths multiples of full-sized blocks where possible. This eliminates the necessity of block cutting. The standard nominal

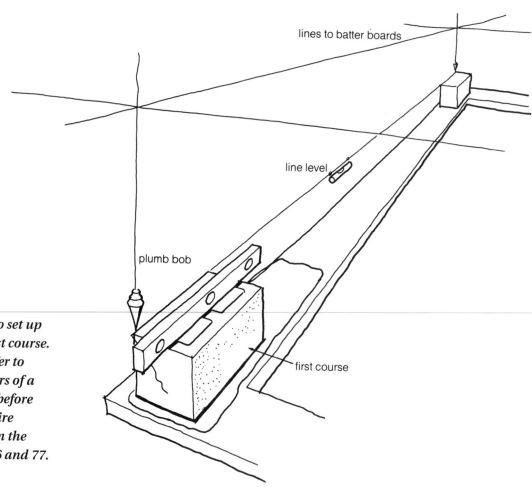

5-40. Here's how to set up blocks for your first course. Some masons prefer to build up the corners of a block foundation before completing an entire course, as shown in the photos on pages 76 and 77.

block size is 8x8x16 inches. The actual size is 8x7⅝x15⅝ inches, allowing for a ⅜-inch mortar joint between blocks and between rows.

Block walls can be strengthened by cementing reinforcing rods vertically inside the block cores or by adding galvanized hardware cloth or metal tie bars to the top of a *course* or horizontal layer. Always add a layer of reinforcing mesh on top of the second-to-last course to strengthen the top of the wall and to suspend the mortar for the anchor bolts, placed every 4 feet for block walls. Insert them in mortar 15 inches into the wall so that 3 inches are left to hold the sill plate.

Mixing Mortar

As with concrete, you can mix mortar in either a wheelbarrow or a power mixer. Type M mortar, used for foundation walls, is a mix of 1 part Portland cement, ¼ part hydrated lime, and 3 parts clean sand. Eighty-pound bags of mortar mix are also available from building suppliers. When mixing mortar, always remember to dry-mix it first, just like concrete. Then add just enough water to give it a

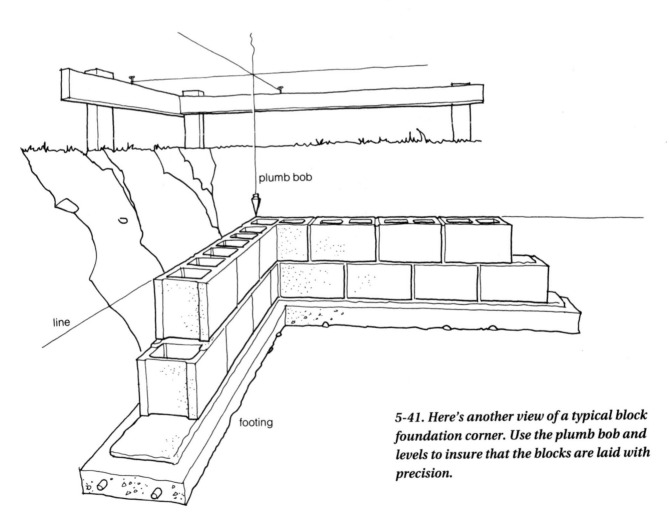

plumb bob

line

footing

5-41. Here's another view of a typical block foundation corner. Use the plumb bob and levels to insure that the blocks are laid with precision.

75

pliable consistency and test it for proper stiffness with a hoe. If it is pliable, yet pulls up easily with a hoe and stays without slumping, it is mixed properly.

Laying Blocks

After completing the footings, put your building outline strings back on the batter boards to reestablish the outside corners of the foundation. At the point where the strings intersect, drop a plumb bob to find the exact corners of the foundation wall. Place one block at each corner, then position strings from the top outside corner of one block to another. This will give you a straight and level line for laying the first row of blocks.

Mix only enough mortar to use before it hardens (about an hour) and begin at two of the marked corners. Trowel about ½ inch of mortar under the two corner blocks and, using the string and a line level, adjust them until they are straight and level with each other. After leveling the blocks you should have a ⅜-inch mortar joint beneath each block. It's also a good idea to level the individual blocks with a small wooden mason's level both lengthwise and crosswise.

Smear mortar on the end of a new block with a trowel and position it against the corner block, again making sure it is level with the first block and straight along the string. Continue laying the first course in this manner. The last block may have to be cut if your joints are too thick or the wall dimensions uneven with multiples of 8 inches. If so, use a mason's hammer and brick chisel to score the block

5-42A. Stretch a mason's line from corner to corner, and, to complete the first course, lay the top outside edge of each block to this line.

5-42B. When installing the last block, in each course, butter all edges of the opening block with mortar and all four vertical edges of the closure block, then carefully lower closure block in place.

5-42C. Handling the blocks correctly is important. By tipping the block slightly toward himself, the mason is able to see the upper edge of the course below, so he can place the lower edge of the block directly over the course below.

76

on both sides until it cracks along the desired line. Be sure to use safety glasses when scoring the blocks.

After finishing the first course, position your strings for the second course and proceed as with the first. Blocks at the corners of the walls should overlap, if using a running bond, so that the vertical block joints do not line up. Use a small level as you build up the corners to ensure they are plumb.

Before the mortar dries, use the trowel to strike off excess mortar from the joints. Then smooth the mortar with a jointing tool and create concave recesses along the joint lines. This action packs the mortar joint and helps create a strong, secure, water-resistant wall.

After laying a block foundation, waterproof it with tar just as you would a poured concrete wall. In northern climates, the walls should also be insulated, and this can be done in two ways. Either apply rigid foam to the outside of the wall, or pour loose fill vermiculite insulation down the block cavities.

5-42D. If work is progressing rapidly, mortar scraped from joints may be applied to face shells of block just laid. If there are delays, mortar should be reworked.

Grading

The final step in constructing a foundation is to backfill the foundation trenches or holes and grade the building site. These are important steps for a long-lasting foundation. Proper backfill material and grading will keep water away from the foundation and minimize the possibility of frost cracks and heaving.

Backfilling is simply a matter of placing excavated dirt back against the foundation of the wall. But do it carefully to avoid damaging the foundation or insulation. Large rocks should be placed, not thrown, against the wall. This is especially important if the concrete is still green. Use only clean fill. Wood and other organic matter will decompose and leave pockets. If drainage is a problem or you have heavy clay soils, backfill with sand or gravel to protect the wall from freeze expansion.

After the walls have been backfilled and tamped down, do the final site grading. To enhance drainage, make the ground slope away from the foundation. If this is impossible because of an uphill slope on one side, dig a drainage trench two feet out from the foundation wall to catch surface water and carry it away. During the final grading, use the topsoil that was set aside during excavation. Spread it evenly, seed it with grass, and cover it with a thick layer of hay mulch to prevent soil erosion. The grass will grow through the mulch.

CHAPTER 6

Framing

6-1. Several different types of framing techniques can be used to construct barns and outbuildings, but the most common is stud framing using dimension lumber.

Most barns, sheds and outbuildings constructed today utilize one of three principal construction methods: stud, post and beam or pole framing. Stud framing consists of two styles, platform or balloon, and is the most common modern method. Pole barn framing, however, has fast become popular due to the increasing availability of pressure-treated poles, the ease of construction, and the relatively low cost. Stud framing uses dimensional lumber, ranging in size from 2x4s to 2x12s, regularly spaced to frame walls, floors, and roofs. Post and beam framing is an older, traditional method of barn construction dating back to medieval times. With this method large beams and posts, such as 6x6s and 8x8s, form the frame. Pole framing is actually the oldest method, dating back to the Stone Age, and widely used by Indian cultures in North and South America. Today, pole building is simplified by machine-cut and pressure-treated poles or posts. Also, it is usually combined with platform framing of doors, floors, and roofs.

PLATFORM FRAMING

Platform framing is the most widely-used method of wood construction for small barns, sheds, and shelters for several reasons. It requires less labor and materials than post-and-beam framing and its construction details are simpler. Because the individual framing members are relatively light and small, one person can erect an entire building alone. Kiln-dried lumber for platform framing is fairly expensive, but sometimes this high cost can be avoided by using native rough-sawn lumber from local sawmills.

The accompanying figures show details and terminology for conventional platform framing suitable for barns and outbuildings with and without ground-level floors. In either case the framing begins with sill plates laid on the foundation.

If a floor is going in, joists are laid on top of the sill plates and then boxed in with joist headers. For larger buildings, a girder is used to support the flooring joists at their midpoint. Girders are usually 6x10s or larger beams. The beams and joists form a deck on which subflooring (usually plywood) is laid. Sole plates are nailed to the subflooring and studs are nailed on top of the plates to form the vertical support elements of the walls. Double top plates are nailed on top of the studs and then the whole process starts again for the second floor. Rafters set directly on top of the topmost wall plates and extend up to a ridge plate to form the top of the roof triangle.

BALLOON FRAMING

Although seldom used as much these days, balloon framing is one method of framing a two-story structure. The primary difference is the studs run continuously from the foundation to the rafter plate. A ribbon support cut into the studs is used to support the ends of the second-floor joists which are also nailed to the studs. In some cases of barns and buildings where the inside won't be finished off, the support boards can simply be nailed to the studs, although cutting them in place does add more support. Firestopping blocks are cut and installed between the studs at the floor levels. One common use of balloon framing is a building that features a story and a half, such as a barn with a loft.

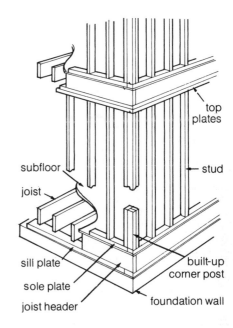

6-2. Platform framing is the most common stud framing style. It consists of a platform upon which the walls are fastened. The upper story also rests on a platform.

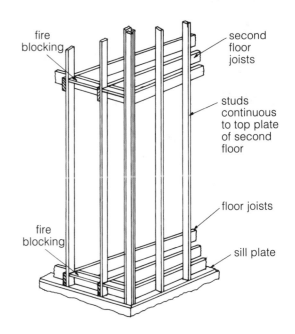

6-3. An alternative two-story construction is the balloon style. Studs run continuous from the bottom floor to the top plates of the second story. The floors are supported on ribbons or supports fastened to the studs. Often these supports are cut into the studs. Fire blocking must be placed on the walls between the studs.

POST-AND-BEAM FRAMING

6-4. Post-and-beam structures are distinguished by heavy beams, typically 6x6s or 8x8s, that provide strength and durability.

While post-and-beam framing requires more materials and skill, it has its advantages. Post-and-beam structures are extremely strong and durable. If properly joined, framing members can support tremendous loads. Also, they are more fire-resistant since it takes a long time to burn through a solid beam. Another advantage to post and beam framing is its flexibility. Large windows and doors can be framed easily into walls without worrying about headers and other supports because the entire load is carried by the individual posts.

Finally, post-and-beam construction lends itself to traditional barn raising. After the timbers have been cut to a length and notched to fit together, the entire structure can be assembled in one day by a gathering of friends and neighbors. The sill beams, usually 8x8s, are the first members to be put in place in post-and-beam framing. Girts are notched into the sills and serve the same purpose that girders do in platform framing. Joists between the girts support the flooring. Wall girts joining the posts at the second floor level are braced diagonally to make the structure rigid. The rafter plate supports 4x6 or larger rafters and 4x4 purlins set between the rafters support the roof decking. As in platform framing, collar ties keep the rafters from sagging in the middle or spreading out at the bottom. For more information on this type of framing, see the discussion of "Timber Framing" on p. 105.

POLE FRAMING

Pole framing is much like post-and-beam construction, using poles or posts set on concrete piers or footings. The beams, however, are replaced by nailed-on or bolted girts and girders of dimensional lumber. Floors and roofs are usually framed with dimensional lumber just as in platform framing, although a popular alternative is to use trusses either purchased or constructed on the spot. The latter are quite common on larger buildings, as they offer more free-span space without interior dividers. For more information on pole framing, refer to p. 105.

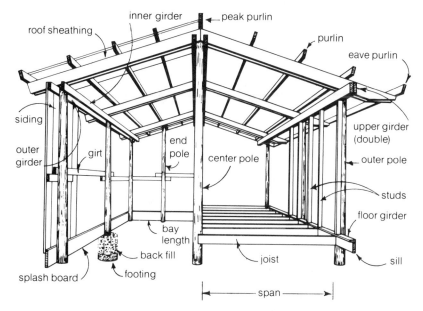

6-5. Pole buildings may be framed two ways. The left side shows a widely-used method for barns and outbuildings. The right side is for more finished buildings with suspended floors.

Platform Framing

Since platform framing is a particularly suitable building method for the owner-builder, it is discussed here in detail. Many of the construction details of platform framing may be used in conjunction with post-and-beam and pole framing, which are covered in general terms at the end of the chapter.

Sill Plates and Beams

After the foundation is in and the concrete has cured for several days, the building frame is started by setting the sill plate or beams. The sills are the building's connection to the concrete foundation.

Use pressure-treated 2x8 stock for the sill plate on top of full wall foundations. The 2x8 should be flush with the outside of the concrete wall and secured every 4 feet with anchor bolts. To set the sill plates accurately, first cut enough 2x8s to cover the top of the foundation and then hold them in place alongside the anchor bolts. The placement of the anchor bolt holes can then be marked accurately using a framing square to find the distance of the bolts from the outside edge of the wall. The

6-6. Typical platform framing for barn or small outbuilding with a "hay loft."

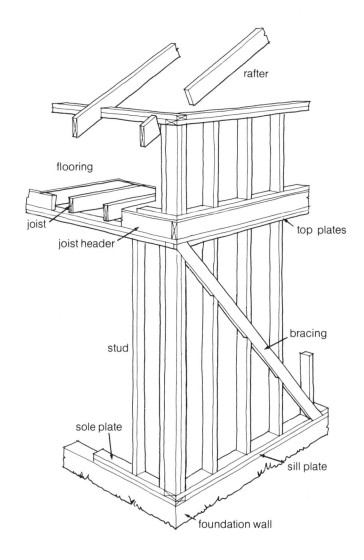

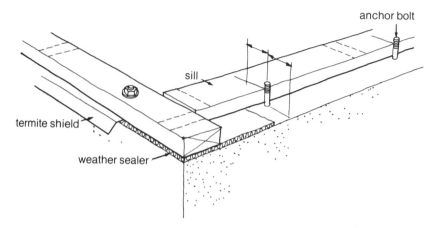

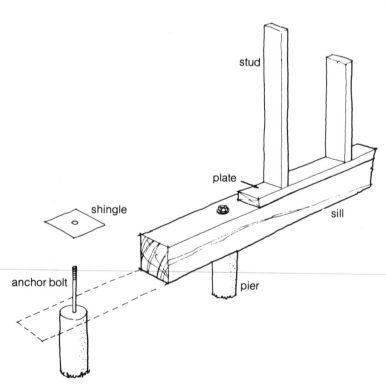

6-7. Use a framing square to accurately mark anchor bolt holes on sill plate. Dotted lines indicate stud positions.

holes should be the same distance from the outside edge of the plate. Drill the holes the same size or slightly larger than the anchor bolts and secure the plates with washers and nuts.

For a barn built on a pier foundation, either a built-up 8x8 beam or a solid 8x8 is used for the sill to carry the building load between piers. A built-up beam is easier to work with and, if properly constructed, is almost as strong as a solid beam. To assemble a built-up 8x8, first attach a 2x8 plate to the top of the piers and secure it with the anchor bolts. Nail together five layers of 2x8s, the length of the building. Make sure, in succeeding layers, to overlap the joints or places where the 2x8s butt together end to end. This way the beam will be stronger, because one layer's joint will not fall directly above another's. Nail each layer on with three 10d nails every 12 inches.

The built-up 8x8 can then be tipped on edge and toenailed into the bottom plate with 8d nails to hold it securely on top of the piers. Where the anchor bolts stick up, the built-up beam must be notched to fit over them. Simply drill holes with an electric drill to form pockets for the bolts.

It's a good idea to separate wooden sills from the concrete foundation walls or piers with a moisture-proof material. Concrete absorbs water and will pick up moisture that will eventually rot the sills. Special foam insulating material is made for residential construction. It separates the plate from the concrete and seals any air gaps as well. For outbuilding con-

6-8. Asphalt shingles help prevent moisture from a concrete pier from seeping into a heavy-beam sill.

struction, I use old asphalt shingles or strips of tar paper; these provide an excellent and rugged moisture barrier when placed under the sills.

Flooring

Flooring Joists

If the barn is to have a wooden floor, the next step after setting the sill plates is to lay out and attach the floor joists. (If you plan to have a dirt, gravel, or concrete floor, you may overlook the following sections on flooring and proceed to "Wall Framing.") The size and spacing of the floor joists

NAILS, NAILS, NAILS

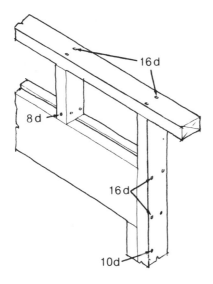

6-9. Select these nails for framing at points shown.

Choosing the proper-sized nail in wood framing is very important. If the nail is too small, it will not hold properly; if it is too large, it may split the wood. Blunting a nail be setting it head down on a firm surface and tapping the point with hammer will minimize the chances of splitting.

For exterior nailing or working with wet, green lumber, always use galvanized nails coated with zinc to prevent rusting.

Galvanized nails also have superior holding power, so use them wherever there is a chance regular nails might pull out. For pole building, there are special grooved nails that have even better holding power.

TOENAILING

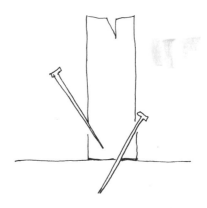

6-10. Proper nail angles for toenailing; use four 8d nails.

Toenailing is one of the hardest things for the beginning carpenter to learn. This is because toenailing involves holding a piece of wood in place while starting a nail and driving it at an angle. To toenail a stud, first start an 8d nail about 1½ inches from the bottom at approximately a 60-degree angle. Position the stud on the plate and gently but firmly drive the nail into the stud until the nail contacts the plate. If the stud moves, reposition it and then sink the nail fully into the stud and plate. Once the first toenail is in place, the others are much easier to nail since the stud is secured.

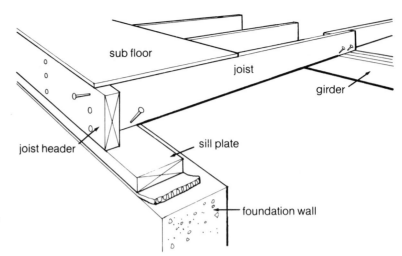

6-11. Attach floor joists so their tops are the same height as the joist header. Use a girder to support wide floor spans.

depend on the amount of weight the floor will carry and the span of the individual joists. Table 6-1 lists the recommended joist sizes and spacing for various spans and loads. Normally, floors are built to carry a "live load" of 40 pounds per square foot. For heavy equipment, this figure should be increased, or you should consider a concrete

TABLE 6-1

Span of Joists

Span calculations provide for carrying the live loads shown and the additional weight of the joists and double flooring

Size	Spacing	20 lbs. Live Load	30 lbs. Live Load	40 lbs. Live Load	50 lbs. Live Load	60 lbs. Live Load
2" x 4"	12"	8'- 8"				
	16"	7'-11"				
	24"	6'-11"				
2" x 6"	12"	13'- 3"	14'-10"	13'- 2"	12'- 0"	11'- 1"
	16"	12'- 1"	12'-11"	11'- 6"	10'- 5"	9'- 8"
	24"	10'-8"	10'- 8"	9'- 6"	8'- 7"	7'-10"
2" x 8"	12"	17'- 6"	19'- 7"	17'- 5"	15'-10"	14'- 8"
	16"	16'- 0"	17'- 1"	15'- 3"	13'-10"	12'- 9"
	24"	14'- 2"	14'- 2"	12'- 6"	11'- 4"	10'- 6"
2" x 10"	12"	21'-11"	24'- 6"	21'-10"	19'-11"	18'- 5"
	16"	20'- 2"	21'- 6"	19'- 2"	17'- 5"	16'- 1"
	24"	17'-10"	17'-10"	15'-10"	14'- 4"	13'- 3"
2" x 12"	12"	26'- 3"	29'- 4"	26'- 3"	24'- 0"	22'- 2"
	16"	24'- 3"	25'-10"	23'- 0"	21'- 0"	19'- 5"
	24"	21'- 6"	21'- 5"	19'- 1"	17'- 4"	16'- 0"

This shows maximum safe spans for high-quality wood joists. Values vary for different species of lumber; always check local buidling codes for exact data.

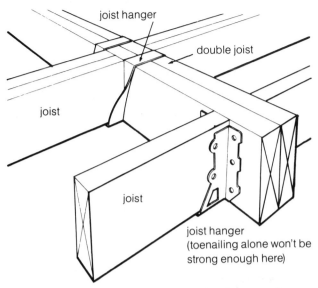

joist hanger

double joist

joist

joist

joist hanger
(toenailing alone won't be
strong enough here)

6-12. In framing situations like this, or when attaching joists to a beam, use joist hangers.

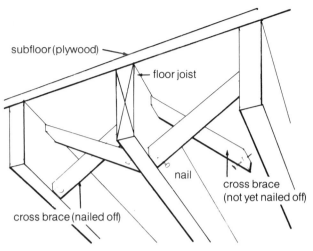

subfloor (plywood)

floor joist

nail

cross brace
(not yet nailed off)

cross brace (nailed off)

6-13. Cross bracing or blocking is recommended for long joist spans. Braces may be cut from 1x3 strapping. Blocks may be cut from scrap lumber.

floor. For more information and assistance, consult your nearest Extension Service office.

Using a tape measure and a combination square, mark the position of the joists on the sill plate. The joists will be either 16 or 24 inches on center (o.c.) from the end of the wall. This is a modular spacing that allows you to use 4x8 sheets of plywood for the subflooring and have a nailer every 4 feet from the edge of the sheets. If you are careful in your layout, the joints of the plywood will land exactly halfway on a joist, giving you a ¾-inch nailing surface for each end. If you are sloppy, the plywood will run off the joists and you will have to cut each piece individually, thereby wasting a lot of plywood and time.

After making the sill plates, cut the joist headers to length, mark the joist positions on them with a framing square, and toenail them in place. Then cut the joists to fit between the headers. Nail the joists with three 16d nails driven through the header and several 8d nails toenailed into the sill. Use the marks you put on the joist header to ensure plumbness of the joists. Always make sure the crown or bow in a joist faces up so the weight of the floor straightens it rather than causing it to sag.

If the building is too wide for the joists to span the entire width, then put in a girder, a solid or a built-up beam that supports the joists at their midpoint. Girders can be installed on top of the foundation wall. In this case, the joists must be the same width as the girder and fastened to it with metal joist hangers. The girder should be supported by concrete footings or piers placed inside the building.

If you are using a pier foundation with large sill beams, attach the joists with metal joist hangers just as they would be to a girder. Simply lay out the joist marks on the beam, set the hangers and slip the joists into them. When you are nailing the hangers in place, make sure they will hold the joists flush with the top of the beam. Use a small sample block of joist material to simplify accurate placement of the joist hangers.

For joist spans of 12 feet or more, it is often recommended that you brace the joists with blocking or wooden cross braces, which keep the joists from twisting. Metal cross bracing is also available that goes on very quickly. Braces or blocks should be set in the middle of the joist span, taking care to preserve the 16- or 24-inch spacing of the joists. Also, if you anticipate having a partition that will run parallel to, but not directly over a joist, now's the time to add blocking that will later be nailed into when the partition is secured to the floor.

Subflooring

After nailing the joists in place, lay down the subfloor. Subflooring can be either plywood, 1-inch boards, or 2-inch planks. Plywood is most often used because it is stronger, lighter, and goes on faster than boards or planks. Use ⅝-inch plywood for floors that will be covered with another layer of wood for finished flooring. Use ¾- to 1-inch plywood if it is to be the finished flooring.

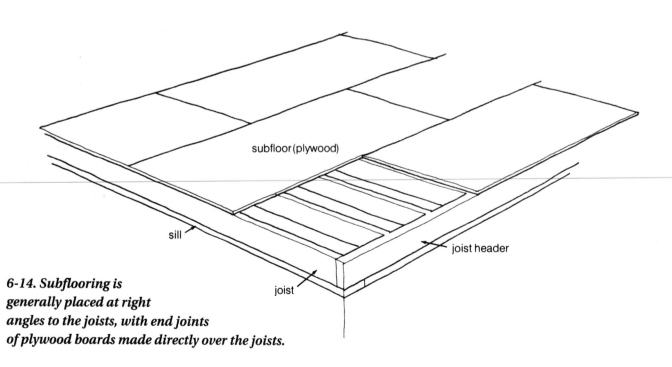

subfloor (plywood)

sill

joist header

joist

6-14. Subflooring is generally placed at right angles to the joists, with end joints of plywood boards made directly over the joists.

You can also use rough-sawn lumber for subflooring, a much less expensive choice than plywood. Rough-sawn, 1-inch boards laid across the joists for subflooring and then 1-inch boards laid perpendicular to these for the finished flooring make a durable and economical barn floor. For floors that must support heavy loads, use 2x6 planks for the subflooring.

Wall Framing

With the floor deck completed, the walls can be framed and raised in place. If the outbuilding has only a dirt or gravel floor, or a poured concrete slab, the walls are built directly on top of the bottom plate which is then called a sole plate.

Walls consist of bottom sole plates, studs, and doubled top plates. Rough openings for windows and doors are framed with headers and jack studs or trimmer studs. Walls can be built with 2x4 studs spaced 16 inches or 24 inches on center (o.c.). Or 2x6 studs can be spaced 24 inches apart. I prefer 2x6 studs for some larger barns because there is less labor involved in cutting and nailing, the walls can be better insulated, and the extra cost of the lumber is negligible.

The first step is to find two very straight pieces of lumber for the top and bottom wall plates. If the plates are crooked, your wall will be, too. Cut the plates to the length of the wall, or, if the wall is over 16 feet long, cut two or more sets of plates and build the wall in sections. A 16-foot wall, 8 feet high, is about the biggest section two men can lift in place.

Put the plates side by side and, starting from one end, mark the plates for studs either 16 or 24 inches on center depending on your stud size. One of these plates will be the bottom and the other the lower top plate. Because they are symmetrical, they can be laid out together side by side. With a combination square, go back and make lines across both plates ¾ inch on either side of the center marks. These mark the outside edges of the studs. (A framing square has one small tongue just the thickness of a standard stud, and can be used to make these as well.)

Rough openings for windows and doors must also be laid out on the plates. These are called rough openings

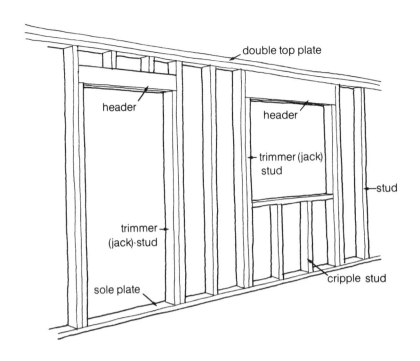

6-15. Wall framing includes special studs and headers for windows and doors.

6-16. To mark locations of studs on bottom and top plates lay plates side-by-side on foundation or slab. Measure and use small tongue on framing square (1½" wide) for making location of studs.

87

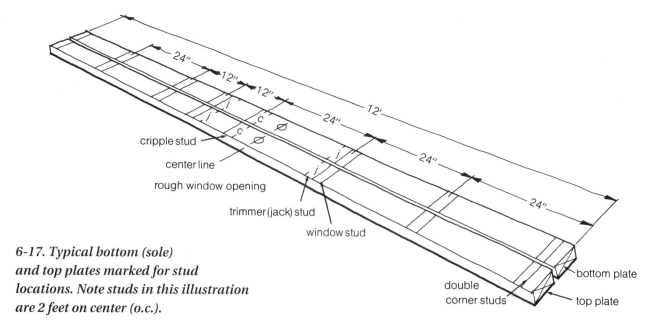

6-17. *Typical bottom (sole) and top plates marked for stud locations. Note studs in this illustration are 2 feet on center (o.c.).*

because they are wider than the actual window or door. This extra space leaves room for jambs, the pieces of wood that encase a window or door. Usually rough openings are 1¼ inches larger than the window or door on all four sides to allow for a ¾-inch-wide jamb and ½ inch for shims. The shims make the jambs perfectly level and plumb. For preassembled windows and doors that come with jambs attached, leave only ½ inch on all sides for shimming. Because of the need for accurate rough openings, always select windows and doors and calculate their rough openings before wall framing begins.

USING NATIVE LUMBER

If you decide to use low-cost native, rough-sawn lumber, the first problem you'll encounter is uneven dimensions. One 2x6 may be 5¾ inches wide and the next 6¼ inches. If you use this lumber without taking these differences into account, you are likely to end up with uneven walls and floors.

For wall framing, the solution is to line up all the studs flush with the outside of the wall plates. This will give you an even exterior wall surface for attaching siding. The inside line of the wall may be a bit wavy, but for barns this is usually not a problem since the studs will be left uncovered. If the interior wall surface is to be finished, it can be evened by attaching horizontal strapping that is shimmed out with shingles.

Laying floor joists is a little more complicated. Always select pieces of oversized stock for the joist headers that are as wide or wider than the rest of the joists. By doing this, the joists will always be the same size or narrower than the headers and won't stick up above them and interfere with the flooring. Simply nail all the joists in place flush with the top of the header. If they are too narrow to sit on the sill plate, slip a shim shingle underneath them until they are supported firmly. This will give you a smooth and level floor deck.

LAYING OUT WINDOWS AND DOORS

The position of windows and doors should be marked clearly on a building's blueprints or working drawings. Normally, the centerline of the unit and its rough opening are indicated on the drawings. If a rough opening is not indicated, you must get this information from the manufacturer or building supplier. If you are custom-building the windows and doors yourself, prepare a detailed drawing showing the construction of the unit, its overall dimensions, and how it is to be mounted so you can determine its rough opening.

Rough openings for site-built doors and windows are normally 2½ to 3 inches wider and higher than the unit itself. This leaves room for installing and leveling ¾-inch jambs that form the box the window or door sits in. For large doors that require greater support, 1¼-inch jambs are often used, and this must be taken into account when framing the rough opening. The thickness of the thresholds (if used) must also be considered when figuring the rough-opening height for doors. In general, leave ½ inch more on all sides than the width and height of the unit with jambs attached. This will give you room to adjust the unit with shim shingles until it is level.

To lay out rough openings, locate the center of the window or door on the wall plates that are being marked off. Take half the width of the rough opening and measure back on each side of the centerline and make a mark. This marks the inside of your jack studs. Mark another line 1½ inches outside of this and another 1½ inches from that. These lines mark the placement of your jack and window or door studs. The area between the lines for the jack studs should be marked with a J to remind you that this is not a full stud. Any of the regularly spaced stud marks (either 16 or 24 inches o.c.) that fall within the rough opening of these windows should be marked with a C to remind you that these are cripple studs that only extend up to the windowsill. It is important that these cripple studs be placed at the regular stud interval to serve as nailers for the wall sheathing.

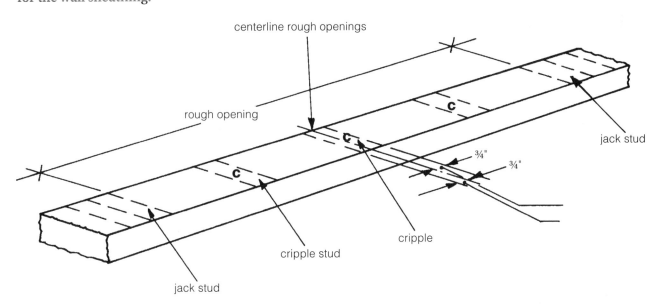

6-18. Typical locations of jack and cripple studs on bottom plate. Cripple studs are consistent with layout of wall studs, whether 16 or 24 inches o.c.

6-19. *Walls are constructed on the flat ground outside or inside building or on slab if building is supported on a concrete slab.*

6-20. *Standing on boards helps hold them in place while driving nails.*

6-21. *Headers are constructed of doubled 2x's with a ½ inch plywood spacer between, then fastened in place between the studs.*

Stud Assembly

After laying out the plates, build the walls on the platform or ground and then lift them into position. Cut the studs to the correct length and then secure them with two 16d nails through the top and bottom plates. The rough openings should be framed with: headers, which carry the load over windows and doors; jack studs, which support the header; and cripple studs, which support the rough sill. Headers can be made from two pieces of 2x6 or 2x8 lumber, depending on the load to be carried. They are nailed together with a spacer between. One piece is nailed flush with the inside of the wall and the other with the outside of the wall. Nail the headers well into the top plate and the two side studs with 16d nails. Then support the headers with jack studs.

When the framing is complete, brace the wall before lifting it into place. All stud walls must be braced to keep the studs at right angles to the plate and to make the wall a rigid structure. If the walls are going to have plywood sheathing or plywood or hardboard siding, this can sometimes be installed on small walls before erecting the walls and will act as a brace. If the walls won't be covered with plywood or hardboard siding, diagonal 1x4 bracing can be nailed across the studs.

First, square up the wall section using a tape to measure the diagonals. Use a sledgehammer to tap the corners of the wall until the diagonals are equal. Be sure the top and bottom plates are perfectly straight. If you use plywood sheathing, attach this to the outside of the wall to brace it. Remember to leave enough overhang on the bottom of the wall so the plywood will hang down and cover the top of the foundation by at least an inch. The upper end should also extend 1½ inches to cover the upper plates that will be installed later. Except for very

90

6-22. Small walls, such as this 14 footer can be raised by one person if prop boards are first nailed in place as shown. As wall is lifted up, boards slide along and help brace the wall.

6-23. Wall is squared, plumbed, then braced in place.

6-24. Long walls are built in sections, splitting the top and bottom plates over the center of a stud.

small buildings, I've found it easier to erect the walls without the siding.

If you are using diagonal 1x4 braces, nail these directly to the inside of the wall so they won't interfere with the exterior siding. Or, if both sides of the wall are to be sheathed, cut notches in the studs for the 1x4s so they are recessed. Purchased metal bracing strips can also be installed on the outside of the walls. These will hold the walls square until the siding and top plates are installed.

Erecting Walls

Once braced, the wall is raised into position. Usually this is a two-or three-person job, since the wall must be held plumb while temporary braces are set to hold it upright. Line up the bottom wall plate with the outside edge of the building and secure it with 16d nails. Drive these through the plate and into the subflooring and joists or sills below. While one person holds the wall plumb using a carpenter's level, another can attach 2x4 braces to the end studs of the wall. Arrange the braces so they run out at a diagonal to stakes in the ground. This temporary bracing holds each wall section plumb until all the sections are built and tied together with the top doubled plate.

Long walls are often built in sections by cutting the top and bottom plates to fall on the center of a stud, and then

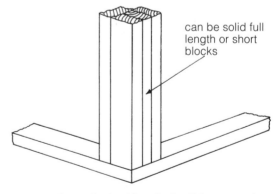

can be solid full length or short blocks

6-25. Where the inside of a building is to be finished, blocking is used to provide a nailing surface at the inside corners. Blocking may be solid or the center blocking may be short lengths added in place.

6-26. Once all walls have been erected (including partition walls that join the exterior walls,) the upper top plate is nailed in place. Make sure the walls are straight before nailing down.

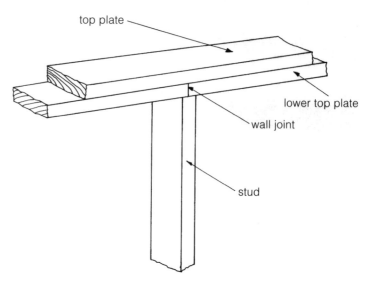

top plate

lower top plate

wall joint

stud

6-27. Upper top plate must overlap joints where two wall sections meet to provide strength.

extending the wall by placing the next section butted against the first with its top and bottom plates nailed to the same stud.

When installing a wall on a concrete pad, or in some cases a foundation, the bottom plate also acts as a sill plate. In this case, the bottom plate must first be bored to accept the anchor bolts. The wall or wall section is then lifted up over the bolts and placed in position. The wall is thoroughly braced with boards running out to stakes, then the anchor nuts and washers are used to secure the wall in place with the bottom plate.

Once the first wall is in position, build, raise, and brace the other wall sections in the same manner. When laying out the side walls, be sure to take into account the width of the front and back walls that the side walls butt against. Take into account the width of the front wall when marking off the side wall plates so that the second stud from the end is 16 or 24 inches o.c. from the *outside* of the building, not the end of the plate. This will keep the sidewall studs on center for plywood sheathing. In conventional residential framing, two studs and blocking are often used for corners. One advantage of this design is to provide needed interior nailing surfaces for drywall. However, this is usually unnecessary for small barns and outbuildings that will not have an interior wall covering.

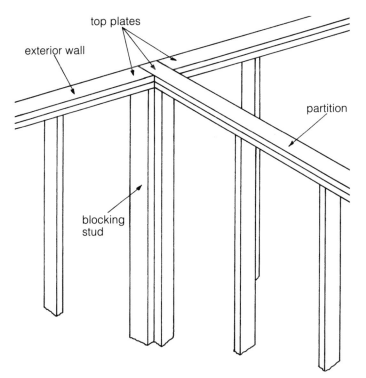

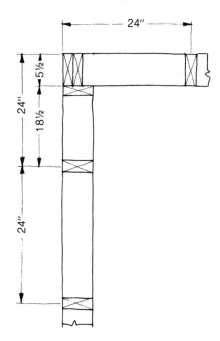

6-29. Sidewall layout for 2x6 studs (1½x5½ inches actual dimensions), spaced 24 inches on center (o.c.).

6-28. Where a partition meets the exterior wall, the upper top plate extends over the partition and across the bottom top plate of the exterior wall.

When all the wall sections have been built and braced, attach the top double wall plate. This top plate should overlap any joints in the lower plate and overlap the corner joints. This will tie the wall sections together securely and give a rigid box structure that can carry the second-floor or roof rafters.

Partitions

Normally, interior wall partitions are made of 2x4 stock and built in the same manner as outside walls. The plates are laid out, the studs nailed in place on the ground, and then the wall raised into position. If the wall runs at right angles to the floor joists, simply nail the plate into these through the subflooring. If the wall runs parallel to the joists, but not directly on top of it, it is necessary to block between the joists under the wall to support the plate. This of course, must be done before the subflooring is put down. Where a partition joins an exterior wall the upper top plate runs across the partition and the lower top plate of the exterior wall, locking them together. If the inside is to be finished off, blocking studs are positioned in the exterior wall on either side of the edges of the partition wall to provide nailing surfaces.

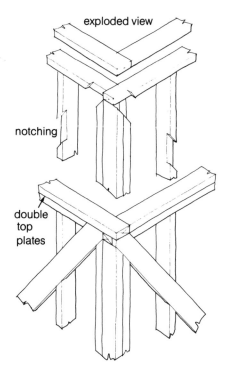

6-30. Diagonal bracing may be notched into studs to add strength. This is recommended for walls without plywood siding. Purchased metal bracing strips may also be nailed in place over the exterior of the walls, then siding applied.

6-31. Temporary bracing may also be used on small buildings until siding is applied.

Partitions for pole barns or barns built without a floor can be erected in at least two ways. Either poles can be driven into the ground to support a fence-like wall or concrete piers can be poured inside the barn to support load-bearing partitions.

With the first-story walls up, the second-floor joists and subflooring can be built if there is to be a hayloft or storage area above the first floor. If the barn has only one story, then the roof rafters are framed on top of the doubled plate.

THE RIGHT ROOF

The choice of roof style depends on the nearby architecture, the local climate, and the building's function. The most common type is the gable roof, which is widely used because of its simplicity and adaptability to any climate. A *gable, or pitched roof* has two equal pitches that meet at a center ridge. The *shed, or lean-to roof* has a single continuous slope from the front of a building to the back. It is most often used for animal stalls, woodsheds, or other narrow buildings that require high front and low back walls. The *salt box* is basically a gable with two unequal roofs or pitches. The *gambrel* utilizes two separate pitches for both sides of the roof. Salt box and gambrel roofs are two designs common to barns and outbuildings. The gambrel provides more useable space than a shed or gable roof. The salt box design is popular in New England because the long, sloping roof provides a lot of protection from winter storms. The *hip roof* has four sides, all sloping toward a center line of the building. Corner rafters run diagonally up to meet the centerline. Roofs may cover more than a simple rectangle or box, and where the two building sections meet the roof forms a valley. This may occur with either a gable roof or hip roof building.

The most important factor in roof design is pitch. The *pitch* of the roof is the ratio of its vertical rise to its horizontal width or "run." The pitch determines how much room is available for storage and how well the roof sheds rain and snow. Thus, if a gable roof rises 8 feet vertically from the top of the wall plate to its peak and is 12 feet wide from the outside wall to its centerline it is called an 8 in 12 roof. For northern climates, an 8 in 12 pitch is a good angle to keep excessive snow loads off the roof. For southern climates, lower pitches are acceptable. The minimum pitch

6-32. Four common roof styles are the gambrel, gable, shed and saltbox.

94

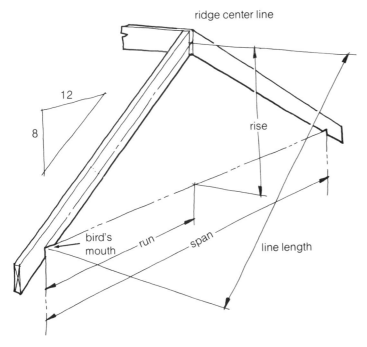

ridge center line

12

8

rise

bird's mouth run span line length

6-33. The roof pitch is expressed as a ratio of the vertical rise to the horizontal run; in this case, the ratio is 8:12. The line length is the distance from the center of the ridge to the outside edge of the bird's mouth cut, in this case, also a plumb cut.

that can be covered with asphalt shingles or corrugated metal roofing is 3 in 12. At lower pitches than this, continuous roll roofing or built-up tar roofing must be used to keep it waterproof.

Gable Roof Framing

To frame a gable roof, you must first determine the length of the rafters. This is often done by a method called "stepping off" or using a rafter square to mark the units of rise and run on a rafter. With this method, you can quickly and accurately determine the length of the rafter and the position of the rafter cuts. There are three special cuts: the *plumb cut* at the top of the rafter that rests against the ridge plate; the *bird's mouth* that allows the rafter to sit on top of the double top plate; and the *tail cut* that forms the outside edge of the building's eaves. The tail cut can either be a plumb cut or a square cut depending on how you want the eaves to look. The plumb cut is used most frequently and is necessary for buildings with eaves.

Before cutting the rafters, decide how large the building's eaves or roof overhangs are going to be. The primary purpose of eaves is to protect the siding of the building and the foundation from rain coming off the roof. For this purpose, they should extend at least 1 foot from the building. Eaves can also screen out the summer sun, but let the winter sun at its lower angle shine through windows. To achieve this, build the eaves larger than normal; extend them 2 or 3 feet out from the building. The optimum distance depends on the local sun angles.

6-34. Framing square is used to layout rafters.

95

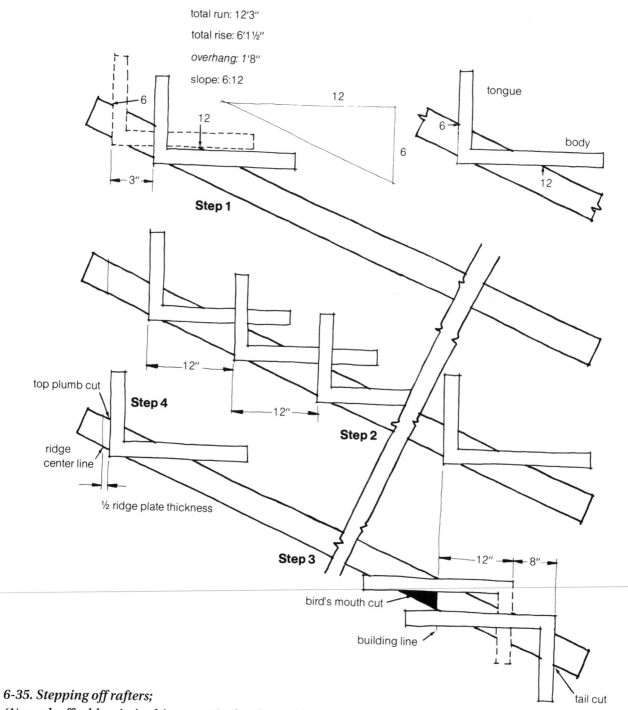

total run: 12′3″

total rise: 6′1½″

overhang: 1′8″

slope: 6:12

12

6

6

12

tongue

body

6

12

Step 1

3″

top plumb cut

Step 4

12″

ridge
center line

12″

½ ridge plate thickness

Step 2

Step 3

12″   8″

bird's mouth cut

building line

tail cut

6-35. Stepping off rafters;

(1) mark off odd unit, in this case, 3 inches from ridge center line;

(2) step off 12-inch increments until building line is reached;

(3) reverse square to mark bird's-mouth cut, overhang, and tail cut; and

(4) return to top of rafter to make top plumb cut.

After you determine the roof pitch and overhang, cut a pattern rafter to use as a template for cutting the others. Test this pattern rafter by holding it against a temporary ridge plate held at the proper height, to see if the length is right and that all the cuts fit. (The temporary ridge plate need only be a short piece of 2x8 material held by hand at the center of the wall.) If the rafter does not quite fit squarely against the ridge plate or the wall plate, adjust the cuts until it does by a process of trial and error. When you are certain the pattern is correct, cut four rafters from it.

Also cut a section of ridge plate from the same width stock as the rafters. The ridge plate must extend the full length of the roof line, but it is easiest to put it up in 10- or 12-foot sections.

STEPPING-OFF RAFTERS

The most accurate way to mark rafters for cutting is to use a rafter square and "step-off" the rafter. This means using the square to represent the rise and run of the roof pitch. Since the square forms a right triangle, like the imaginary rise and run of the roof, you can use the scale on the small blade, called the *tongue*, to represent the roof's rise and the scale on the large blade, called the *body*, to represent the run.

For this example, we'll mark off a rafter with a total run of 12 feet 3 inches, a rise of 6 feet 1½ inches, and an overhang of 1 foot 8 inches. Dividing the rise by the run, we see that this is a perfect 6 in 12 roof.

First select the straightest piece of rafter stock you have to use for the pattern. With the rafter up on sawhorses, locate the 12-inch mark on the body of the square and position it against the edge of the rafter at one end. Locate the number for the unit of rise, in this case 6, on the tongue of the square and position it against the edge. Draw a line along the back of the tongue. This marks the top plumb cut at the centerline of the ridge. When this is cut you will have to subtract half the thickness of the ridge plate from this line to compensate for the thickness of the plate.

To begin stepping-off the rafter, first measure off the odd unit of run, in this case 3 inches. With the square in the original position, measure off 3 inches on the body and make a mark on the rafter. Slide the square down the rafter, holding the 6 in 12 position, until the back of the tongue is on the 3-inch mark you made. Now mark a line along the back of the tongue and body. Once you have marked off the odd unit of run, you are ready to move the square down the rafter in full 12-inch increments to mark off the remaining 12 feet of run. Simply move the square down the rafter in the 6 in 12 position until the 6-inch mark on the tongue lines up with the previous 12-inch mark on the body. Step the rafter off 12 times in this manner to measure the 12 feet of run. Your last mark will indicate the entire length of the rafter from the centerline of the ridge to the outside of the wall plate.

To mark the bird's mouth cut that will sit on the wall plate, turn the square upside down and position the 12-inch mark on the body at the top of the last plumb line and the 6-inch mark on the tongue on the rafter's upper edge. The horizontal line along the bottom of the body marks the bird's-mouth cut.

Finally, two more steps can be made with the square in its upside-down position, one full 12-inch step and another 8-inch step. This will give you the 1-foot, 8-inch overhang for the eaves. Make the tail cut, then return to the ridge and make the top plumb cut. Remember to shorten the rafter by one-half the thickness of the ridge plate.

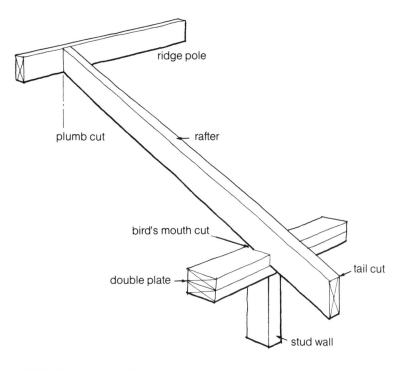

6-36. Three special cuts are made in rafters; the plumb cut; the bird's mouth; and the tail cut.

TABLE 6-2

Rafter Lengths*

Size	Spacing	Live Load p.s.i.	Groups A & B	Groups C & D
2" x 4"	16"	20	9'- 9"	7'-10"
		40	7'- 4"	5'-11"
	24"	20	8'- 0"	6'- 5"
		40	6'- 0"	4'-10"
2" x 6"	16"	20	14'-11"	12'- 0"
		40	11'- 4"	9'- 1"
	24"	20	12'- 4"	9'-11"
		40	9'- 4"	7'- 6"
2" x 8"	16"	20	19'- 8"	15'- 9"
		40	15'- 0"	12'- 0"
	24"	20	16'- 4"	13'- 1"
		40	12'- 4"	9'-11"
2" x 10"	16"	20	29'- 6"	19'- 9"
		40	18'-10"	15'- 1"
	24"	20	24'- 7"	16'- 5"
		40	15'- 7"	12'- 6"

* National Building Code
Maximum allowable lengths for rafters that are sloped greater than 3 in 12.
Length is the distance from the plate to the ridge.

Next, mark positions for the rafters, every 16 or 24 inches o.c. on the ridge plate and on the top of the wall plates starting from the outside edge of the building. Mark these with double lines, 1½ inches apart, just as you would wall studs.

Assembling the Roof

When both plates have been marked off, attach the two end rafters to the ridge plate on the ground. Use 16d nails through the plate to attach the first rafter and then toenail the second onto the plate with 10d nails. Adjust the spread of the rafters until they span the width of the building exactly and lock them in this position using a collar tie nailed across them temporarily. Then with the help of two other people, hoist the rafters on top of the wall plate and fasten the ridge plate level with a support pole.

Toenail the end rafters to the end of the wall plate using 16d nails. With the ridge plate level, attach a set of rafters to its other end. The ridge plate is then self-supporting and the rest of the rafters can be cut and nailed in place. Follow the same procedure for the next section of ridge plate until the entire roof is framed.

If you don't have time to put the roof deck on immediately, brace the rafters with 1x4s. Simply tack long strips of 1x4 or other scrap lumber diagonally across the roof rafters on both sides. If metal roofing is planned, you can install 2x4 nailers instead of temporary bracing. Install the nailers horizontally across the rafters, spaced 24 inches o.c.

As soon as the rafters are set and before any weight is put on them, attach collar ties to keep the weight of the roof from bowing the rafters in or pushing the building walls out. Usually, collar ties are cut from 1x6 or 1x8 stock and attached to at least every third pair of rafters. Don't use permanent collar ties on the gable ends.

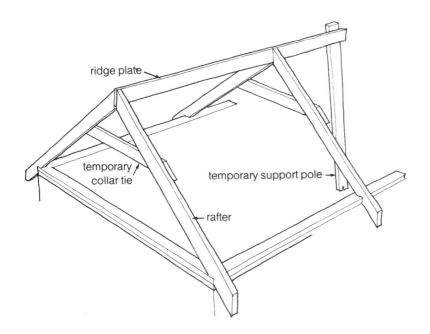

ridge plate

temporary collar tie

temporary support pole

rafter

6-37. Use a temporary support pole to hold up a ridge plate section and rafters. When all rafters are in place, remove the temporary collar tie at the gable end and replace it with gable end framing.

To finish off the gable-end of a small barn, shed, or shelter, remove the temporary collar ties and install vertical studs. The accompanying illustration shows two ways to fit the studs to the gable end; notching and cutting the stud at an angle to fit beneath the rafters. I prefer notching as it provides more nailing area and a sturdier support.

Shed Roofs

A shed or lean-to style roof is built in much the same way as a gable roof, except it has only a single pitch. Two bird's mouth cuts are made in each rafter so it fits over the front and back wall plates. Eave headers are sometimes attached to both ends to lock the rafters together and to finish off the overhangs. This type of roof is much simpler to frame and install than a double-pitched roof, but be-

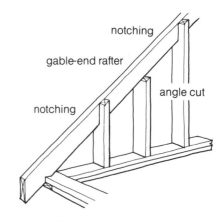

notching

gable-end rafter

notching

angle cut

6-38. Gable-end framing with vertical studs cut or notched to fit beneath rafters.

6-39. Rafters must be braced in place until all are erected and roof decking is installed.

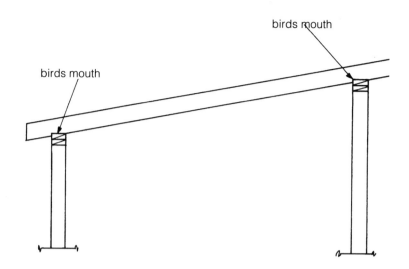

6-40. Shed roof framing can consist of rafters simply nailed in place with metal hangers, or bird's mouth cuts can be made on both ends of the rafters.

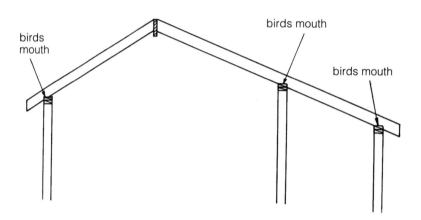

6-41. Saltbox roof consists of unequal rafter lengths. Long rafter often has support in the middle. Note the two bird's-mouth cuts on long rafter.

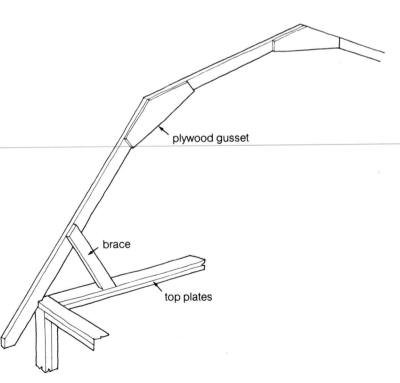

6-42. Gambrel roofs have two sloped pitches and framing members held together with braces and plywood gussets.

cause of its often low pitch and design limitations, it is normally suitable only for small sheds and outbuildings.

Salt Box Roofs

Salt box roofs are also similar to gable roofs in a structural sense. However, there are differences: the salt box has short rafters for one roof pitch and long rafters for the other. Usually, the long rafters are formed by overlapping two long boards and are often supported at their midpoints by girder or interior walls.

Gambrel Roofs

Gambrel roofs are often selected for barns because they provide more head and storage room. They have two sloped pitches; usually, the top is about 30 degrees and the bottom 60 degrees to the horizontal. Each is framed with two members held together with braces and plywood gussets. A different type of roof construction is often used to frame small gambrel buildings. In this case, the rafters meet at both an upper and two secondary "ridge" boards instead.

Trusses

When optimum utilization of space immediately beneath the roof is essential, you can use roof trusses. Trusses are constructed in several different patterns depending on the span and are quite strong. With the correct truss design, spans up to 60 feet are possible. It's a bit harder to construct a ceiling on trusses than one using rigid frame construction. Without a ceiling, however, you will often have problems with birds roosting on the trusses, particularly in open buildings.

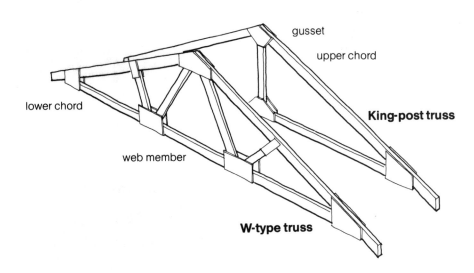

gusset

upper chord

lower chord

King-post truss

web member

W-type truss

6-43. Truss roofs allow for clear coverage of wide spans without interior support. Two common types are King-post and W. Trusses can be fastened together with glue and nailed plywood gussets or metal fastening plates.

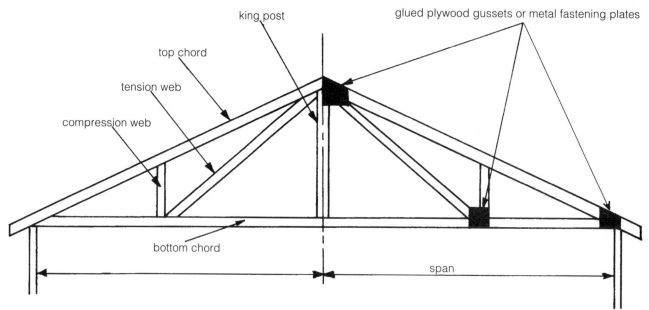

king post

glued plywood gussets or metal fastening plates

top chord

tension web

compression web

bottom chord

span

6-44. Prat web truss is another popular style. Shown are the different parts of a truss.

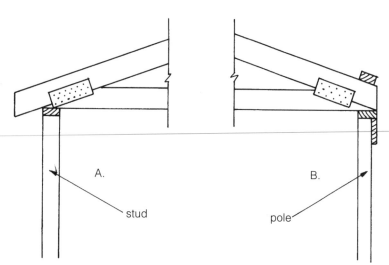

A.

stud

B.

pole

6-45. Trusses may extend past building sides to create an overhang (as in A), or end flush with building (as in B) when used with metal coverings and finishing trim.

You can order factory-built trusses to fit almost any roof pitch and design from most building supply dealers. These pre-assembled units can cut roof framing down to just a few hours. With assembled trusses and a little help on hand you can easily frame the roof of even a large building in one day. You can also build your trusses fairly easily, but you must follow an engineered design that will support anticipated roof loads for your area. Most building supply dealers can offer advice on the best type of design for your building, climate, etc. Excellent information on 10,000 specific truss designs is also available in *Designs for Glued Trusses,* published by Midwest Plan Service (see listing in the Appendix) or from the Extension Agricultural Engineers from several universities.

You will need a large flat surface for assembling the trusses. A plywood floor or area where you can lay out plywood flat and nail blocks to form a jig is your best bet, but is not always available. The first step in assembling your own truss is to lay out and cut the pieces to the correct size and shape, then assemble a pattern truss. Do this by first cutting the bottom chord or chord pieces to length. The bottom chords should have a slight camber or bow upwards at the center. This will take care of the slight settling that occurs without allowing the truss to sag. Then cut the top chords to length, using the proper angle cut at the top. Make sure the bottom chords have a scarf cut so they properly meet the top chord and sit flat on the top wall plates, allowing the angled top chord to hang below it creating an overhang as shown in the accompanying illustration.

Lay these pieces in position, then measure and make sure they are the correct size and cuts for your building width, have the correct roof slope, etc. This may take some cutting and recutting to get the first truss exactly right. Once these "pattern" pieces are cut correctly use them to cut a second set. Assemble the bottom and top chords of the truss with glued and nailed gussets or nailing plates. If using glued and nailed plywood gussets apply plenty of Resorcinal glue with a brush or paint roller and space nails properly.

Cut and add the rest of the truss components, fitting them into the first assembled truss and using these pieces as well for patterns. Once a truss is constructed, leave it lying in place and merely construct the next truss on top of it, using the bottom as a pattern to position pieces where they belong. Once you have an assembled truss and are sure it's correct, along with the pattern pieces, it's simple to mass-produce all the pieces, cutting all one size and shape, then going to the next. With all pieces cut, proceed to assemble the rest of the trusses.

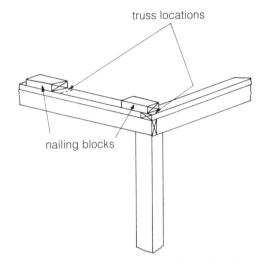

6-46. Trusses can be fastened in place in several ways. Nailing blocks shown here can also be used to "position" trusses as they are erected.

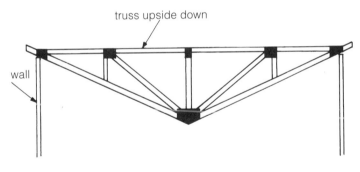

6-47A. To erect truss place ends on side walls, slide over against nailing blocks, then using a push pole to which top of truss is fastened securely with chain or heavy rope, swing truss up into position.

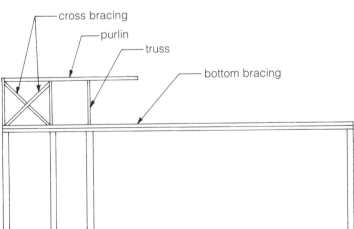

6-47B. Trusses should be braced with purlins, cross bracing and bottom braces as they are erected.

103

6-48. *On roofs with trusses spaced 48" and closer where shingles are to be used, plywood decking is often used to tie the trusses together and provide a nailing surface for the shingles.*

Erecting Trusses

The major disadvantage of trusses is their weight. Even small trusses are heavy, and a 40-foot truss is a monster. Erecting it can be quite dangerous. Smaller trusses can usually be lifted up by hand, placing first one end on the top wall plate, then lifting the other end in place. Be careful the first end doesn't slide down off the wall while lifting the second end in position. A hoist must be used to lift heavier trusses in place, or a tractor lift can sometimes be used. I like to nail "stop blocks" in position at each truss location. Temporary blocks are used on the outside walls to prevent the trusses from sliding off the ends. Permanent blocks are used on the inside trusses as nailing surfaces, although the trusses must also be anchored with additional means.

Once both ends of the truss are resting on the walls, with the top hanging down use a push pole to push and swing the truss up in position. (I use a long, sturdy 2x4 with a small chain bolted to it. The chain is wrapped around the top chord of the truss to prevent the truss and pole from separating.) With the truss upright, slide it over against the stop blocks and have a helper on each end ensure that the truss is properly positioned across the width of the building and against the stop blocks. Nail the truss to the stop blocks, then brace it in place. (Note: the first truss must be braced to the outside and inside of the building. Second and additional trusses are then braced to the first and to each other.) Make sure you put in stiffeners, purlins, and bracing as the trusses are put up to prevent the possibility of wind damage and collapse during construction and before the roof covering can be applied.

The type of roofing applied determines the next step. On buildings with trusses spaced 48 inches and closer, plywood sheathing is sometimes used, although it should be ¾- or ⅞-inch. In many instances of truss-constructed buildings, metal roofing is commonly applied over purlins. With larger buildings utilizing truss spacing over 4 feet, and where a shingle roof is desired, purlins are commonly applied at 24 inches o.c. over which sheathing is then applied. Purlins are commonly 2x4s and may be laid either flat or on edge depending on snow loads, truss spacing, etc. It's a good idea to check this with local building supply dealers or your local

6-49. *On metal roofed buildings nailing purlins are used instead. Here nailing purlins are being installed on a rafter roofed building.*

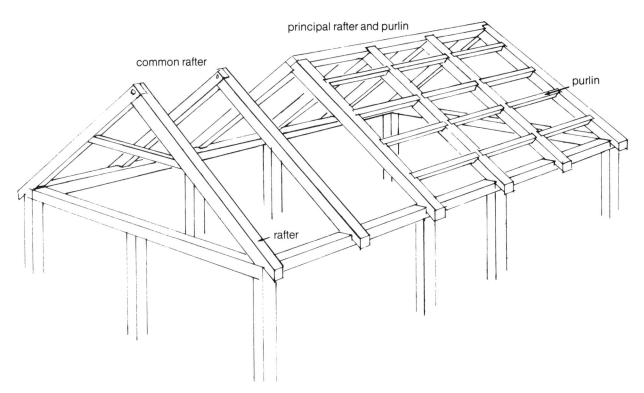

common rafter

principal rafter and purlin

purlin

rafter

Extension Agricultural Engineer. Purlins should be over-lapped at the ends. An alternative tactic is to use purlin hangers to install purlins between the trusses instead of on top of them.

6-50. In timber framing, two basic roof systems are used: the common rafter and the principal rafter and purlin.

TIMBER FRAMING

Post-and-beam framing or *timber framing* as it is commonly called, has been the traditional method for building barns for centuries. Only in the past 100 years has it been replaced by stud framing because of rising material and labor costs. In many parts of the country, however, timber framing is making a comeback due to the availability of low-cost native timbers and a renewed interest in craftsmanship. Probably the major disadvantage of this centuries-old method is acquiring the heavy timbers needed. This often necessitates cutting your own. These timbers can be hand-hewn, using old-time methods, or you can use a sawmill such as the Foley/Belsaw circular or bandsaw mills, or the Wood Miser bandsaw mill from Laskowski Enterprises. A brief introduction to timber framing follows.

The type of wood selected for timber framing varies from one part of the country to the next. Oak is the popular hardwood, valued for its strength and durability. It also has excellent working qualities for joinery. Spruce, hemlock, and pine are the three most widely used softwoods.

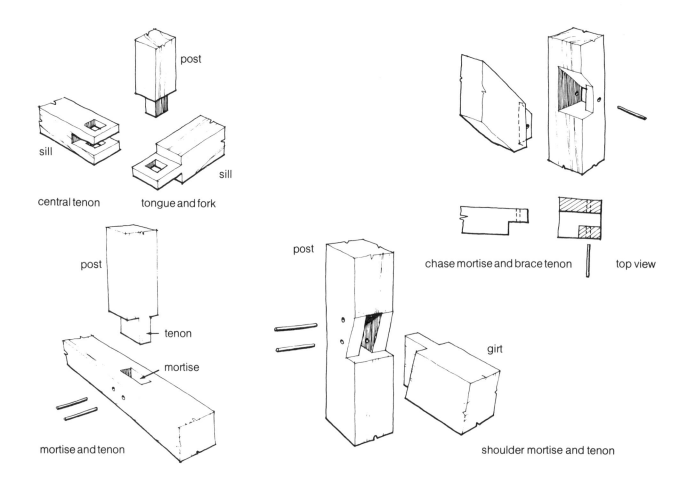

post

sill

central tenon

sill

tongue and fork

post

tenon

mortise

mortise and tenon

post

girt

shoulder mortise and tenon

chase mortise and brace tenon

top view

6-51. Four principal joints are used in post-and-beam construction.

They are easier to cut and chisel than oak, but not as strong. Often other types of wood are chosen simply because they are locally available at low cost.

The principal challenge of timber framing is the joinery of posts and beams. Only standard woodworking tools such as saws, chisels, and drills are necessary to make the joints, but to do the job properly requires patience and attention to detail.

Four principal joints are used in timber framing, though there are many variations of these for special situations. A tongue-and-fork with a central tenon is the simplest way to lock adjoining sill plates and the corner posts together. A mortise-and-tenon is the standard joint for setting posts into the sill. A shoulder mortise-and-tenon is used to fasten the girts. It is important to note that the shoulder carries the weight of the girts with the full thickness of the beam while the tenon anchors it without undue stress. A chase mortise and brace tenon is used for corner braces to keep the post and beam structure rigid.

The actual process of timber framing is much like platform framing. Once the sills are in place on top of the foundation, the walls (or *bents*, as they are traditionally called) are lifted into place as a unit. Girts are set across the first- and second-story floors, and then rafter beams are

MORTISE AND TENON JOINERY

The mortise-and-tenon joint is a standard timber-framing joint that can be made with four basic wood-working tools: a hand saw, chisel, drill, and square.

To cut the mortise for a post-and-sill-beam joint, first use a combination square to lay out where the post and its tenon will sit on the beam. Usually, the tenon runs the width of the post and is 1½ inches or more in thickness. The tenon always runs parallel with the beam so pegs can be driven through both to lock them together.

Drill out the mortise using an electric drill or hand auger with a bit that is slightly smaller than the width of the tenon. Be careful to drill squarely into the beam so that the walls of the mortise are plumb. Use a stop guide on the drill bit to accurately measure the depth of the mortise. It can be a little deeper than the tenon, but it can't be shallower.

After removing most of the waste from the mortise with a drill, use a sharp chisel to even out the sides of the housing and to scrape the bottom clean. Don't overdo it. The tenon must fit snugly into the mortise and a few too many passes with the chisel can make for a loose joint. Once the mortise is finished, drill two holes in the side of the beam through the mortise for the wooden pegs. These holes should be at least 1½ inches up from the bottom of the mortise.

Now cut the tenon. Measure the depth of the mortise, then mark a squared line around the post at that measurement. Then mark the outline of the tenon on the end of the post and running back up the sides to meet the square lines. Use a circular saw or hand saw to cut along the squared lines that run with the tenon. Cut only to the depth where the tenon begins. Make parallel cuts 1 inch apart and to the proper depth across the two faces of the beam that will be removed to expose the tenon. The waste wood can now be removed with a wide chisel using the depth of the saw cuts as a guide.

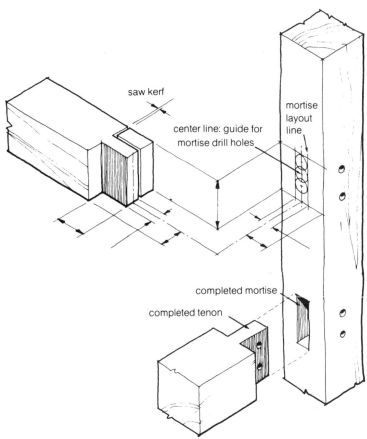

6-52. Use a square to mark beams for tenon and mortise; note corresponding dimensions shown here.

After chiseling out the tenon, test-fit it in the mortise and make small adjustments as necessary to get a good fit. With the tenon in place, mark the peg holes on the tenon with a pencil stuck through the mortise holes. For barn construction, the normal peg size is ¾ to 1 inch in diameter. For a tight fit, the holes in the tenon and mortise are slightly offset so that the peg draws the tenon into the joint. This is referred to as "drawboring." The tenon should be drilled ⅛ of an inch off center toward the top of the post so that the pegs will be forced in a shallow S curve that helps pull the tenon into the mortise.

set as pairs. The two principal differences from platform framing are that the beams and their joints are precut and fitted on the ground and then the entire frame is assembled before any subflooring or platforms are put down.

Two types of roof systems are used in timber framing, the common rafter and the principal rafter-and-purlin. The common rafter system uses 4x6 or larger rafters set 3 to 6 feet apart with the roof sheathing laid perpendicular over the rafters. The principal rafter-and-purlin system uses larger rafters, 6x6 or larger, spaced 6 to 9 feet apart with 4x4 purlins running between them. The roof sheathing is then laid parallel to the rafters over the purlins.

After the timber frame has been assembled, usually with a small army of people, the walls are studded in and the floor joists set using regular dimensional lumber. This is necessary to hold the siding and the floorboards and results in the additional materials required for timber framing. Windows and doors are framed in the same way as platform framing except that load-carrying headers are

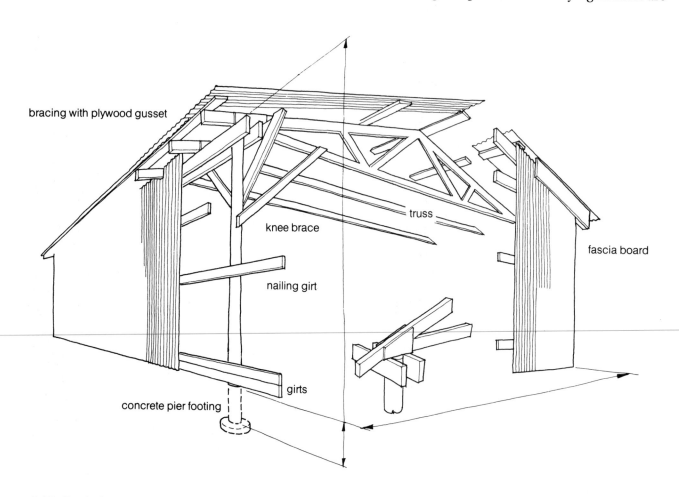

bracing with plywood gusset

truss

fascia board

knee brace

nailing girt

girts

concrete pier footing

6-53. Basic framing pieces of a pole building are shown in this cut-away view. Although this framing method is known as pole building, posts are often substituted for poles.

not required since the building load is completely carried by the wall girts. All that is necessary is a single plate above and below to frame the rough opening.

POLE FRAMING

Pole framing is also fast becoming popular across much of the country due to its simplicity and economy. Another advantage is the speed at which a building can be erected. Unlike other types of construction, pole framing is basically a do-it-as-you-like-it process. You don't need a whole work crew and can simply work on the building as you like, as long as erected poles are firmly braced in place. This type of framing is ideally suited for small barns and outbuildings, as well as larger livestock confinement and grain- or equipment-storage buildings, especially those that don't have to be completely closed off.

Poles are available in many sizes and lengths, ranging up to 60 feet or more. Common building poles are under 30 feet in length and range from 4 to 8 inches in diameter. Poles may be round or dimensionally sawn to square or rectangular shape with the latter 4x4s and 4x6s the most common sizes. If the poles are to be set in the ground, they must be pressure-treated to last more than a few years. Pressure-treated, southern yellow pine, dimensionally sawn poles are readily available at building supply centers in all parts of the country.

You can also sometimes acquire second-hand poles from utility companies and salvage operations at a lower cost. Or you can cut and treat your own, although the latter chore is hard and somewhat dangerous.

Pole Construction

The first step in pole building is to lay out the building and locate the position of the poles. It's extremely important that pole positions are correct. Then set the poles in the ground on concrete footings or on concrete piers with anchor brackets. In many instances the footings are actually *punch pads,* a pad of concrete poured in the hole and allowed to harden before the poles are set in place. Once the poles are in position, additional concrete can be poured around them to further anchor them in place. Poles can be spaced from 4 to 16 feet apart, depending on their size, the wind load the roof must carry, etc. Check with the agricultural extension engineer in your area for exact pole sizes and spacing for your particular building. Most knowledge-

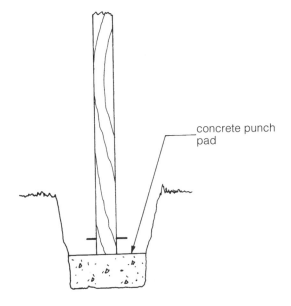

concrete punch pad

6-54. *Posts are often supported on concrete punch pads, then additional concrete poured around the posts once they are plumbed and braced.*

6-55. *These posts have been plumbed, braced, and concrete poured around them.*

109

6-56. Once all poles have been erected, nailing girts or horizontal nailing boards are fastened across them at regular intervals.

able building supply dealers can also provide this information.

Normally, the poles will extend from the foundation right up to the roof line. Large poles can either be raised by hand with several people or raised with a tractor. Once a pole is up, it must be held plumb while braces are attached to keep it in position.

When all the poles have been raised, girts are nailed onto them horizontally every 2 feet to carry the siding. These are normally 2x4s or 2x6s depending on the span between poles and the weight of the siding. Bottom girts in contact with the soil should be pressure-treated or coated with a wood preservative.

The top members that carry the rafters are usually doubled girts, made from 2x8s or 2x10s placed on the inside and outside of the poles. These are bolted together through the poles or nailed in place with 20d spikes. The height of the top members is measured from either the top or bottom of the bottom girt. A second method providing top wall support is also shown in the accompanying drawing. The height of the top members is measured from either the top or bottom of the bottom girt.

Wall bracing is absolutely necessary in pole buildings. Braces are run from the upper girt to the poles and are

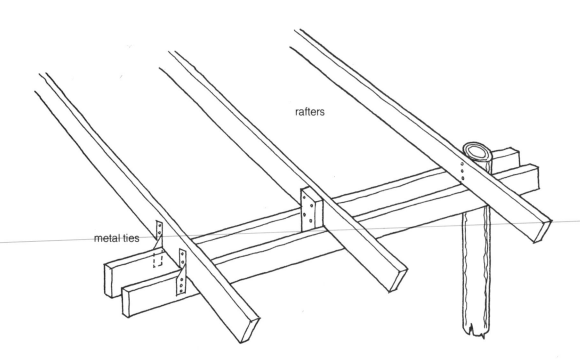

rafters

metal ties

6-57. You can attach rafters to girts with metal ties or wooden blocks. Sometimes it's convenient to simply attach a rafter to a pole.

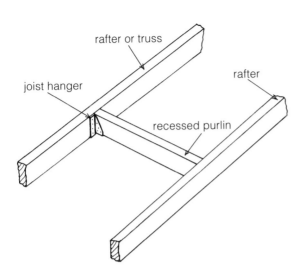

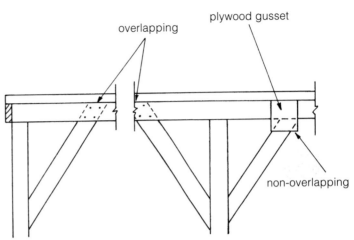

6-58. Nailing purlins are often placed over the rafters. Joist hangers can also be used to recess the purlins between the rafters.

6-59. Pole buildings must be laterally braced. Plywood gussets can be used for added strength in anchoring the non-overlapping bracing in place.

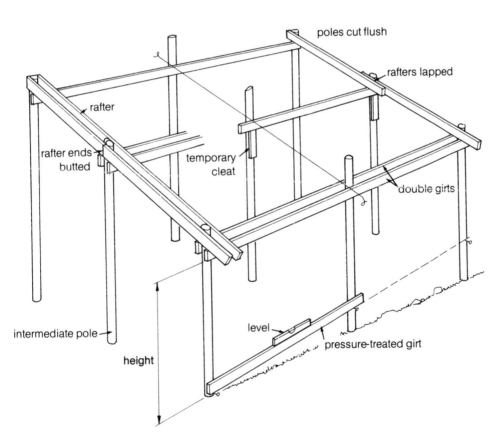

6-60. Make sure all nailing girts are level. Use the first girt as a guide for measuring the height of the building. Line up girts on intermediate poles with a string. It may prove helpful to rest girts on temporary cleats before securing with nails. Compound rafter may be butted or lapped as shown.

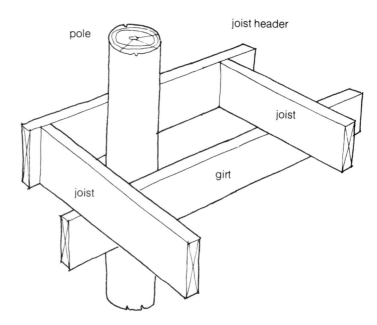

Labels in figure: pole, joist header, joist, girt, joist

6-61. To install a floor at any level of a pole barn, simply support the floor joists on the wall girts.

often reinforced with plywood gussets. These braces are the only things that give the walls any lateral strength if metal or board siding is used. Plywood siding helps brace the building, but is not a substitute for adequate diagonal bracing.

With the walls framed and braced, you can assemble roof rafters or trusses on top of the double girts. You can cut and assemble rafters in the same manner as described in the section on platform framing. Or, to simplify things, omit the bird's-mouth cuts, and, after placing the rafters on top of the double girts, secure them with wooden or metal ties. Whenever possible, secure them to adjacent posts. If you use rafters rather than trusses, be sure to attach collar ties too, so the outside walls will not be pushed out.

You can build floors in a pole barn at any level by using wall girts to support the joists. Attach joist headers to both the ends of the joists and poles.

CHAPTER 7

Siding

7-1. Any number of siding materials can be used for covering small barns, sheds and outbuildings. Plywood siding shown here is quick, easy.

Many siding materials are available for barn and outbuilding construction. When you make your choice, consider not only the cost, durability, and appearance, but also the all-around suitability of a particular material. For instance, galvanized-steel siding might be the cheapest and most weather-resistant siding you can find. For animal housing, however, it might be a poor choice because it is so easily dented.

Or, you might consider native board and batten siding as an ideal, low-cost solution for your garage. But your house and all the others in the neighborhood are covered with painted clapboards. The juxtaposition of the two different sidings would be unsightly and might detract from the value of your home. Despite the added cost, it would probably be better to use clapboards on the garage to match the style of the house and neighborhood.

Two other important items to consider before installing siding, especially if you are planning a livestock barn or outbuilding, are ventilation and insulation. Discussion of these topics follows in Chapter 9 on Windows.

Rough-sawn, native lumber is an attractive, low-cost siding material. It may be installed vertically to achieve one of four principal styles. *Board and Batten,* the traditional barn siding, consists of 6-inch or wider boards butted edge to edge and 1x 2-inch battens which cover the gaps between the boards. In a variation of this style, *Batten and Board,* the battens are placed behind the boards, not over them. Another variation, *Board on Board,* dispenses with the battens altogether; the boards are simply overlapped. Dimensional lumber may also be used for this type of siding, although the cost will be greater. Then there's *tongue-and-groove* and *shiplap* or *half-lap* siding, which are installed vertically. Tongue-and-groove has ¼-inch to ⅜-inch tongue-and-groove joints while *half-lap* has ¾-inch half-lap joints. When joined, both form a fairly watertight seal. Both are available commonly in 6- and 8-inch-wide boards.

Pine, spruce, and hemlock are three softwoods commonly used for board siding. Pine and spruce are prefer-

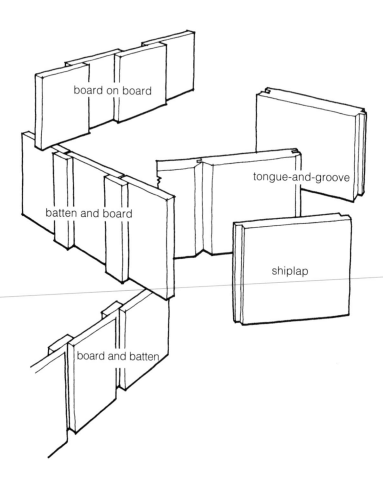

board on board

tongue-and-groove

batten and board

shiplap

board and batten

7-2. Native-lumber and dimensional sawn solid siding may be installed in several styles, including batten and board, board and batten, board on board and shiplap.

able because of their better weathering characteristics. Oak, a hardwood, is occasionally used for siding, especially in the South, where it is plentiful. If oak is used, it must be installed while green. Once it has dried, it's nearly impossible to drive a nail through it. Alaskan or white cedar, usually available with one side planed and the other rough-sawn, is another excellent siding material. Both it and redwood have the best weathering characteristics of commonly available wood. Unfortunately, they're both fairly expensive. See Table 7–1 for additional wood weathering characteristics. These boards are commonly available in 11½-inch-wide boards.

Rough-sawn boards can be milled to any width and length. Normally, mills turn out 6-, 8-, 10-, and 12-inch-wide boards; lengths are in increments of 2 feet, starting at 8 feet. You can also buy boards in random widths, sometimes at a savings over dimensional lumber. In this case don't purchase boards narrower than 8 inches, or it will require more time and effort in covering the wall. Some native materials when milled to wider widths, such as 12 inches, may also warp or crack.

TABLE 7-1			
Weathering Characteristics of Wood*			
	Resistance to Cupping 1 = Best 4 = Worst	Conspicuousness of Checking 1 = Least 2 = Most	Ease of Keeping Well Painted 1 = Easiest 4 = Hardest
SOFTWOODS			
Cedar, Alaska	1	1	1
Cedar, white	1	—	1
Redwood	1	1	1
Pine, eastern white	2	2	2
Pine, sugar	2	2	2
Hemlock	2	2	3
Spruce	2	2	3
Douglas fir	2	2	4
HARDWOODS			
Beech	4	2	4
Birch	4	2	4
Maple	4	2	4
Ash	4	2	3
Chestnut	3	2	3
Walnut	3	2	3
Elm	4	2	4
Oak, white	4	2	4

* From *Wood Handbook*, Agriculture Handbook No.72, United States Department

INSTALLING BOARD SIDING

Horizontal wall framing members must be present to provide a nailing base for installing vertical board siding. On pole construction the wall girts provide the nailing surface to carry the siding. For stud construction, nail 2x4 blocks between the wall studs or 1x3 strapping horizontally across the studs. I prefer to nail strapping across the studs every 2 feet o.c. because there is less cutting and the strapping goes faster than individual blocks. If the building is to be insulated, plywood wall sheathing should be applied over the wall framing and the siding applied over that.

Where a single board will not reach the top of the wall, such as on the gable end of a barn, another board must be installed. There are three ways of joining the boards together. The simplest is to use metal flashing. Or you can cut a 45-degree bevel joint so the top board overlaps the

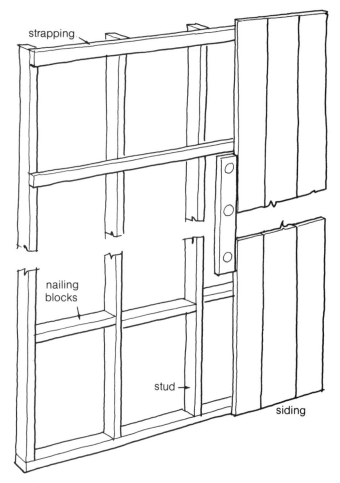

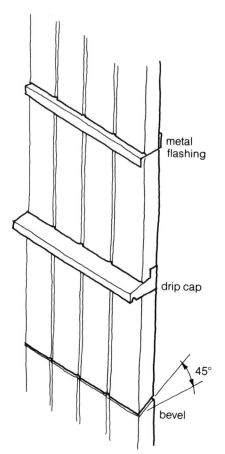

7-3. When installing vertical board siding, make sure each piece is plumb. With standard platform framing, you can nail siding to blocks or strapping. With pole barns, nail the siding to girts.

7-4. Three ways to join siding boards are with metal flashing, with a drip cap or with a 45-degree bevel joint.

bottom one. This ensures that gaps for wind and water won't develop between the ends of the two boards. Always make this joint on top of a horizontal nailer so the ends of both siding boards can be fastened securely. Another way to join boards is to use a drip cap, which overhangs the bottom boards and sheds water from the joint.

BEVEL SIDING AND SHINGLES

Two horizontally installed wood sidings are bevel siding and shingles. Bevel siding boards, usually cut from spruce or cedar, come in widths of 5 inches and larger. When overlapped in horizontal rows, they form a weathertight seal. Shingles, also made from spruce or cedar, are overlapped in the same manner, but in wider rows.

116

Both bevel siding and shingles require a nailing base of either boards or plywood sheathing. Combined with the higher cost of the siding, this makes for an expensive and labor-intensive protective covering. Bevel siding and shingles are extremely durable, however, and are still widely used in residential construction. These sidings are also often selected for outbuildings where architectural standards or the severity of the climate dictate their use.

Install both sidings from the bottom up, overlapping each course by about half the width of the pieces. For example, install bevel siding with 3 inches exposed "to weather." I recommend using 6d galvanized siding nails

ESTIMATING SIDING

Estimating the amount of siding material you need is not always easy. You must calculate the square footage of wall area to be covered, adjust for differences in nominal and actual dimensions of the siding boards, account for window and door openings, and adjust for waste and errors.

For plywood siding, the task is fairly straightforward. Simply measure the square footage of your walls by multiplying their length times height, then add up all the surfaces. Divide this figure by 32, the number of square feet in a 4x8 sheet of plywood and you'll have the number of sheets that you need. To calculate the area of the triangular gable ends, multiply the length of the end by the height of the triangle from the top wall plate to the peak and divide by 2. Don't subtract any square footage for windows or doors, as these will usually be covered with plywood and cut out later. Large garage doors or other openings should be subtracted, of course. When you arrive at a final figure for the number of sheets you need, add 10 percent to cover for waste in cutting and errors.

To calculate the number of boards you will need for vertical board siding, use the same method. First, find the wall area and divide this by the number of square feet an individual board will cover. For example, if you are using 1x8 rough-sawn siding 8 feet long, each board contains 5.3 square feet. To cover a 30x8 wall you would need about 45 boards (240 divided by 5.3). Make sure you use the actual width of the board and not its nominal size. A rough-sawn 1x8 is actually 8 inches wide, but a planed, kiln-dried 1x8 is only 7¼ inches wide. When you arrive at a final figure for the number of boards you need, always add 10 to 15 percent for waste and bad wood.

Estimating bevel siding is perhaps the most difficult task. You buy bevel siding by the nominal board foot, so you must convert this figure into square-foot coverage. First, determine the overlap on the siding. If the siding is to have 3 inches exposed to weather, then each row will only cover a 3-inch-wide section of wall. The minimum width siding you could use for this would be 6-inch nominal-width siding that would actually measure 5½ inches wide. Because each row will only cover half the width of the siding, 2 board feet of siding will be needed to cover one square foot of wall area. A formula for figuring this would be:

$$\frac{nominal\ width\ of\ siding}{inches\ to\ weather} \times \text{sq. ft. of wall} = \text{bd. ft. of siding}$$

Add ten percent to this figure for waste.

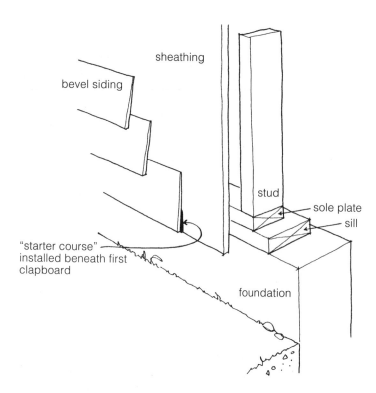

to fasten the siding to the sheathing, nailing through the siding's thick bottom edge. Align the nailing pattern vertically, so the nail penetrates into the stud behind the sheathing.

In addition to wooden bevel siding, manufactured hardboard and plywood sidings are also available in strips 12 to 16 inches wide. These go on like bevel siding, but are much less expensive and require less labor. Hardboard manufacturers recommend leaving an ⅛-inch gap between butted siding, as well as between the end of a piece of siding and window, door, or corner trim. Failure to leave this gap often results in buckled siding once it is installed and exposed to the weather. Flexible caulk is used to fill the gap.

7-5. Bevel siding and shingles, whether wood or hardboard, are installed similarly. Note that the plywood sheathing overhangs the foundation slightly to create a tight seal.

PLYWOOD SIDING

Plywood siding is popular because it is fast and easy to apply, fairly economical, and comes in several different textures and patterns. T 1–11 is a commonly used type that has a rough-sawn surface and grooves every 8 inches to imitate rough-sawn boards. Another choice, regular A/C exterior glue plywood, has a smooth surface for painting and is less expensive. (For more on plywood characteristics, see Table 2–1 in Chapter 2.)

Plywood siding goes on quickly and is very strong and light; you can apply it directly to the studs without any other sheathing, strapping, or bracing. Because of this, cost is its main advantage. Manhandling a 4x8 sheet of plywood into place is a tough chore, but it can be done by one person, if that person is fairly strong. Two people can make the job fast and easy.

The normal size of plywood is 4x8 foot panels, although it can be ordered in longer lengths. For most outbuilding construction, plywood siding should be ½ or ⅝ inch thick. More economical grades are available ⅜ inch thick and will suffice on small buildings that don't get a lot of abuse. Normally, plywood is installed vertically to eliminate horizontal seams. It should be secured with 8d galvanized or ring-shank nails spaced 12 inches apart along the studs and 6 inches apart around the perimeter of the sheets.

If you wish to decorate plain plywood siding or cover

7-6. Plywood siding is quite popular because it is fast and easy to apply.

118

7-7. Plain plywood siding can be made more decorative by nailing 1½-inch battens over the plywood.

7-8. One fast way of installing plywood is to fasten it right over window and door openings, then cut out the openings with a circular saw, chain saw, or reciprocating saw.

the vertical seams, you can nail 1½-inch battens over the plywood either every 8 inches or every 4 feet at the seams. Horizontal seams, such as on the gable end of a building where a second row of plywood must be installed to reach the peak, should be flashed with aluminum drip cap. The aluminum flashing keeps water from entering the joint as it drips down the siding.

When installing plywood sheets around windows and doors, don't bother to precut the sheets for the rough openings. Simply nail the siding over the opening, then cut out the opening with a saw. A small electric chain saw is ideal for cutting out the opening from inside the building. Use the rough opening frame as a guide. A circular saw will also work, but first mark the area to be cut on the outside of the plywood since the saw must be used from the outside. To do the marking, drive nails through at the corners from the inside, then use a chalk line to mark the outline of the opening.

HARDBOARD SIDING

Hardboard panel siding is also an extremely economical, fast and easy way of covering a building. It is available in a wide variety of textures, and in prefinished panels which eliminate the need for priming once installed. Hardboard siding is also available in 4x8-foot and longer sheets and is applied in the same manner as plywood. If using prepainted sheets, color matching nails are used.

7-9. Hardboard panel siding is also fast, easy and comes prefinished so you don't even have to paint or stain it.

119

Corrugated sheets of aluminum or galvanized steel are popular barn and shed sidings. Both metals are long-lasting and require almost no maintenance. Galvanized steel is available in a full range of colors and with different protective coatings. Metal's biggest disadvantage is that it can be dented easily. If used for large-animal housing, the insides of stalls and walls must be protected with wooden rails to keep animals from kicking or rubbing against the siding.

Two common widths for galvanized siding are 32 and 38 inches. When the pieces are overlapped, these two widths cover 30 and 36 inches respectively. A standard length is 8 feet, although the material can be ordered in any length in 1-inch increments.

Before installation, always store color-coated and galvanized metal panels in a dry place. Do *not* store outdoors, or in any indoor area subject to moisture. Stand panels on end and fanned out at the bottom to provide air circulation, or stack them flat in piles on blocks or racks to protect the bottom panels from ground moisture. Don't slide the panels when unpiling, or you may scratch the coating on the bottom panel.

Galvanized siding is fastened in place with hot-dipped, galvanized, ring-shank nails with lead or neoprene washers. Galvanized hex-head screws are also used. These can be driven with an electric drill with a hex-head bit. Aluminum siding must be fastened with aluminum ring-shank nails, not galvanized nails. Dissimilar metals will react chemically with each other, causing corrosion and oxidation.

Start installation on the edge away from the prevailing wind, from either the right or left edge. It is extremely important that the first panel be plumb to ensure straight alignment of the entire row of panels. Take plenty of time to make sure you get the first panel installed correctly.

Temporarily nail a 2x4 horizontally as a guideline about 3 inches below the top of the baseboard or foundation to set panels on and ensure they all line up evenly. Make sure

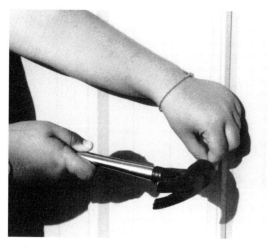

7-10. Corrugated metal siding is extremely popular and also easy to apply.

7-11. Metal siding overlaps to create a standard width. Shown here is a 36-inch cover width siding.

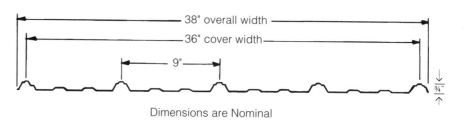

38" overall width

36" cover width

9"

¾"

Dimensions are Nominal

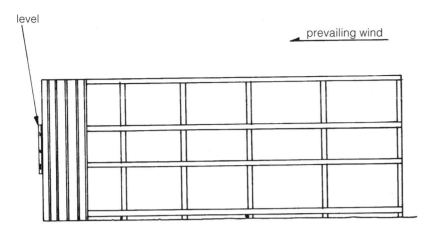

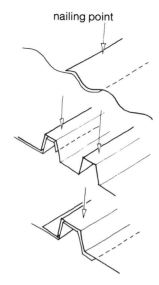

7-12. Metal siding must be started on the side away from the prevailing wind, from either the right or left edge. Make sure first panel is plumb and square with the building.

7-13. Install the next sheet over the first, overlapping over the corrugated feature.

this support is level before nailing it in place. After the siding has been installed, remove the 2x4 support.

Temporarily fasten the first panel at the starting edge. Always tack each sheet in position with just a couple of nails and check it for squareness before driving all the nails. Once driven home, the nails are very hard to remove without damaging the siding.

Install the second panel next to the first, side-lapping over the corrugated feature. Temporarily fasten all the panels in place, then permanently fasten them when they are all in line and correctly positioned. Make sure you do not temporarily or permanently fasten the underlap rib of any panel until it is side-lapped by an adjacent panel. To permanently nail the siding, space nails every 8 inches into the girts or strapping set 2 feet o.c. Prevent the panels from permanently contacting the ground.

Make sure you select nail lengths for the siding that do not go completely through purlins or girts. Do not over-drive nails, as this tends to damage washers and overcompresses the steel. On the other hand, if not driven in far enough, the siding will be loose and rattle in the wind; also water will get in under the washers. If a nail misses a nailing support, pull out the nail and fill the hole with a galvanized sheet-metal screw or sealant. Make sure the nails are driven in perpendicular to the center surface of the rib and are located in the center of each main rib.

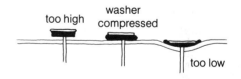

7-14. Metal siding is fastened in place with galvanized roofing nails with rubber washers or hex-head screws. Note the nails or screws must be driven in just the right amount. Follow manufacturer's directions.

Cutting Corrugated Metal

Despite the fact that it is metal, aluminum and galvanized steel siding and roofing can be cut easily on the job. All you need is a circular saw with a metal-cutting blade,

7-15. Metal siding can be cut with shears, tin snips, or with a portable electric circular saw fitted with a special metal-cutting blade.

7-16. Special trim pieces can be used to trim and finish off the building around corners, edges, windows, doors, etc.

which can be bought at any hardware or building supply dealer.

When cutting metal, **ALWAYS WEAR SAFETY GOGGLES** to protect your eyes from metal bits thrown out of the saw. Also, leather gloves are a good precaution since freshly cut edges can be ragged and sharp. Some form of hearing protection should also be used, as cutting metal panels with a circular saw is extremely noisy.

To mark for a cut along a piece of metal siding, use a chalk line. With it you can snap a colored line that is easily visible. Pencil marks are nearly invisible on the gray surface of sheet metal or darker-colored panels.

When cutting metal, hold back the blade guard on the circular saw with your left thumb so that it doesn't catch on the corrugations. Slowly, but steadily, move the saw, taking care not to force it as you cut. The blade will send sparks flying, so don't cut around flammable materials or gases. After cutting for a couple of hours, you'll notice that the blade seems smaller than when you started. If you have an abrasive blade, you're right — it *is* smaller. The blade cuts by abrasive action and grinds itself away with every revolution. I prefer a more durable, non-ferrous, metal-cutting blade.

Trim pieces are available for finishing the corners of the building and around the windows, doors, and eaves. Install these according to each manufacturer's instructions.

OIL STAINS AND PAINTS

All wood siding should be protected from the weather. Paint, a time-tested preservative, requires periodic and expensive maintenance. Every few years or so, it must be scraped, washed, and replaced with a new coat. Clear linseed oil and oil-based stains can often serve as replacements for paint, and they require far less labor to apply and maintain. Stains protect wood just as well as paint and can be reapplied without scraping and preparation.

CHAPTER 8

Roofing

8-1. Roofing also comes in a wide variety of materials and applications. Masonite Woodruf is shown being applied here.

Like siding, roofing materials come in many shapes and sizes. Galvanized-metal roofing is perhaps the most common choice for barns and outbuildings because it is durable, inexpensive, and easy to apply. Asphalt or fiberglass shingles are also used where it is important for appearance to match existing shingled roofs, or on buildings that are to be insulated. Shingles, however, require more labor and must go on over a roof deck of plywood, which adds to the expense. On roofs with extremely low pitch, asphalt roll roofing or built-up tar roofs are used to form a watertight seal.

Metal Roofing

Like metal siding, metal roofing consists of long, rectangular panels installed vertically. Nailing purlins, 2x4 strapping run across the rafters every 2 feet, are used to support the metal sheets. The first row of purlins or strapping should be even with the bottom edge of the rafters and overhang the ends of the building by the width desired for eaves. In many instances, metal-covered buildings don't have eaves, instead finishing off the roof and wall with an edge trim. If eaves are desired, they are framed in and the fascia boards attached as previously described, then the

123

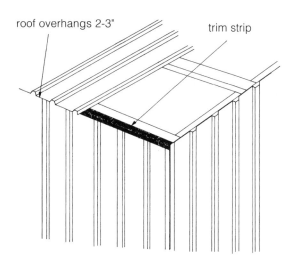

roof overhangs 2-3"   trim strip

8-2. In many instances metal covered buildings are not finished off with eaves, but have the rafters or trusses flush with the sides. Roof edge can protrude slightly over building side, or be flush and covered with metal trim.

sheets are nailed in place.

Starting at the bottom corner of the roof, at the edge away from the prevailing wind, place the first sheet and make sure it is square with the rafters and plumb. This ensures straight alignment of the entire row of panels. Stretch chalk lines between nails driven into the ridge and end of each edge rafter to serve as a guideline to align panels. If an eave is desired, make sure the bottom edge of the panel overhangs the fascia boards by at least 1½ inches on the sides and bottom. This overhang will keep water from dripping onto the fascia boards. Then complete the bottom row of roofing, overlapping the sheets according to the manufacturer's directions and keeping a constant 1½-inch overhang on the bottom. (**Note:** The sheets are cut and installed with nails or screws just as with the metal siding previously described.)

If more than one row of sheets is required to reach the peak of the roof, simply overlap the rows like shingles. For steep-pitched roofs (6 in 12 or more), a 4-inch overlap will seal out water. For lower-pitched roofs, use an 8-inch overlap. Or follow the manufacturer's specifications.

ESTIMATING ROOFING MATERIALS

Roofing materials, except metal sheets, are measured by the square. A square is a 10×10-foot area or 100 square feet.

Asphalt shingles are sold in bundles, each containing enough shingles to cover ⅓ square. Thus nine bundles would cover 3 squares or 300 square feet of roof area. Shingles are also measured by their weight, the heavier ones being more durable: 235-pound shingles are common for most roofing applications, meaning a square of these shingles weighs 235 pounds. When ordering shingles, simply measure the square footage of both sides of the roof and divide by 100 to determine the number of squares you need. Add about 5 to 10 percent to this figure for starter courses, ridge caps, and waste.

Double-coverage roll roofing comes in 3-foot-wide rolls that are 36 feet long. When installed with a half-lap, each roll covers a net area of 50 square feet or ½ square.

Some other materials you will need for shingling are tar paper and roofing nails. Fifteen-pound tar or "felt" paper comes in 3-foot-wide rolls that are 144 feet long. One roll will cover 4 squares. Galvanized roofing nails, 1¼ inches long, are used for new roofs, and you'll need about 2½ pounds per square of roof area.

Galvanized metal and aluminum roofing come in different widths from different manufacturers, but two common widths are 32 inches and 38 inches. When properly overlapped, these widths will cover a net width of 30 and 36 inches, respectively. To figure the square-foot coverage of metal roofing, simply multiply the net width by the length of the rafters.

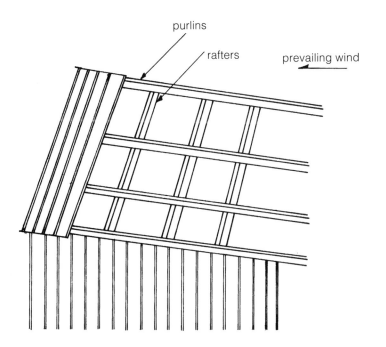

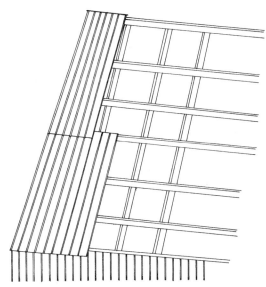

8-3. *Metal roofing is started at the bottom of the roof and on the edge away from the prevailing wind.*

8-4. *Where more than one row of roofing is needed to reach the peak, the bottom sheets are installed first, then the top sheets overlapping them according to the manufacturer's instructions.*

When both sides of the roof have been covered, nail a ridge cap over the peak to seal the joint there. Metal ridge caps are adjustable to conform to any roof peak.

Corrugated fiberglass panels are also available that you can use in conjunction with metal roofing to let light into a building. Install these just like the metal sheets, with an overlap on each edge. Though these panels are fairly expensive, they allow natural daylight into the barn, thereby cutting your electrical lighting bill.

PLYWOOD DECKING AND EAVES

Before the roofing such as shingles go on, the roof deck must be put down and the eaves framed in. (Eaves and overhangs are desirable to help protect barns, sheds, shelters, and outbuildings from the weather. However, fully-enclosed eaves are not always necessary; fascia boards, soffits, and other finishing details may be omitted when air tightness is not required.) For shingles and roll roofing, a full plywood or flakeboard deck is needed. Extend the roof deck beyond the two end rafters, the width of the eaves.

Getting plywood onto a roof can be a difficult task, particularly in windy conditions. On two-story roofs, plywood is usually passed up on the inside of the building,

125

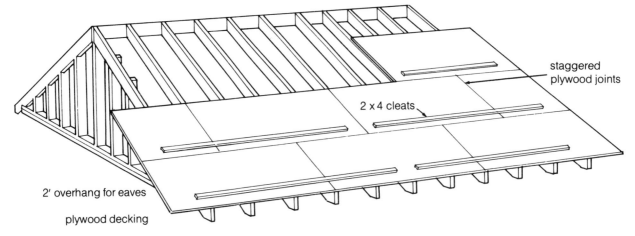

staggered plywood joints

2 x 4 cleats

2' overhang for eaves

plywood decking

8-5. Roofing such as asphalt, fiberglass or wood shingles require plywood decking to be installed first. Plywood must be staggered at the joints and overhang the eaves. Note the 2x4 cleats used for help in working on steep pitched roofs.

through the spaces between studs and rafters.

Here's how to put down a plywood deck. Starting at the bottom, run 4x8 plywood sheets horizontally across the rafters for greatest strength. Make sure the plywood is flush with the bottom of the rafters and overhangs the end walls of the building by the amount desired for the eave. Stagger the joints on the next row of plywood so they don't all line up on the same rafter. The first row of plywood is set using ladders and standing on the top of the wall plate. On shallow-pitched roofs you can sometimes stand on the lowest plywood row to install the one above it. On steep-pitched roofs nail 2x4 cleats across the plywood and into the rafters to provide footholds as you work your way up the roof. For added safety on steep roofs, use roofing brackets and planks (also known as *ladder jacks*) instead of cleats. Use ring-shank nails spaced 12 inches on rafters and 6 inches along plywood edges. "H-clips" should be used with rafters more than 16 inches o.c.

After the plywood has been nailed in place, the gable-end eaves must be framed to support the overhanging deck.

First, cut two sets of fly or "hanging" rafters, which form the outside edges of the two gable-end eaves. Fly rafters are exactly like the normal rafters, except they don't have bird's-mouth cuts and they are ¾ inch longer because there is no ridge plate to butt against. Normally, fly rafters are cut from 2x4 or 2x6 stock depending on the width of the fascia board that will cover them.

With the fly rafters cut, haul them up on the roof. With one person on the roof and another on a ladder, hold a fly rafter in position under the edge of the decking. Drive 16d nails through the decking to secure the rafter in place. The fly rafters should butt against one another at the roof peak and be even with the tail cut of all the other rafters.

Once all four fly rafters are nailed to the decking and

126

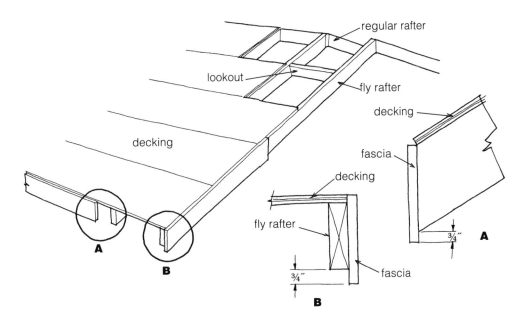

suspended temporarily, they should be blocked against the building every 2 feet o.c. These blocks, called *lookouts,* should be toenailed into the end rafters and nailed to the fly rafters with 10d nails. When the blocking is in place, the eave overhangs will be strong enough to support the roofing and to work on. The fascia boards can now be applied to the fly rafters and over the ends of the main rafters. The top of the fascia boards should be flush with the top of the roof decking. Also, they should be wide enough to hang down ¾ inch below the soffit boards that will be nailed up under the lookouts. Once the fascia boards are on, you are ready to start roofing.

8-6. After decking has been applied, fly rafters, lookouts and then fascia boards may be installed.

Asphalt or Fiberglass Shingles

Asphalt shingles come in several different styles and a wide variety of colors. They must be installed over a plywood or flakeboard deck covered with 15-pound tar paper and edged with galvanized-metal drip edge.

When the plywood deck is in place, frame the gable eaves and attach the fascia boards as previously described. After the fascia boards are on, nail galvanized drip edge to the edges of the roof deck. Be sure upper sections of drip edge overlap lower ones along the gable ends, and that the gable end drip edge overlaps the drip edge along the bottom edge of the roof. The drip edge, an important feature, keeps water off the fascia and gives a straight line for the roof edge.

Next, staple 15-pound roof felt (tar paper) to the ply-

8-7. Before installing asphalt, fiberglass or hardboard shingles, first attach a drip edge, then lay down tar paper.

wood. On steep roofs start at the top of the roof, not at the bottom as is commonly recommended, so you can use the 2x4 cleats to work on. Roll out the first course at the peak and staple it down except along the bottom. Move down 3 feet and roll out the next row, slipping it up under the top row to give a waterproof overlap of about 4 inches. Most paper has chalk lines printed on it for easy alignment. This "upside-down method" requires a little more patience to "under-lap" the tar paper, but it saves the time and work of having to first remove all the cleats, reattach them as you install the tar paper, and then remove them again. If you install the tar paper from the top down, you only need to attach the cleats once and remove them once as you work down the roof.

With the tar paper installed, you're ready to shingle. Some tools you will need are a straight edge (a rafter square is good for this), a utility knife for cutting shingles, and a chalk line for making shingle courses. (These days most shingles come with alignment lines printed on them, alleviating the chalk line step. On larger roofs, though, my experience is that it is still necessary to snap a chalk line every five or six courses to keep the courses straight.) For steep roofs you will also need roof jacks, which are wooden brackets for holding the staging if the roof has more than a 4 in 12 pitch. You can attach these brackets to the roof without damaging the shingles to create a secure work platform. You can usually rent these from local tool rental shops or building suppliers.

To install asphalt shingles, follow the manufacturer's instructions. The first course of shingles is called the starter course. Install this course of shingles with the tab edges pointing up toward the peak of the roof. One-sixth of the first shingle must be cut off so there won't be an overlap of

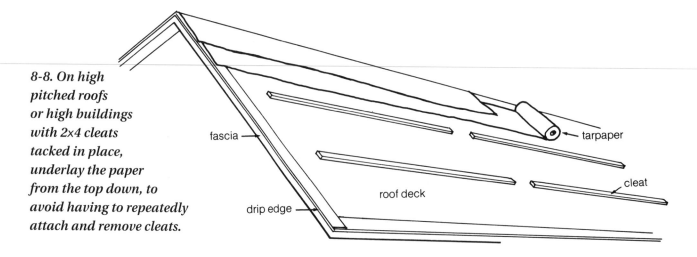

8-8. On high pitched roofs or high buildings with 2x4 cleats tacked in place, underlay the paper from the top down, to avoid having to repeatedly attach and remove cleats.

fascia

tarpaper

cleat

roof deck

drip edge

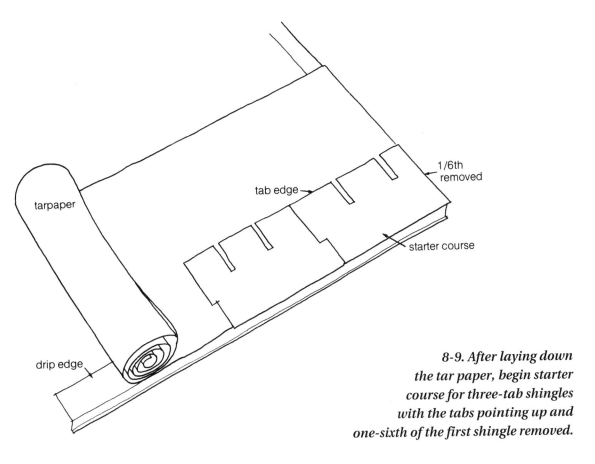

8-9. After laying down the tar paper, begin starter course for three-tab shingles with the tabs pointing up and one-sixth of the first shingle removed.

the tab slits of the next course. When cutting shingles, always cut from the back so the knife blade won't be dulled by the pebbled front surface. Nail the shingles in place with four 1¼-inch galvanized roofing nails above the line marking the bottom of the next course of shingles as shown.

Nail the first "exposed" course directly over the up-side-down starter course, using full shingles. To start the next course, use a shingle with one-sixth (6 inches) of one end removed. The bottom of the shingles should be even with the top of the tab slits on the underneath course. For the next course, cut off one-third (12 inches) of the first shingle. Then cut off half (18 inches) of the shingle that starts the next course and so on until you reach the last one-sixth (6 inches) piece. Butt and nail shingles in place against these starter courses until you complete them across the roof, then start the process over again. When you reach the end of the shingle course you will have to cut

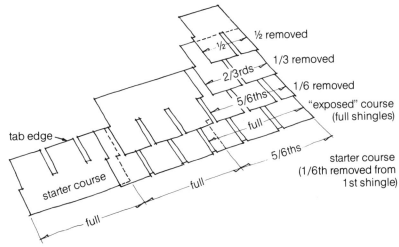

8-10. When installing asphalt shingles, follow manufacturer's directions, or trim shingles in a sequence like this to avoid overlapping slits.

129

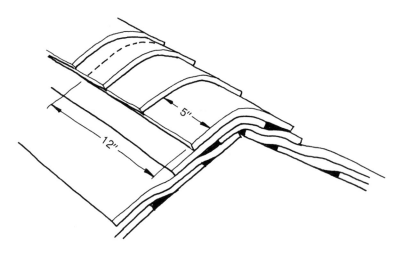

8-11. At the ridge, the last course of shingles are bent over so they overlap, then a "cap row" of 12-inch shingle pieces is placed as shown.

the last pieces, reversing the process in one-sixth to complete the starter courses. As you move up the roof, use a chalk line and tape measure to keep the rows straight and horizontal.

When you get to the peak, don't cut the last course off even with the peak. Bend it over to the other side of the roof and nail it down. Once both sides have been shingled, a ridge "cap row" is nailed in place. This consists of shingles cut into three 12-inch pieces. These are folded over the ridge sideways, leaving the normal amount "to weather." Work along the ridge toward the direction of the prevailing winds, so the overlaps face away from the wind and the driving rain. "Blind nail" or nail so following overlapping shingles cover the nails.

ROLL ROOFING

For shallow-pitched roofs with less than a 4 in 12 pitch, you can use asphalt roll roofing applied on top of 15-pound felt (tar paper). Staple the tar paper to the roof deck.

Roll roofing comes in 3-foot-wide rolls that cover 50 square feet each, when applied as double coverage (with a half-lap on each row). Unroll horizontally across the roof and apply strips from the bottom up. Overlap each strip half its width by the row above it and fasten with roofing nails along its top and bottom edges. Spread a sealing compound called "blind nailing cement" under the bottom edge of each row to seal the nail holes where they penetrate the roofing.

For extra protection on shallow-pitched roofs, tar the roll roofing with asphalt roof coating. Apply the coating with a tar brush. The coating hardens into an impervious surface within a few days.

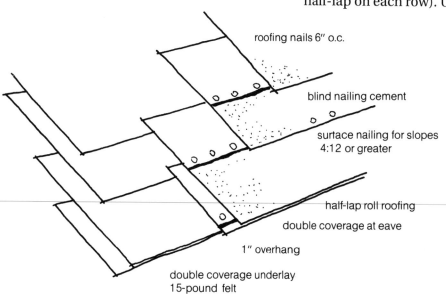

roofing nails 6" o.c.

blind nailing cement

surface nailing for slopes 4:12 or greater

half-lap roll roofing

double coverage at eave

1" overhang

double coverage underlay 15-pound felt

8-12. Roll roofing is installed with double coverage at eaves. And beneath the roofing is a double layer of 15-pound felt (tar paper).

Hardboard Shingles

One extremely effective method of acquiring the look of wood shingles, without the hassle and skills needed when installing individual wood shingles, is to utilize hardboard shingle sheets such as the Masonite Woodruf. Available in 4-foot lengths, they are somewhat harder to install than asphalt but much easier than individual wood shingles.

8-13. Masonite Woodruf, simulates wood shingles but without a lot of the hassle of installation.

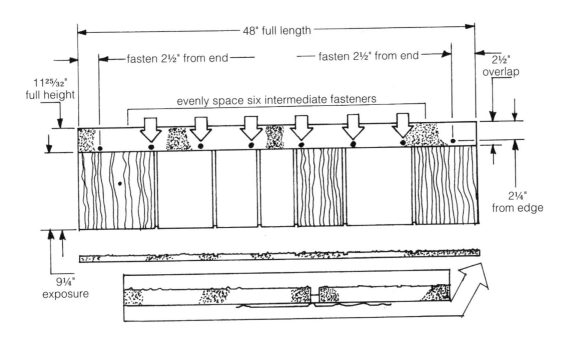

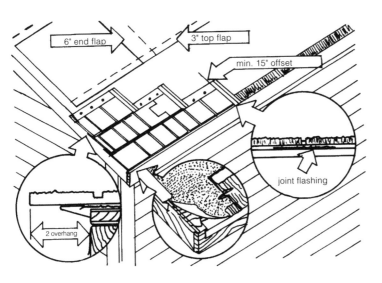

8-14. A 3-inch wide starter strip is applied at the eave over the roofing felt and metal drip edge. Then start laying shingles at the bottom of the roof with a minimum of 15 inches offset. Joint flashing is placed between the sections.

131

CHAPTER 9

Windows, Doors, and Finishing Details

9-1. Once the building has been roughed-in then the fun begins as you close it off with windows, doors and add trim and decorative effects.

Usually there is a big rush to "rough-in" a building, or get the building framed, roofed, and the siding installed. This provides some amount of protection from the weather to the building and building materials. Once this step has been achieved you can usually slow down the pace. This change in pace also occurs quite naturally, because the installation of windows, doors, soffits, and other finishing details is more time-consuming and exacting work than the actual framing and covering of the building.

The type of windows and doors you select for your building will depend on your budget and carpentry skills. Windows and doors for barns and outbuildings can be anything from a hole cut in the side of a building to special solar-operated panels that open and close automatically to provide ventilation in confinement hog barns. The "style" is normally dependent on what the building is to be used for.

Factory-built, pre-hung windows and doors are available that can be simply nailed in place. These are easy to install and weathertight, but fairly expensive. If you have a limited budget, you may want to make your windows out of fixed panes of glass, and your doors from plywood or solid stock. One method often used is to simply fasten aluminum storm windows or doors over the opening, although this doesn't provide a secure means of locking the building. Another alternative is to recycle doors or windows from a building that has been torn down. Home-made and recycled doors and windows do require a great deal more work, tools, and carpentry skills. They can, however, save as much as $50.00 or more per unit.

WINDOWS

Four principal types of windows are used for small-barn construction. The most common is a double-hung window with two sashes that slide up and down in tracks. On old windows the sashes are often divided into "lights" (small panes) by wooden muntins. On modern double-hung windows the windows may be left plain, or the muntins may be made of wooden or plastic grills mounted over a large single pane of glass to give the impression of small panes. This allows for easier cleaning of the glass and the use of double- or triple-glazed windows for greater energy efficiency.

Awning windows have only a single sash that opens out at the bottom. These are usually less expensive than double-hung windows and more energy-efficient. They come with either a roller opening mechanism or sliding friction hinges to hold them open at the desired angle.

Sliding windows are available that have either two moveable sashes or one fixed sash and one moveable one that slides to the side over the fixed sash. These are available with both wood and aluminum frames. These units are ideal where wall height is a limiting factor or where windows must be placed high on a wall to keep them away from animals.

Barn sash is another type of window widely used for outbuilding construction. It is simply a fixed pane of glass in a frame that is mounted in a fixed position.

Windows should be concentrated on the south side of a building to capture solar heat and to avoid winter winds. While the primary function of windows is to let in light, they serve other purposes as well. At strategic points, windows should be operable to help with summer cooling and cross-ventilation. Not all windows need to

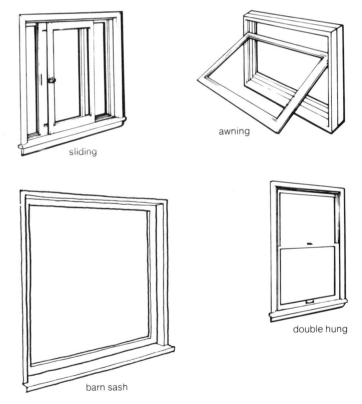

9-2. Windows typically used for small barn and out-buildings include the sliding and awning type, the double hung window and ordinary barn sash.

133

be operable, however. Fixed panes of glass and barn sash are less expensive, easier to install, and more energy-efficient than operable windows.

Installing WIndows

Pre-Hung Units

Pre-hung windows are installed after the sheathing is on, but before some types of siding are installed. If there is no sheathing (if for example the plywood beneath board siding has been omitted) the windows are simply installed over the siding.

An important measurement of windows is the width of the jambs. The jambs must be wide enough to extend from the outside of the siding to the inside of the finished wall. This will allow window casing to be nailed onto the edge of the jambs to finish the window trim. You can order windows with different jamb widths to fit different walls. Or, you can order or make jamb extensions to extend the inside of the jamb. If the inside of the building will not be finished with wall paneling and trim, then the width of the jamb is not as critical.

To install pre-hung windows, first staple 15-pound building felt paper (tar paper) in a strip about 12 inches wide around the window rough opening and over the sheathing, making sure upper felt paper pieces overlap lower ones. (Omit this step if the window is being installed on siding and there is no sheathing.) Then run a bead of caulking on the tar paper or siding, under the places where the window casings will be nailed to the sheathing. This will help minimize air infiltration. Hold the unit tightly against the building and have a helper wedge shim shingles under the window from the inside of the building until the window is level and at the proper height. When the unit is level, adjust the side jambs with shims until the window frame is perfectly square. You can check this either by using a rafter square on the inside of the jambs or by checking the diagonals of the window opening to make sure they are equal.

When the unit is level and square,

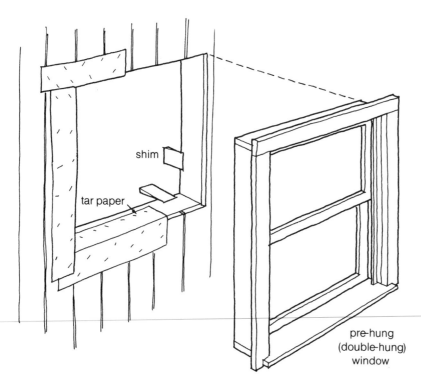

9-3. When installing windows, especially pre-hung units, use tar paper to reduce air infiltration. (Illustration shows tar paper partially installed.) Shims can hold the window so it is perfectly level and square.

shim

tar paper

pre-hung (double-hung) window

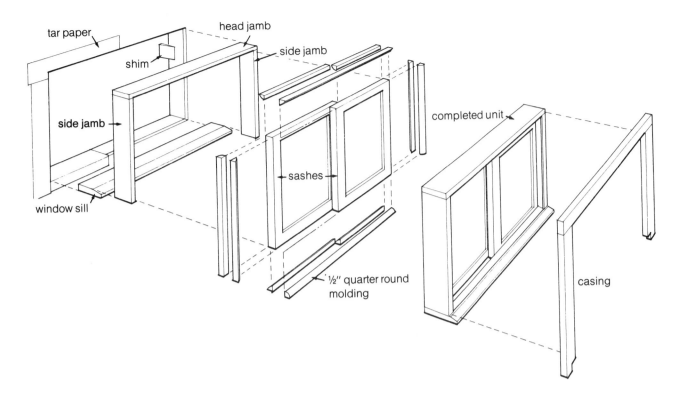

Labels in figure:
tar paper
shim
head jamb
side jamb
side jamb
window sill
sashes
½" quarter round molding
completed unit
casing

drive a nail part way into each corner of the casing to hold it in place. Check again for squareness, and, if it hasn't moved, nail the window permanently through the top and side casings. Use 12d galvanized casing or finishing nails driven securely into the framing of the rough opening.

9-4. Recycled window sash is made into a new window in this sequence: build frame of sill and jambs; secure sashes with quarter-round molding; install completed in rough opening; and finish off with casing.

Recycled Windows

Recycled window sashes are ideal for barns and outbuildings. You can rebuild sashes into double-hung, awning, or fixed windows at a fraction of the cost of factory-made units. It is best to try to get a set of similar windows from a single building so they will all match. If the units are in good shape, you can remove entire frames in one piece from the old building and simply reinstall them in your rough openings. If the frame is rotted or cracked, you can still salvage the sashes and either rebuild the frame and tracks or install the sashes as fixed units.

A horizontal sliding window is the easiest operable type to build from old sashes. The first step is to build and install a window frame that will fit in your rough opening. The frame also must have the proper dimensions to hold the sashes. To make the frame, use a clear piece of 2x8 or 2x10 for the windowsill and 1-inch stock cut to the proper width for the jambs.

The first step is to cut and install the sill. It should overhang the siding by 1½ inches so that it projects out beyond the window casing and allows water to drip free of

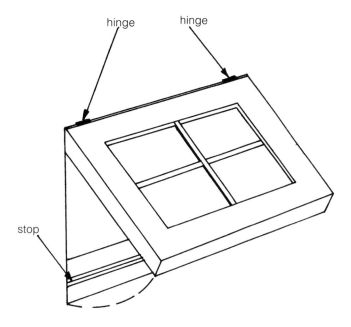

9-5. Awning windows can also be made quite easily.

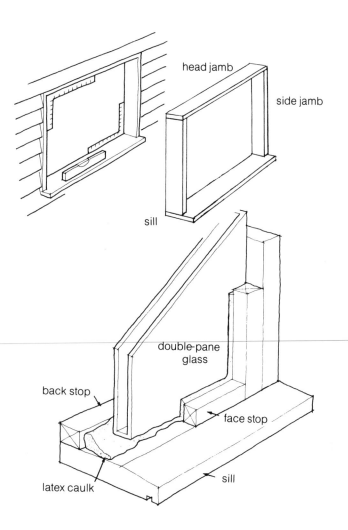

the siding. Bevel the part that extends out beyond the siding at a slight angle so that it sheds water. Level the sill and nail it in the rough opening with 12d galvanized nails (about 8 inches o.c.).

Cut and nail together the side jambs and head jamb to complete the frame. (This part is assembled and then installed in place on top of the sill and in the rough opening.) The finished height of the window frame should be ⅛ inch taller than the height of the sashes so they can slide easily. The width should be 1 inch less than the combined widths of the sashes so that they overlap. Use shims between the jambs and studs to get the proper dimensions and to make sure the frame is square.

When the frame is nailed securely in place on the sill, mount the two window sashes. To do this use ½-inch quarter-round moldings to form tracks for the sashes to slide in. Lay the tracks so the inner window sash is fixed and the outer one slides by it.

Awning windows are also simple to make from an old sash. First, a sill and jambs are installed as outlined above to make a window frame. Make the frame exactly to the size of the window sash. The sash can then be hinged either to the inside or the outside edge of the top (or bottom) jamb so it swings either in or out. Use pieces of ½-inch quarter round for stops to keep the window in position when it is closed. A variety of arrangements can be used to hold the window open. A short stick, a hook and screw eye attached above the window, or friction side hinges are all possibilities. All that you need to keep the window closed are a hook and screw eye.

All barn windows do not have to be operable. After you have provided adequate ventilation with moveable windows, use fixed panes or barn sashes, which are cheaper and easier to install. Barn sash can

9-6. Fixed glass is easy to install in a rough opening; simply follow the sequence of steps shown on page 138.

simply be mounted in a window frame with quarter-round molding and caulking to make it airtight.

If you desire large areas of glass for either solar heating or more light, consider using custom-made pieces of Thermopane glass. Thermopane is two layers of plate glass fused to a thin metal frame. It is also called insulated glass and is readily available from glass suppliers, made to your specifications. Thermopane units can be mounted directly in framed window openings just as barn sash is.

If you are considering passive solar heat for your livestock barn, an inexpensive option is to use replacement glass panels for sliding doors. This is sometimes called patio glass. This glass comes in ⅝-inch Thermopane and in several standard sizes including 34x76 inches. An extra advantage of these units is that the glass is usually tempered to resist breakage. On a square-foot basis, they are inexpensive, energy-efficient, and easy to install.

Other options for solar heating are fiberglass and plastic glazing panels. Most of these are not transparent so you cannot see through them, but they let in almost as much sunlight as clear glass and can be bought at a fraction of the cost. They are also stronger than glass. Solar hardware suppliers have complete listings of these glazings along with installation information.

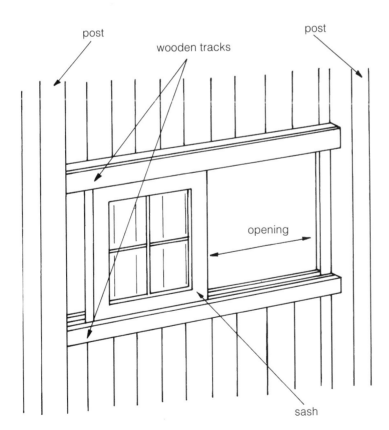

9-7. Another excellent way of creating a window in a pole or post-and-beam barn is to slide a sash in tracks between the poles.

137

INSTALLING FIXED GLASS

Fixed glass is double-pane or insulated glass, set in a casing to make a window. "Fixed" means the window is not operable.

Here's how to install fixed glass:

1. Build a frame for the glass that is ⅛-inch wider and higher on its inside dimension than the piece of glass. Make the sill from 2x8 stock with a beveled outer edge to shed water. Make the side and head jambs from 1x6 pine.
2. Install the frame in the rough opening, making sure that the sill is level and that the frame is perfectly square. Use shim shingles driven between the frame and rough opening to adjust its squareness. Secure the frame with 12d galvanized casing nails driven through the jambs.
3. Attach a back stop to the rear of the sill and jambs. The stop can be made from a small strip of wood or quarter-round molding and nailed with 4d finish nails.
4. Apply a good bead of latex caulk or neoprene gasket strip around the window frame against the back stop. This should form a weathertight seal and expansion joint for the glass.
5. Place the glass pane or insulated glass unit against the back stop. Press it inward until it makes a tight fit along the back stop.
6. Nail the face stop in position to secure the glass. Be careful not to angle the nails into the glass.

DOORS

Several types of doors can be used for outbuilding construction. The standard side-hinge door is commonly used for both interior and exterior doors that need only be large enough to allow a single person, or small equipment through. These are available in solid wood, insulated steel (exterior doors), and in economical, lightweight, hollow-core doors commonly used for interior-doors. Normal dimensions for exterior hinged doors are 36 inches wide by 80 inches high. Interior doors are usually narrower. Side-hinge doors may be purchased alone or as a pre-hung unit. The latter are more expensive but a great deal easier to install, and usually result in a better-fitting door.

The Dutch door and double door are two variations of the side-hinge door. The Dutch door has a separate top and bottom panel so that either can be opened while the other is closed. These are used with horse stalls so the horse can have light and air through the opened upper part while remaining confined by the lower panel. The double door is simply two hinged doors mounted side by side to form a large opening. Usually these are made on the site out of boards or plywood and

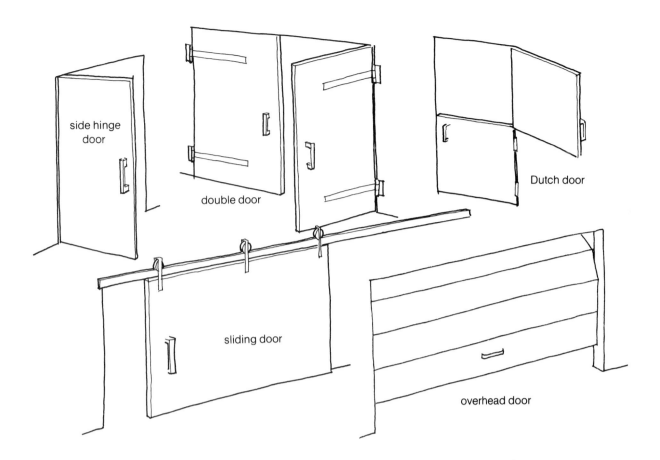

mounted with large T or strap hinges.

Sliding doors are often used for large barn or shop openings to accommodate tractors and other oversized equipment. These doors are usually made on site from dimension lumber and covered with metal, plywood, or siding to match the building. They are hung on an overhead track with rollers so they can slide to one side of the door opening. An extremely large opening may require two doors sliding in opposite directions.

Overhead doors or garage doors are sometimes preferred to sliding doors because they take up less room and are out of the way when open. They are made of hinged steel, plastic, or wood panels that are spring-mounted on a set of curved steel tracks. Electric garage door openers add to the convenience.

Installing Doors

Pre-hung purchased hinge doors, complete with jambs, casing, and threshold are installed in the same manner as pre-hung window units. Prepare the rough opening with 12-inch strips of tar paper (building felt) and a bead of caulk around the outside. (Again, omit the tar paper if the door is to be installed on siding and there is no

9-8. Doors suitable for small barns and outbuildings include the side hinge door, double door, dutch door, sliding door and overhead door.

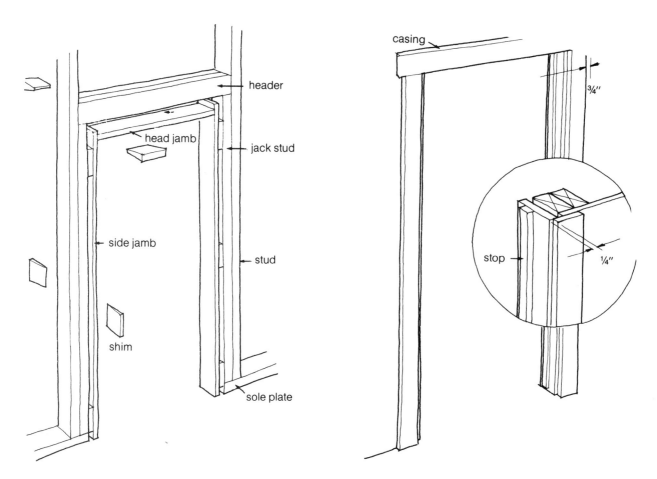

9-9. Major parts of the door frame are the side and head jambs. Thresholds are often omitted for small barns and sheds.

sheathing.) Tilt the door unit in place, plumb and square it using shim shingles, and drive 12d casing nails through the side jambs to secure the door.

If you are using old recycled doors or new ones that are not pre-assembled, you will have to frame in the rough opening and hang the door yourself. The door frame is built just like a window frame except a door threshold, if used, replaces the windowsill. Normally, thresholds are made of oak to withstand the constant wear of boots and shoes. These can be purchased pre-shaped from building suppliers, or you can make them from 1½-inch stock. Preformed aluminum thresholds with built-in rubber weather strips are also available. Install the threshold first and then the side and head jambs. As with the window-frame units, it's best to assemble the frames outside the opening then place them in the opening. Or you can make an entire pre-hung unit, building the frame, then installing the door in the opening of the frame, and finally installing the entire unit in place.

If you are creating a frame without pre-hanging the door, the first step of course is to build and install

140

the frame. Mount the door hinges about 6 inches from the top and bottom of the door. On extra-heavy doors a third hinge is often added in the middle. Install the leaves on the door first. A sharp chisel is used for this job, although a router and mortising template can also be used if you have several doors to install. It is extremely important to have the surfaces of the hinge leaves flush with the surface of the jambs and doors. The leaves on some hinges are held with a removable pin. Removing the pin makes it easier to install the individual leaves of the hinge.

Next, place the door in the frame in its closed position. Place a ¼-inch shim under the door for clearance when it swings. Mark the position of the hinge on the side of the jamb. Remove the door from the opening and chisel out the mortises for the other leaves. Then mount them on the jamb. The leaves of the hinges are mortised into the side jamb and into the door so the door fits tightly against the jamb when it is closed. If you have measured accurately, you can now hold the door in place, slip the hinges together and reinsert the pins. Rarely is it quite this easy. In many instances a slightly out-of-square frame will require adjusting the hinges or planing off some portion of the door surface.

After the door is hung, apply stops to the side and head jambs to keep the door from swinging too far when it is closed. You can buy stops in a variety of styles with different edge moldings. For outbuildings you can also easily make the stops from 1x stock. With the door in the closed position, nail the stops in place so they make a tight fit against the door. On exterior doors, you can add weatherstripping to the stops to create an airtight seal.

To finish off the door, nail casing around the inside and outside of the door to cover the jamb and frame installation. Normally, casings are made of 1x4 or wider pine stock or molding nailed over the edge of the jambs and the siding. The casings are set back from the inside edge of the jambs by ¼ inch on all sides. You can also set the casings flush with the jambs.

Fit all doors and windows with drip caps to keep water from getting behind the casings. Make *drip caps* from a piece of crimped aluminum or a beveled piece of wood set on top of the window or door casing. Even a large bead of caulk is often satisfactory if the window or door is partially protected by the eave overhang.

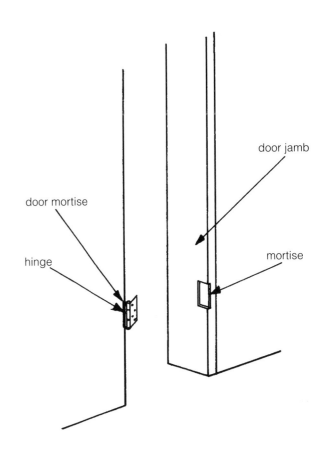

9-10. When butt hinges are used they must be mortised into both the door jamb and the door.

141

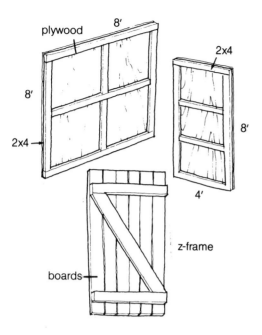

9-11. Simple home-made doors can be fabricated from 2x4's and plywood or rough-sawn boards backed by a Z-frame.

Homemade Doors

An inexpensive alternative to buying solid-wood doors is to make your own. These can be solid plank, plywood, or doors of plywood or siding to match the building, mounted on a 2x4 frame. You'll need a good flat work surface to assemble the doors. To create the plywood 2x4 frame doors, cut the plywood outer skin to the correct size first. Then cut the 2x4 bracing members to size. (**Note:** These should be installed flat rather than on edge to create a door 1½ inches thick plus the siding or plywood.) The frame should have cross members at least every 4 feet to support the plywood skin. Make sure the plywood skin is cut square, then place the 2x4 bracing members on a flat surface, lay the plywood skin on top, and fasten to the 2x4 framing members with 4d galvanized nails. Turn the door over and fasten the wooden framing 2x4s together with either corrugated fasteners or metal braces. Nailing 16d boxing nails through the ends can also add more support. The outside of the door can be decorated with additional 1x trim to add design decor. This can simulate a Dutch door, or the cross-bracing often seen on doors.

Doors for stalls and pens for small livestock can also be made from single sheets of exterior plywood mounted on hinges. The minimum thickness of plywood that can be used for this without undue warping is ¾-inch.

Doors can also be made of solid stock, either kiln-dried or treated dimension lumber, car siding (tongue and groove), or even rough-sawn native materials. Z-bracing is used to fasten the boards together. Make sure the top angle of the Z is placed on the hinge side to add support.

Regardless of their construction, doors must be sturdy enough to hold up to the rough wear and tear an outbuilding receives. A flimsy door, or one that is hung on undersized hinges, won't last long.

Sliding and Overhead Doors

Sliding doors are good equipment entrance doors for barns. These can be constructed on a wooden frame assembled with metal joining plates or plywood glued gussets and covered with either metal, wood, or plywood. The door has rollers bolted to its top that slide in an overhead track.

When designing a sliding door, consider its size and weight carefully. Most overhead tracks are designed to carry a maximum weight of 300 pounds. A 10x8 sliding door made with rough-sawn framing and covered with 1-inch boards might put this carrying capacity to the test. It would also take three people just to stand it up and put it in place. One way to minimize the door weight is to use kiln-dried framing members and plywood or metal skins.

Sliding doors are installed either inside or outside. Probably the biggest problem with large, horizontal sliding doors is figuring out where they will slide when they are opened. Keep in mind that a 10-foot sliding door mounted inside a 20-foot-wide barn won't open all the way. By the same token, a large door mounted near the edge of a building where the door must slide past the building edge will require an additional post to support the track and door. When a door slides along the inside of the wall, this wall surface must be free from obstructions and partitions. One way to minimize the area large sliding doors take up is to break them into two pieces that either slide open in opposite directions or overlap and slide to the same side. When built with two halves, sliding doors are also easier to mount and operate. Obviously, doors mounted on the outside would be encumbered by fewer obstructions. But to prevent rusting, their mounting hardware should be protected from the weather, or you should use galvanized metal mounting hardware.

If the barn layout interferes with or restricts the use of sliding doors, overhead garage doors are an alternative. Because of the complicated construction of the door panels and runners, it is not practical to build these yourself. Overhead doors can be purchased from most building supply dealers and installed yourself, or you can contract with a local overhead door company to install a completed unit.

9-12. One extremely popular door for larger sheds and barns is a sliding door.

FINISHING DETAILS

After the windows and doors are installed, it's time to tackle the rest of the finish work. In the case of barns that are primarily used for storage, or as open livestock shelters, very little is usually left to do. In many such cases the rooflines do not have fascia boards, nor are the eaves closed off. This does allow for more ventilation. For livestock barns or other structures that must be kept warm, however, the eaves should be closed in. (In southern climates this may not be necessary. In any case, be sure to provide adequate ventilation.) Closing the eaves off on

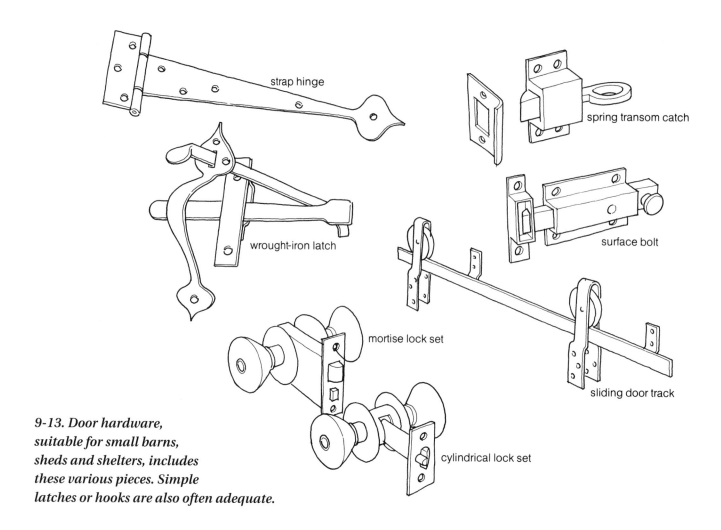

strap hinge

spring transom catch

wrought-iron latch

surface bolt

mortise lock set

sliding door track

cylindrical lock set

9-13. Door hardware, suitable for small barns, sheds and shelters, includes these various pieces. Simple latches or hooks are also often adequate.

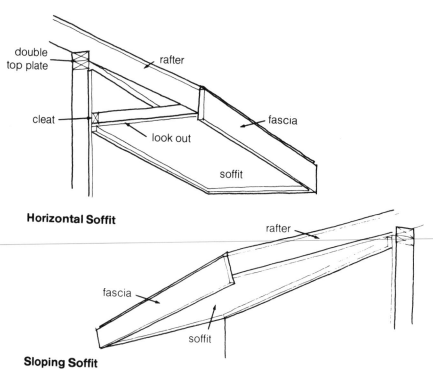

double top plate

rafter

cleat

fascia

look out

soffit

Horizontal Soffit

rafter

fascia

soffit

Sloping Soffit

9-14. If you plan to finish off the eaves of your building you can do so with either horizontal or sloping soffits.

144

storage buildings will also help keep out pests such as squirrels, birds, and insects.

Once the fascia boards have been installed, by nailing to the rafter ends and to the hanging rafters, all that is left to do is attach the soffit boards. Soffit boards, either single pieces of ¼- to ⅜-inch plywood, special hardboard soffits, or 1x6, or 1x8 boards can be used for the soffits. These can be attached as a horizontal soffit or as a sloping soffit. (In case added air circulation is required, you can install soffits with vents.)

If the soffit is to be sloping, simply attach the soffit board(s) to the underside of the rafters, butting the edges tight against the fascia board in front and the barn siding in the back. Secure the soffit with 8d galvanized nails. On the gable ends of the building, nail the soffit up under the lookouts.

If you are using the horizontal soffit detail, you need to attach lookouts to the ends of the rafters as horizontal nailers for the soffit. The lookouts can be 2x4s nailed into the side of each rafter, flush with the bottom of the rafter and extending horizontally back to the siding. Nail them into the rafters with 10d nails, and with 8d nails toenail the other end to a cleat mounted on the siding. The cleats attached to the siding must also run the full length of the building. Then attach the soffit boards underneath the lookouts.

A horizontal soffit requires an eave return to join the two different soffit planes where they come together at the corners of the building. Eave returns can be ornate, complicated structures, especially on older residential houses, but for a barn they can be quite simple. Your basic task is to join two boards that meet at different angles.

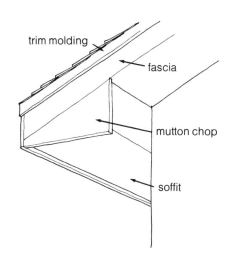

9-15. Mutton chop or eave returns are names for the end pieces that join the horizontal soffit and vertical soffit board. They can be plain or decorative.

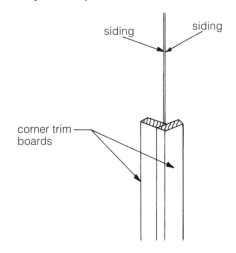

9-16. In some cases corner trim boards are used to finish off a building.

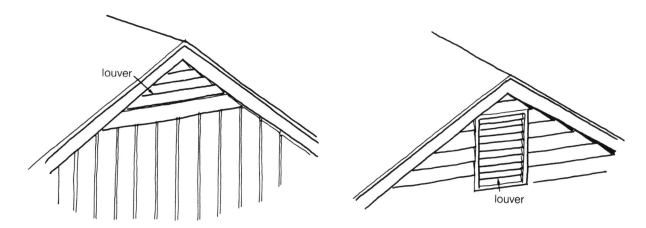

9-17. Gable-end vents such as these assist air circulation in small barns and livestock buildings. Vents should be screened inside to keep out insects.

145

9-18. One popular method of providing ventilation is the cupola.

In some cases corner trim boards may also be required to "finish" off or dress up a building.

Ventilation

A controlled environment is important to the health and productivity of farm animals. Adequate ventilation can help remove summer heat, moisture, dust, and toxic gases; for, when these accumulate, mold, dampness, and disease are the result.

The simplest way to ventilate a barn is to have well-placed windows that provide cross-ventilation. Consider prevailing winds carefully when laying out the windows and interior spaces of a barn. But do not locate windows so that livestock receive direct drafts. Drafts on confined horses and cows will cause pneumonia.

Rising hot air also induces ventilation. By installing air vents in the roof, such as cupolas or gable-end vents, hot air can be exhausted naturally from the building. As the hot air escapes, it pulls in fresh cooler air through the

VENTILATION FOR LIVESTOCK

Ventilation in livestock barns is critically important to the health and well-being of the animals. Ventilation is necessary for cooling, to remove disease pathogens, and to control moisture.

A common rule-of-thumb for livestock buildings with average animal density is to provide for four air changes an hour during the winter and 40 air changes an hour during the summer. Thus for a 30x20 barn with an 8-foot-high ceiling, you would have to change 4,800 cubic feet of air every 15 minutes in the winter. This would require a ventilation system capable of handling 320 cubic feet per minute (cfm) of air. In the summer you would need ten times this rate, or 3,200 cfm.

To meet the dual requirements of summer and winter ventilation, you can either buy a single exhaust fan with variable speeds or buy two fans, a small one for winter and a larger one that can pick up the extra summer load. A fan is rated by the cfm so that you can determine how much air it will move. Never buy a fan solely based on its horsepower or blade diameter; these features may have little relation to its air-moving capacity.

The fan size must be matched to the air inlet's area if the fan is to work at maximum efficiency. A good rule-of-thumb is 18 square inches of air inlet for every 100 cfm of summer fan capacity. This will give you an air velocity of 800 cubic feet per minute. During the winter, part of the air inlets can be closed off to compensate for the reduced air flow.

Allowing four air changes an hour during the winter can put a real strain on your heating system and pocketbook, especially in northern climates. One possible solution is to install an air-to-air heat exchanger. A heat exchanger works by extracting the heat value of the air that is being exhausted from the building and transferring it to cool, fresh air that is entering the building. Heat exchangers are widely used in residential construction today, and for livestock operations in areas with large heating requirements and high fuel costs, this may be an economical solution.

ground-floor windows. This process, called *thermosyphoning,* occurs naturally whenever the air inside the barn is heated.

Natural ventilation can also be assisted by electric exhaust fans placed in a cupola or wall vent. Fans can assist cooling by pulling cool night air through a building and lowering its temperature. In the morning, shut off the fans and close up the building to keep out the hot daytime air.

Insulation

Heated livestock barns, tool shops, or garages should be well insulated to maximize comfort and minimize fuel bills. Insulation is easy to install, and it is one of the best investments you can make. (For more information on insulation, see Chapter 2.)

The minimum amount of insulation recommended for heated buildings is 3½ inches of fiberglass in the walls and 6 inches in the ceiling. For northern climates, the minimum standard is 6 inches in the walls and 12 inches in the ceiling. Keep in mind that the entire building doesn't always need to be insulated; only those rooms that are going to be heated or cooled to maintain fairly constant temperatures.

Insulation value is measured by a number called the R factor. The higher the R factor, the better the insulation. Fiberglass, 3½ inches thick, has an R of 11; 12 inches has an R of 38. Fiberglass comes in batts or rolls, 16 or 24 inches wide, to fit between studs and ceiling joists. It also comes with an aluminum-foil backing, a kraft-paper backing, or unfaced. I would recommend using kraft-faced or unfaced fiberglass for most outbuildings as it is the most economical per R value. Don't bother with rigid foam insulation, except for insulating concrete walls and foundations, as it is very expensive.

Always cover insulation with a vapor barrier to keep moisture out of it. A good vapor barrier is 4 mil polyethylene plastic that comes in 8-foot and wider rolls. This is stapled over the inside of the wall studs and ceiling joists to keep moisture that originates inside the building from entering the walls. Never put a vapor barrier on the outside of a building or insulation; this would trap moisture in the

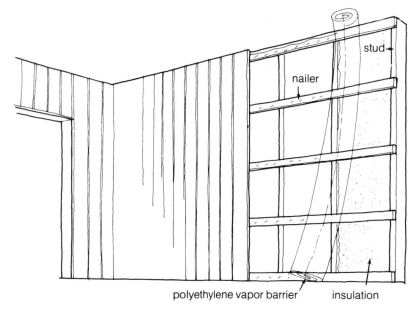

stud

nailer

polyethylene vapor barrier   insulation

9-19. Place the vapor barrier behind the paneling to prevent moisture from entering the walls and damaging the insulation.

147

walls and rot them. Exterior building surfaces should be able to "breathe." Even a layer of tar paper (once considered a good practice) over the wall sheathing is a bad idea as it will trap moisture in the building.

To protect the insulation and vapor barrier from puncture, some type of interior wall sheathing is necessary. One possibility is building board, an inexpensive wall covering made from compressed cardboard. It is available in several different brands and thicknesses. It will not, however, hold up to much abuse. Plywood is a more rugged alternative, but also more expensive. These days one low-cost covering that can be utilized is "chip" or "flake" board. Made of compressed wood pieces glued into sheets, it is also available in a variety of thicknesses. Interiors of stalls and other livestock holding areas require heavier materials, and the best choice is pressure-treated 2x materials or ¾-inch plywood.

If you are building an insulated and heated barn, ventilate the roof to avoid ice buildup and moisture problems. If there is an attic space that has insulation in the floor and is unheated, install gable-end vents at both ends of the building. These vents allow air to flow beneath the roof, keeping it cool and venting moisture.

If the rafters are insulated, then install soffit vents and ridge vents; these will allow air to flow up between the rafters and out the peak of the roof. Leave at least 1 inch air space between the insulation and the roof deck so the air can travel freely. Pre-made aluminum soffit and ridge vents are available that provide the correct amount of ventilating area.

CHAPTER 10

Wiring

The biggest question of whether or not to do your own wiring is whether it is legal in your area. Some towns, counties, or municipalities do not allow wiring to be done except by professionals. In other cases you may be able to do part but not all of the wiring. In most instances, however, wiring a barn or outbuilding is not a particularly difficult job. It does require attention to details, a basic understanding of electricity, and a strict adherence to safety precautions. Regardless, you will probably have to obtain a wiring permit before the work starts and have an authorized inspector approve the work when it is completed. Many communities permit you to install new circuits, but require that a licensed electrician make the final connection at the service entrance panel. This is a good way to have your work inspected for safety; yet you will be saving the electrician's fee for mounting the boxes and routing the wires.

Make sure you check local codes carefully to insure that you follow recommended wire standards and requirements. Wiring in the United States must be installed according to the National Electrical Code. In most cases, this national standard has been incorporated into local

10-1. Most barns and outbuildings require electrical service. In cases where allowed, homeowners can do a great deal of this themselves.

codes. The primary reason for the code is safety, and you should never attempt to do things that deviate from it. A copy of the latest version of the National Electrical Code can usually be purchased at your local building supply outlet. It is also reprinted in various books on wiring.

Don't skimp on materials or time; many barns and outbuildings have been lost because of hasty wiring jobs. A hayloft aflame with animals below is truly a nightmare.

Before embarking on a wiring project, it's a good idea to have a basic understanding of electricity, and the terms used to describe it.

Amperes (amps) is the quantity of current flowing through a wire of a specific diameter. If the wire is too small for the amount of current it must carry, it will overheat and trip the circuit breaker or blow a fuse.

In order for electricity to travel through wiring, it must be under pressure. This electrical potential is measured in *volts*. Most modern homes receive 240 volts of electric power. This allows the installation of 240-volt equipment such as electric water heaters, ranges, and dryers, as well as standard 120-volt lights, outlets, and small appliances.

Wattage is the amount of power flowing at a given point. It is determined by multiplying the amps times the volts. Every electrical device has its rated power usage printed or stamped on it. Some have the wattage used by the appliances on an attached nameplate or tag. By adding up the wattage used by the appliances on a given circuit, you can determine the electrical load of that circuit.

For example, on a 20-ampere, 120-volt circuit you would have a maximum of 2,400 watts available (20 amps x 120 volts = 2,400 watts.) Leaving a safety margin, you would plan to only have 2,000 watts of load on this circuit. If this was to be a circuit for infrared heat bulbs for warming pen areas, you would divide the wattage of each bulb into the circuit load to determine how many lamps could be on this circuit. If the lamps are 300 watts each, you could attach six of them to this circuit bringing the total load to 1,800 watts, well under your design load.

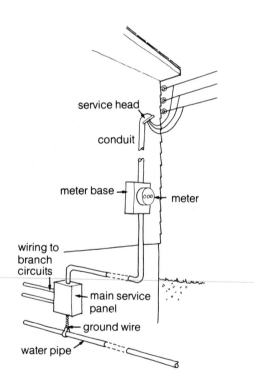

10-2. Wiring can be run directly from the basement main service panel to an outbuilding, or from the main service panel to a subpanel at an outbuilding.

DETERMINING YOUR POWER NEEDS

Before doing any wiring, examine your existing electrical service and determine whether this power is sufficient for your electrical requirements.

Typically a 100- to 150-ampere main service panel is sufficient to service a three-bedroom home, as well as a small outbuilding such as a nearby shed or chicken house. If your main service panel has adequate capacity, but there is no room for a new circuit from the panel, you must install a sub panel for your building. Each outbuilding should be on a separate power take-off from the main service panel.

Many outbuildings, however, especially a machine or woodworking shop, or a livestock building with heat, will require their own main service panel. How the power is supplied to the building depends on the situation and the choice of the utility company. The power may be supplied from a main utility pole, one service entrance and one meter with branch lines off to the different buildings and service panels, or a separate meter may be required. A separate meter is not a bad idea if you have, for instance, a hog farrowing house. The meter readings can be kept separate from your house for tax purposes. If two or more outbuildings with heavy electrical loads are planned, it might be best to bring the power from the utility pole to a centrally located yard pole and meter; then feed the power from there to service panels at the individual buildings. This also gives you the opportunity to place a centrally located barnyard light on the power pole.

Regardless of how the power is supplied, each outbuilding should have its own disconnect switch, so power can be turned on or off at the outbuilding, or at the service pole servicing the building. This is a standard regulation in many areas.

Installation of the yard pole and incoming power lines to the service entrance and meter should be done by electrical contractors or the utility company personnel. If the wiring is done from a yard pole, locate the meter and pole conveniently and centrally between the house and the outbuildings. This routing system is better suited for larger operations with several buildings because it ensures that no building is too far from the main service. Wiring from a main to a sub panel is generally only recommended for a home and single outbuilding.

In either case, the feeder lines to the outbuildings can be located above or below ground. Use underground lines whenever possible, and plan the trench carefully so as not to interfere with underground plumbing, gas, or other lines. Overhead lines are always a problem. They can be damaged by water, ice, windstorms, or falling trees. They're also a nuisance, and provide the danger of snagging with tall farm equipment. In addition, they're unsightly. Make sure you check local codes for the types of underground

10-3. Regardless of how the power is supplied, each outbuilding must have its own disconnect switch. In some cases this is required to be on the main service pole.

151

wiring approved in your area. Bury the lines at least 18 to 24 inches below the surface and enclose them in conduit wherever they may be subject to damage. If installing underground lines in rocky soil place a layer of sand or topsoil in the trench bottom, lay the lines in place, and cover with another 2- to 3-inch layer of topsoil to prevent rocks from working into the lines.

THE SERVICE PANEL

In most instances electric utility personnel will install the electric lines that run from the utility pole to the service entrance on the outside of the building or yard pole. The homeowner must supply and install the service entrance and the feed to the meter socket and service panel, as well as the ground rod installation.

The service panel distributes power to the various branch circuits. It also contains the main disconnect switch for the entire wiring system of that building or buildings, as well as circuit breakers for each individual circuit.

Each circuit includes a series of outlets and/or switches and lights. Each of these circuits is protected by an individual circuit breaker or fuse at the panel. Whenever a circuit is overloaded, the circuit breaker will trip, or the fuse wire will burn. When the problem has been remedied (usually by decreasing the number of electrical devices simultaneously using the circuit), power can be restored by flipping the breaker back on or replacing the fuse.

A three-wire, 30-ampere service panel (7,200 available watts) is usually sufficient for a small barn. This service provides only limited capacity for lighting and a few smaller appliances. Larger barns and confinement buildings require at least a 100-ampere (24,000 available watts) panel; some may require 150- and 200-ampere service for full use of electricity for major tools such as welders and extensive lighting.

Remember when planning your wiring system, extra capacity is important because the size of the service entrance and connecting wires limits the amount of electricity you can use at one time. With extra capacity, you avoid the fear of overloading the wires and tripping the circuit breakers. Ask electric-utility company personnel for advice on the service panel best suited for your buildings.

10-4. The service panel located inside the building distributes the power to the various circuits in the building.

SAFETY PRECAUTIONS

- Don't attach wires to service panels before disconnecting the power.
- Make sure the circuit you plan to work on is dead. Test it before making any repairs or connections.
- Don't wire buildings while standing on wet or damp surfaces. Even though the circuit you're working on is off, you may inadvertently contact another live circuit or wire.
- Be sure to check with local codes before beginning any wiring projects.
- Be sure all equipment and wiring is properly grounded.
- Use pliers and other electrical tools with insulated handles.
- Don't staple or smash through cables.
- Protect underground cable with conduit wherever damage might occur.

THE WIRE

All wiring materials and installations must meet specifications set down in local codes. Most interior lighting and receptacles for barns and outbuildings are installed with plastic-sheathed cable, stapled or strapped every three feet to beams or other frame members.

Type NM (non-metallic, sheathed) cable, commonly called Romex, is suitable for most outbuildings because its plastic sheathing protects it from moisture. Type NM cable is made of a tough outer sheath that covers two or more plastic-insulated copper conductors and a bare copper grounding conductor. The ground wire is essential to your safety. It provides a conductor from the electrical system, either directly or through other conductors, to the ground.

Wire is also rated by numbers; the lower the number, the larger the diameter of the wire and the more current it can carry. To determine the right size of wire to use for a circuit, you must know the length of wire in the circuit and the maximum load from appliances and lights. Table 10–1 gives wire sizes for various loads and distances to the load center.

For example the table shows that No. 12 wire, which has a maximum capacity of 20 amperes can only be used for runs up to 30 feet on circuits that service a 2,200 watt

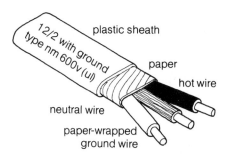

10-5. Twelve-gauge type NM cable with two conductors (12/2) is shown.

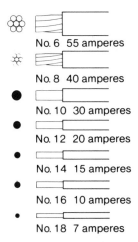

10-6. The smaller the wire number, the greater the diameter and ampacity.

153

load. No. 14 wire, which is sometimes used for small lighting circuits and is rated for 15-ampere service, can only be run 30 feet with a 1,400 watt load.

Cable can be purchased in standard 250-foot rolls or in smaller coils of 25, 50, and 100 feet. Unless the job requires only a few feet of cable, purchase one or two larger rolls rather than several smaller rolls.

GROUNDING

All electrical circuits must be grounded. This means there is a wire connecting the service panel to a rod driven into the earth. In residential areas, this ground wire is usually attached to the water-supply pipe and this in turn leads into the ground.

A grounded system reduces the effects of high voltage

TABLE 10-1

Wire Size Required*

(computed for maximum of 2-volt drop on two-wire 120-volt circuit)

Load/ Circuit	Current 120-volt Circuit	Length of Run (Panel Box to Load Center) — Feet																	
Watts	Amps	30	40	50	60	70	80	90	100	110	120	130	140	150	160	170	180	190	200
500	4.2	14	14	14	14	14	14	12	12	12	12	12	12	10	10	10	10	10	10
600	5.0	14	14	14	14	14	12	12	12	12	10	10	10	10	10	10	10	8	8
700	5.8	14	14	14	14	12	12	12	10	10	10	10	10	10	8	8	8	8	8
800	6.7	14	14	14	12	12	12	10	10	10	10	10	8	8	8	8	8	8	8
900	7.5	14	14	12	12	12	10	10	10	10	8	8	8	8	8	8	8	8	6
1000	8.3	14	14	12	12	10	10	10	10	10	8	8	8	8	8	8	6	6	6
1200	10.0	14	12	12	10	10	10	10	8	8	8	8	8	6	6	6	6	6	6
1400	11.7	14	12	10	10	10	8	8	8	8	8	6	6	6	6	6	6	6	6
1600	13.3	12	12	10	10	8	8	8	8	6	6	6	6	6	6	6	6	4	4
1800	15.0	12	10	10	10	8	8	8	6	6	6	6	6	6	4	4	4	4	4
2000	16.7	12	10	10	8	8	8	6	6	6	6	6	6	4	4	4	4	4	4
2200	18.3	12	10	10	8	8	8	6	6	6	6	6	4	4	4	4	4	4	2
2400	20.0	10	10	8	8	8	6	6	6	6	6	4	4	4	4	4	4	2	2
2600	21.7	10	10	8	8	6	6	6	6	4	4	4	4	4	4	4	4	2	2
2800	23.3	10	8	8	8	6	6	6	6	4	4	4	4	4	4	4	2	2	2
3000	24.0	10	8	8	6	6	6	6	6	4	4	4	4	4	4	2	2	2	2
3500	29.2	10	8	8	6	6	6	4	4	4	4	2	2	2	2	2	2	2	2
4000	33.3	8	8	6	6	6	4	4	4	4	2	2	2	2	2	2	1	1	1
4500	37.5	8	6	6	6	4	4	4	2	2	2	2	2	2	1	1	1	1	1

* Middleton, Roger G., *Practical Electricity* (Indianapolis, Indiana: Audel and Co., a division of Howard Sams and Co., Inc., 1974).

and lightning strikes, and reduces the danger of shock or fire, should some metal be accidentally made "live."

Because of the inherently damp conditions in barns, it is also advisable to have the added protection of a Ground Fault Circuit Interrupter (GFCI). This device can be installed either in individual outlets or in the breaker panel box to protect the entire circuit. The function of a ground fault device is to shut off the circuit whenever it detects stray current that could shock you when using electrical equipment. The ground fault device will shut down the circuit even if the short-circuit in the appliance is not enough to trip the circuit breaker, but enough to give you a good shock.

Outdoor ground-fault-protected receptacles should also be used for such things as water tank heaters during winter months.

Again, check local wiring codes to determine grounding requirements.

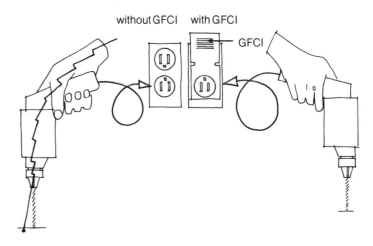

10-7. GFCIs protect people from ground fault currents. Be sure to check codes for grounding requirements.

Planning Circuits, Lights, and Outlets

Before doing any wiring in your building, make up a wiring plan or schematic. This allows you to plan the circuits, so outlets and switches are within easy reach, lighting is adequate for the work you anticipate, and there are plenty of outlets where needed. For example, you may wish to locate switches so that lights can be turned on and off at two convenient locations. Your wiring schematic also makes a handy "shopping guide" when you purchase the electrical wire, fixtures, and other materials.

Following are some suggestions for light and outlet locations:

- **Plan one light in the center of the alley for every two or three stalls. Install one outlet every 10 to 15 feet in the center of the feed alley. You also may need outlets for ventilating fans.**
- **Barns, box stalls, and pens require one light outlet for every 150 square feet of open pen area.**
- **For portable millers, clippers, and other power equipment, you will need outlets every 20 feet.**
- **In the milking room, there should be a light in front of every three cows, with an outlet behind**

every four to five cows.

- For every 100 square feet in the milkhouse, there should be a light and an outlet. Special-purpose outlets also may be needed for fans, coolers, sterilizers, heating equipment, and water heaters.
- For hog and farrowing houses, one light is ample for two hog pens. Provide outlets for heat bulbs and water warmers.
- Poultry-laying houses will need both dim and bright-light circuits.
- Outlets are also needed for motor-driven feeders, and farm shops need outlets for each fixed piece of equipment.
- Special 240-volt outlets are needed for power-driven equipment of ½ horsepower or more and for electric welders. Overloading and poor voltage can damage motors and other electrical equipment.
- Locate lights in feed and tack rooms, in lofts, stairways, and alleyways. For horse and other large animal stalls, use only unbreakable light fixtures, such as those enclosed in wire housing.
- To enhance safety and provide convenience, install floodlights at strategic places outside the building.

INSTALLING OUTLET BOXES

At switch and receptacle locations, mount rectangular outlet boxes. These are available in either metal or plastic, with plastic gradually replacing metal (where legal). Standard boxes are 2½ inches deep, although deeper boxes can be purchased for special purposes. Boxes are also available in single or double, where you wish to provide two switches or outlets at one location. Boxes come with a variety of fastening methods. Some have nailing straps attached to their sides. Some plastic boxes have nails pre-placed in holders in the box, while many of the older-style boxes simply have holes punched in the box for 16d nails to be driven through to anchor them in place.

The boxes are mounted so they will be flush with the finished wall (or studs if there will be no interior wall covering). Mount switches 4 feet from the floor and outlet boxes 13 to 15 inches off the floor, or as required.

Unless you are using fluorescent light fixtures with built-in junction boxes, you will need to mount 4-inch octagonal junction boxes for overhead lights. You will also need these boxes where wire must be joined to provide

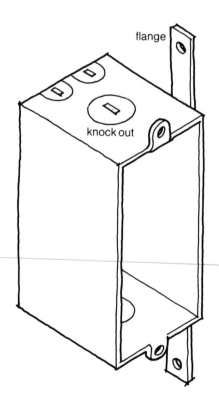

10-8. An outlet box like this may be nailed to a stud by its flange.

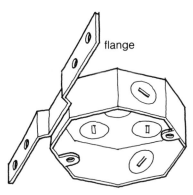

10-9. Octagonal junction boxes are used for overhead interior lights.

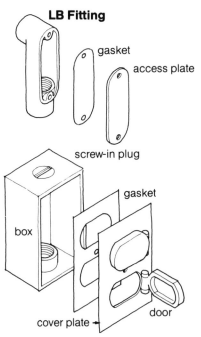

Weatherproof Receptacle

10-10. An LB fitting (L shaped with back conduit opening) protects wire passing out through a wall. A special outdoor cover plate makes the receptacle weatherproof.

power to more than one source, for instance a light plus a line of outlets. Junction boxes are available with side nailers for fastening to joists or with a hanging bar that spans between the studs and allows the box to be positioned at any point between them.

Once the boxes have been mounted, run the circuit wires between them without installing switches or outlets. Don't hook the wire up to the service panel yet. Run the wires by the most direct routes possible, but remember that most codes prohibit running Type NM wire beneath joists or over studs where it could be damaged. Therefore you will have to drill holes through the studs and joists when you have to cross them. When drilling joists or studs, drill in the center of the wood to minimize structural weakening. If you have many wires running across a room, you can run them all together in a ceiling trough (a wooden box) mounted under the joists. When installing the wire into the junction boxes, strip about 6 inches of the plastic sheathing off the wire. Insert the wire through one of the back knockouts and pull it out the front until the plastic sheathing just enters the box. This will give you enough wire to hook up an outlet or switch. If the circuit continues from the box, another 6-inch length of unsheathed wire should also be inserted in the box and the wire clamp (if the box has one) tightened at the back of the box to hold both pieces of cable.

When running wire between boxes, be careful to do a neat job. Run the wire tightly enough to prevent sagging, but not so tight that it is stretched. Staple it every couple of feet with wire staples where it runs along joists or studs. When rounding corners and going through tight spaces, don't crimp the wire, but make easy and gradual bends so as not to damage the copper conductors or the plastic sheathing.

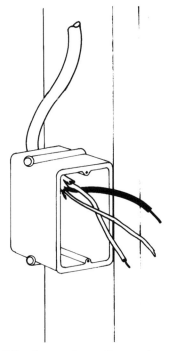

10-11. Wire is run from the service panel to junction or outlet boxes. Strip about 6 inches of sheathing off wire, then insert wire through back of box, pull it out front until plastic sheathing just enters box.

157

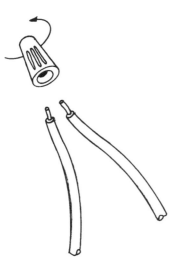

10-12. *One simple method of connecting wires together is using "wire nuts." The appropriate size nuts are simply twisted over the bare stripped wire ends.*

WIRING LIGHTS, SWITCHES, AND RECEPTACLES

After the wiring has been run to all the boxes, but *before the circuit is hooked into the service panel,* install the switches, receptacles, and lights.

In normal, two-wire circuits, there is a black wire that is "hot" and carries current to a fixture, a white wire that is neutral or a ground potential and carries the current back from the fixture, and a bare ground wire that grounds all receptacles (and boxes if metal and installed thus according to some local codes) to prevent shocks. For safety, the white wire and the ground wire should be continuous throughout the circuit. Thus, only the black wire is ever interrupted by a switch.

Several common wiring diagrams are presented to familiarize you with the flow of electricity in a circuit. For further information on more complicated circuits consult a wiring handbook or a competent electrician.

Light Circuit with Single Switch

Power is supplied to the switch box where the black wire is interrupted and attached to the switch terminal. The white wire is connected with a wire nut to the white wire running to the light fixture. The two ground wires are also connected together and joined to a jumper which grounds the switch box (if grounding screw is available). This jumper should be attached to a screw on the back of

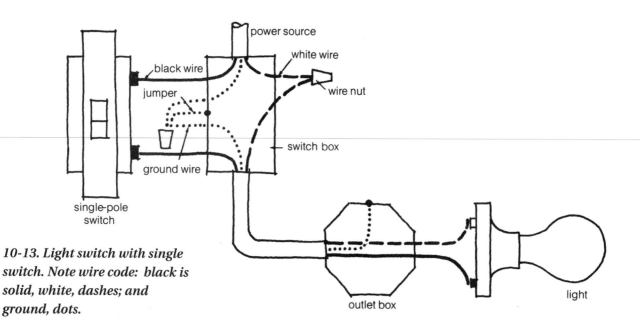

10-13. *Light switch with single switch. Note wire code: black is solid, white, dashes; and ground, dots.*

the box to make a firm mechanical connection. Additional lights could be put on this switched circuit by running a new cable from the junction box, merely attaching the matching wires to the black, white, and ground wires.

Light Circuit with Two Three-Way Switches

A light can be controlled from two locations using this circuit. There are three wires and a ground running from the switches to the light. Three-wire cable must be used; the kind that is commonly available has black, red, white, and ground wires. In this circuit, the power comes to the switch box where the black is interrupted and tied to the "common" terminal of the three-way switch. Make sure you understand which is the common terminal on the switch. The white neutral wire continues on uninterrupted to the light fixture. Red and black "traveler" wires are then attached to the two switch terminals on the switch and are run to the other similar terminals on the other three-way switch. The remaining wire is then attached to the common terminal on the second switch and run back to the light. Notice in the diagram that a black wire is attached to a white wire in the ceiling fixture box. The white wire going to the second switch is actually serving as a black "hot" wire. Whenever this is done for switching purposes, the

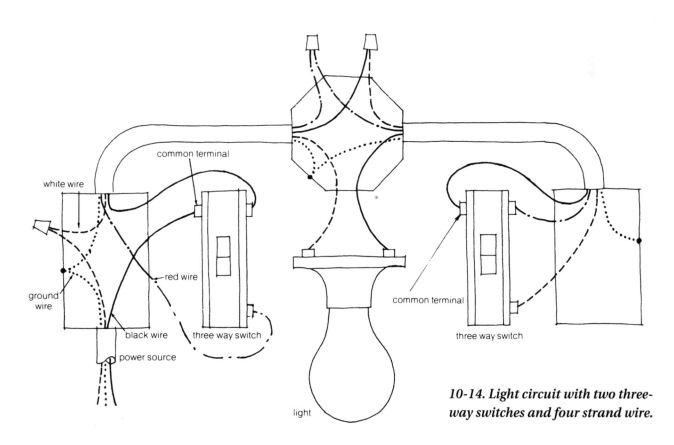

10-14. Light circuit with two three-way switches and four strand wire.

159

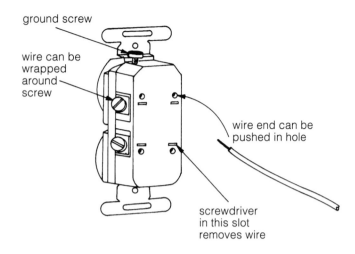

ground screw

wire can be
wrapped
around
screw

wire end can be
pushed in hole

screwdriver
in this slot
removes wire

10-15. These days switch and receptacle wiring is easy. Although appropriate screws are available, most receptacles and screws have holes in the back where the appropriate wires can simply be inserted and are held firmly in place.

ends of the white wire should be painted with a black magic marker to indicate that it is a black wire and, in fact, is carrying current.

Receptacle and Switch Wiring

Attaching wires to receptacles and switches is quite easy these days. Although most receptacles and switches have screws for fastening the wires in place, most also have holes in the back where the appropriate wires can simply be inserted. Strip about ½-inch of the insulation off the ends of individual wires and simply push the wires into the holes until they fit snugly. If you must remove a wire insert a small screwdriver into the slot next to the hole and pull the wire out.

For polarity the white wires must be fastened to silver screws or inserted in the holes in the side of the receptacle with the silver screws. The black wire ends are fastened to the brass screws or inserted in the holes in the brass side. Regardless of whether you're using screws or holes to install the wires, make sure the two wires are fastened directly opposite each other on a receptacle. Receptacles have two sets of screws and holes. To continue power from one receptacle to another simply attach the appropriate wires into the second set of holes or screws. The receptacle has a green screw that is the ground connection. On a single receptacle the bare copper ground wire from the source is fastened to this. A short piece of ground wire is also connected from this green screw to the grounding screw on

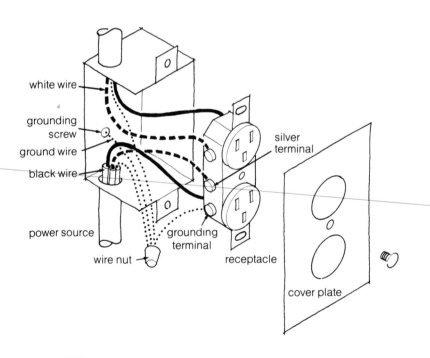

white wire

grounding
screw

ground wire

black wire

power source

wire nut

grounding
terminal

silver
terminal

receptacle

cover plate

10-16. Duplex receptacle, using three-strand wire.

the outlet box (if available). If the power continues to another receptacle, connect the ground wire to either the box grounding screw or to the green screw on the receptacle.

CONNECTIONS TO THE SERVICE PANEL

After all the outlets, switches, and lights have been installed, the final circuit connections to the service panel can be made. In some locations *this must be done by a licensed electrician.* In other areas, it can be done by homeowners on their own property. If you do all the wiring yourself, always have it inspected by an electrician or building inspector *before* your new circuits are energized.

At the service panel, first make sure that the main breaker is off *before starting to hook circuits into the box.* With the main breaker off, you can work safely in the box, but remember that the entrance cable coming into the main breaker is hot and contact with it must be carefully avoided (unless the building has a separate disconnect, which is required in some areas, and it has been turned to the OFF position).

Bring the circuits into the box one at a time through the

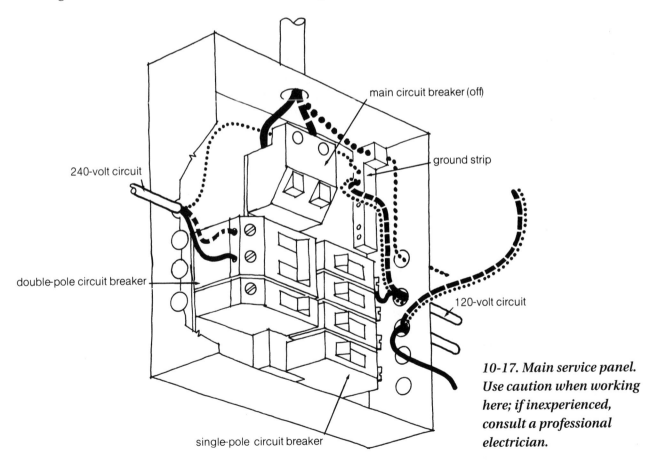

main circuit breaker (off)

ground strip

240-volt circuit

double-pole circuit breaker

120-volt circuit

single-pole circuit breaker

10-17. Main service panel. Use caution when working here; if inexperienced, consult a professional electrician.

161

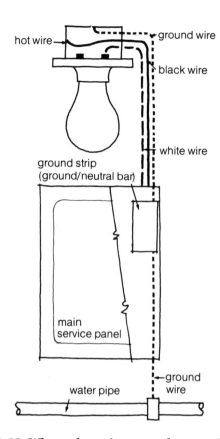

hot wire

ground wire

black wire

white wire

ground strip
(ground/neutral bar)

main
service panel

ground
wire

water pipe

10-18. When a hot wire comes loose and touches the metal box, the ground wire removes this dangerous current, carrying it first to the neutral ground strip, then on to a water pipe (or ground rod) that goes to the earth.

side knockouts, which can be removed with a large screwdriver and a pair of pliers. Romex connectors or "cable clamps" should be inserted in the knockout holes to securely clamp the cable as it enters the box. Remember to leave enough wire so you can run the black lead to the circuit breaker and the white and ground leads to the ground strip. Starting from the top circuit breaker position, attach the circuits in a descending sequence with the wires run neatly inside of the box so they don't interfere with new circuit wires and breakers as they are added.

After all the circuit breakers that you need are in place, prepare the front cover of the service panel to fit over them. The front panel has knockouts for each breaker position, and using a pair of pliers you should remove the ones that correspond to the breakers you have just put in. The unused breaker positions should remain covered until they are needed. Carefully fit the cover over the service panel and label all the circuits you have installed on the circuit directory on the cover door.

A final step before closing up the service box is to ground the panel and circuits. Using No. 4 solid copper wire with a metal casing like BX wire, run a ground cable from the ground strip in the box to either the building's plumbing or a copper rod driven into the earth. Using a copper rod is the surest way to ground the service panel, but if you use the plumbing, make sure the ground connection is to metal pipe that enters the building from underground. Some codes require that, if you can't ground the system right where the pipe enters the building, you should drive a ground rod and attach the wire to it.

CHAPTER 11

Plumbing

11-1. Many outbuildings will require plumbing, and again a great deal of this can be done by the homeowner. In fact, modern materials and methods make plumbing much easier and faster.

The water needs of your outbuildings determine the amount of plumbing required. Large cow barns, milk barns, hog farrowing, finishing, and other buildings will of course require much more plumbing than a simple workshop or general barn. Some, such as a milk barn, will require both hot and cold running water and manure-removal systems, while hog confinement barns will require plenty of water for flushing and cleaning as well as manure-removal systems. These will also require special manure-handling capabilities such as lagoons. The simplest plumbing, on the other hand, is a garden hose run to the building from the main house or an outside freeze-proof faucet. A more convenient system for the small barn is a plastic pipe run underground (below frost line) from the main house or well to the barn and into a freeze-proof faucet to supply cold water at the barn.

In most plumbing systems, other than the simple cold water supply, there are two distinct sets of pipes; the water supply pipes and the waste-water drain pipes. For rural locations the water supply often starts at a well — either a spring box, a shallow-dug well, or a deep-drill well. In some instances the water supply may begin from a pond or holding basin. Regardless of where it begins, the water is

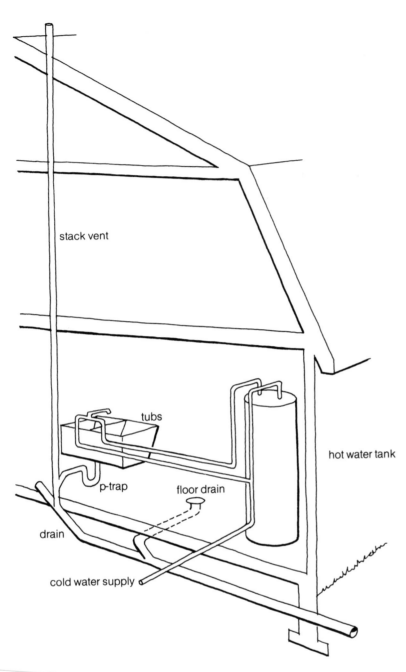

stack vent

tubs

hot water tank

p-trap

floor drain

drain

cold water supply

11-2. A basic plumbing diagram for a barn. Plumbing codes require venting for each plumbing fixture. Usually, hot water is only necessary for livestock operations requiring cleaning and sterilization.

transferred to a storage tank, either by gravity feed from an uphill spring or, more often, by a pump. From the pressurized tank, water is fed through distribution lines called *risers* to various fixtures such as faucets, wash basins, and sillcocks. If hot water is required, an electric, gas, or oil hot water heater is installed with its own separate set of supply lines.

The drainage system starts at the fixtures and carries away waste water through pipes to a septic tank. Traps which hold standing water are always installed directly below sink and floor drains to prevent sewer gas from coming back up the drain pipe and into the building. The drain pipes must also be vented through the roof to allow these gases to escape and to prevent draining water from creating a vacuum in the pipe. At the septic tank, the wastes are partially digested by microorganisms and allowed to settle out. The waste water then travels to a leach field where it disperses into the ground. In the past few years, however, many counties and municipalities have begun requiring lagoons rather than septic systems.

Because of the health hazards associated with improperly installed water supplies and septic systems, plumbing work is strictly regulated by codes. These vary from state to state, especially in regard to who may do plumbing work and what materials are acceptable, but the basic requirements are standard throughout the country. The National Plumbing Code is the basis for all state regulations and its recommendations should be followed; check with your local authorities for specific information on codes and accepted practices.

In the past plumbing required a great deal of tools and special skills. With the advent of plastic pipe and tubing for both supply and drain lines, however, plumbing has become greatly simplified and practical for many homeowners to do themselves. Before you start, however, take advantage of free information and advice that is available on what is required in your area. Contact an Extension Service agent to get recommendations on the size and type of water-supply and waste-removal systems that will meet your outbuilding requirements. Also check with local building inspectors on plumbing codes and health regulations. Building supply dealers can advise you on the best materials to choose. Again, it's a good idea to make up a diagram of the plumbing requirements of your building. This not only guides you in laying out the plumbing, but can also be a shopping list when you go to purchase materials.

SUPPLY PIPES AND FITTINGS

One of the hardest things about doing your own plumbing is knowing what materials and fittings are available and their proper names. A good knowledge of plumbing materials and terms will make your work easier.

Piping for water supply lines is available in galvanized steel, rigid copper, copper tubing, and plastic. These days copper tubing and plastic are the most widely used, with plastic a first choice among do-it-yourselfers for many chores. Because galvanized steel must be cut and threaded with special tools and requires more precision in cutting and joining, it is hardly ever used except where required by the codes.

Rigid copper piping can be cut with a tube cutter or hacksaw and is joined with copper "sweat" fittings, using a propane torch and lead solder. Like rigid galvanized steel pipe, it does require careful measurement and a bit of soldering skill.

Some common fittings used to join rigid copper pipe are: 90- and 45-degree elbows or ells for bends or turns; tees for joining two lines together; couplings for joining straight runs; and male or female adapters to go from a sweat fitting to a threaded one. Unions are also used to allow copper to be taken apart without unsoldering the joints.

Copper tubing on the other hand is flexible, comes in rolls, and can be bent, shaped, or fit into tight places more easily than rigid pipe. It is normally cut

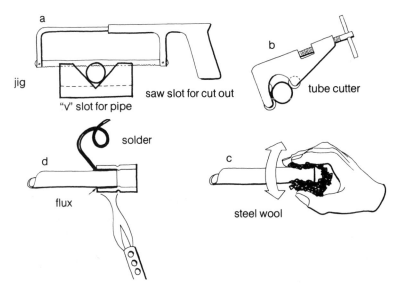

11-3. Rigid copper tubing can be cut neatly with a hacksaw in a jig (a) or with a small tubing cutter (b). Before a connection is made, the outside of the tubing and the inside of the fitting have to be cleaned with steel wool or some fine abrasive paper (c). Surfaces to be soldered should be wiped with "flux" before soldering and after abrasive cleaning. Don't try to heat the solder wire directly (d), instead heat the tubing and fittings and touch solder to them.

165

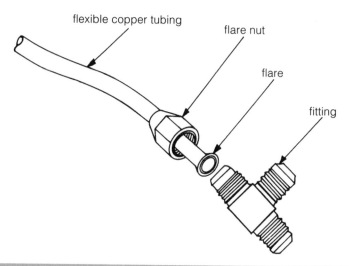

flexible copper tubing

flare nut

flare

fitting

11-4. Flexible copper tubing is extremely easy, even for first timers to work with. It is joined with flare fittings. First a flare nut is slipped over the tubing, then the flare formed on the tubing with a flaring tool. The flare is pushed against the fitting, then the flare nut is turned onto the fitting. Plumbers tape applied to the threads of the fitting prevents leaks.

TABLE 11-1

Pipe Data at a Glance*

Type of Pipe	Ease of Working	Water Flow Efficiency Factor	Type of Fittings Needed
BRASS	Threading required or ask for pre-threaded. Cuts easily, but can't be bent. Measuring a rather difficult job.	Highly efficient because of low friction.	Screw-on Connections.
COPPER PIPE	Easier to work with than brass.	Same as brass.	Screw-on or solder connections.
COPPER TUBING, RIGID	Easier to work with than brass or hard copper because it bends readily by using a bending tool or by annealing. Measuring a job not too difficult.	Same as brass.	Solder connections.
COPPER TUBING, FLEXIBLE	Easier than soft copper because it can be bent without a tool. Measuring jobs are easy.	Highest of all metals since there are no nipples, unions or elbows.	Solder or compression connections.
GALVANIZED STEEL (OR WROUGHT IRON)	Has to be threaded. More difficult to cut. Measurements for jobs must be exact.	Lower than copper because nipple unions reduce water flow.	Screw-on connections.
PLASTIC PIPE	Can be cut with saw or knife.	Same as copper tubing.	Insert couplings, clamps; also by cement. Threaded and compression fittings can be used. (Thread same as for metal pipe.)

with a tubing cutter and joined by using "flare fittings." To join with flare fittings, the fitting is slid over the end of the cut pipe, a reaming tool used to smooth the inside edge of the cut pipe, then a flaring tool used to create a "bell" type of flare on the end of the tubing. The flared nut is then turned over any number of different fittings to join the pipe together or to other fixtures. About the only problem that occurs with tubing is kinking it if you try to bend it too sharply. A tubing bender, consisting of a spring wire slipped over the tubing, will usually solve this problem.

In recent years, plastic pipe has won approval in most states for potable water supply lines. It is cheaper than copper and easier to install. Plastic piping is also less

Manner Usually Stocked	Life Expectancy	Principal Uses	Remarks
12-ft. rigid lengths. Cut to size wanted.	Lasts life of building.	Generally for commercial construction.	Required in some cities where water is extremely corrosive. Often smaller diameter will suffice because of low friction coefficient.
12-ft. rigid lengths. Cut to size wanted.	Same as brass.	Same as brass.	
Three wall thicknesses: K=thickest, L=medium, M=thinnest. 10- or 20-foot lengths.	Same as brass.	"K" is used in municipal and commercial construction. "L" is used for residential water lines. "M" is for light domestic use only: check codes before using.	
Two wall thicknesses: K=thickest, L=medium. 30-, 60-, or 100-foot coils (except "M").	Same as brass.	Widely used in residential installation.	Probably the most popular pipe today. Often a smaller diameter will suffice because of low friction coefficient.
10- or 21-foot rigid lengths. Usually cut to size wanted.	Very durable.	Generally found in older homes.	Recommended if lines are in a location subject to impact.
Rigid, semi-rigid and flexible. Continuous lengths to 1000 ft.	Long life and it is rust and corrosion-proof.	For cold water installations. Used for well casings, septic tank lines, sprinkler systems. (Can be used for hot water in some applications.) Check codes before installing.	Lightest of all, weighs ⅛th of metal pipe. Does not burst in below-freezing weather.

Homeowners Guide to Plumbing (Milwaukee, Wisconsin: Ideals Publishing Co., 1981).

11-5. Plastic supply lines are also easy to work. They can be cut with a hacksaw, and are joined to fittings with plastic cement.

susceptible to mineral scaling. There are two types, CPVC (chlorinated polyvinyl chloride) and PB (polybutylene). PB is the type to use if it is available because it is flexible and can be run around corners without expensive fittings. The plastic casing also provides some insulation for the water line, and it won't burst open even when frozen because it is so flexible. PB is available in rolls up to 1000 feet long and comes in two colors, gray for hot water and blue for cold. It has been approved by the Food and Drug Administration for potable water supplies.

In most instances, however, you will probably only be able to find CPVC or rigid plastic tubing. It is a bit harder to work, only in that it requires more careful measuring and fitting and is a bit tougher to install in tight places. It will also freeze and burst fairly easily and must be insulated.

In general, though, both types of plastic pipe are easily worked, and both can be cut with a hacksaw. The cut ends are smoothed with a piece of sandpaper and then the piping and fittings joined using a special plastic cement.

VALVES AND FAUCETS

Valves are used to shut off water flow in a line when it is no longer needed or when a fixture must be repaired. There are two standard types: the gate valve and the globe valve.

A gate valve does not restrict the flow of water when it is fully open and therefore it is preferable for main shut-offs that are only closed in emergencies. Globe valves do restrict the flow of water when they are fully open, but have the advantage of being able to control the rate of flow from full force down to a trickle. Gate valves don't do this very well. Therefore, globe valves are used in branch lines and other places where the flow of water needs to be controlled or turned on and off often. Also, globe valves are cheaper than gate valves and easier to shut off.

Globe valves are available with a side drain on the non-pressurized chamber that allows you to drain shut-off pipes. These are called stop-and-waste or stop-and-drain valves. These valves must be installed in the proper direction, with the arrow on the body of the valve facing in the direction of the water flow.

The most commonly used fixtures for most outbuildings are various faucets for controlling water at sinks, tanks, and hoses. A *hose bib* is a valve with a threaded spout for drawing off water or hooking up a hose. These

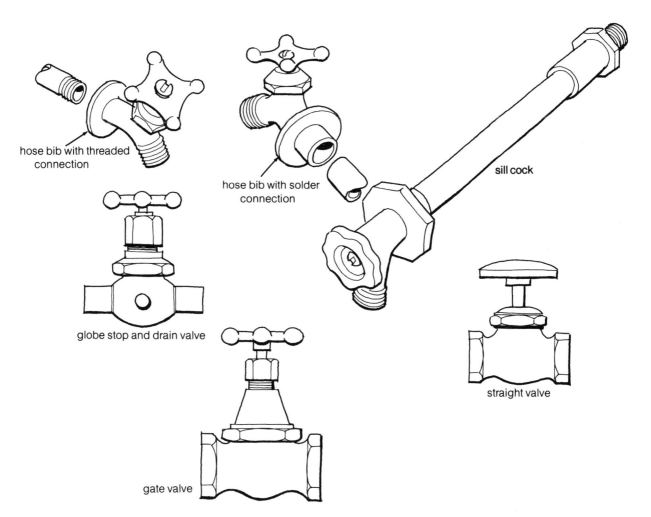

11-6. Valves and faucets suitable for small barns. A frost-proof faucet, known as a sill-cock is located on the outside of the building, but its workings are on the inside where it's warmer.

come with sweat or threaded connections and in different mounting styles. A frost-proof sillcock is a hose bib with a long extended valve. With the valve shut-off located inside a warm building, water freezing is less likely, although it can occur.

There are two practical ways to control the flow of water to livestock tanks. You can mount a hose bib and fill the tank manually, or you can install a valve that opens and shuts when activated by a float. Float valves used in toilets to regulate the flow of water can be modified and installed in livestock tanks. There are also livestock tanks with built-in float valves, or you can purchase an "add-on" unit.

A laundry sink for washing and cleaning is often useful in a livestock barn or equipment shop. These should be equipped with a double-handle faucet (if you have hot water) and a large movable spout. If grease, oil, or hair is likely to get in the drain, install a special drum trap that can be cleaned easily by removing a top cover.

JOINING PLASTIC DRAINAGE PIPE

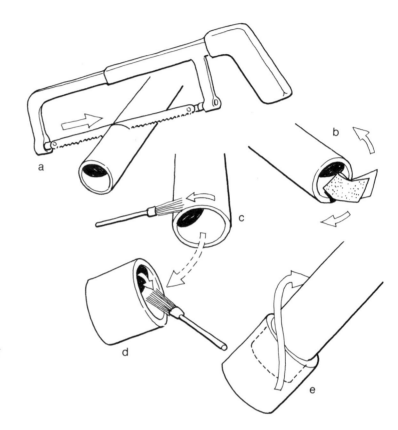

11-7. To join plastic waste pipe cut the pipe to the desired length (a); smooth burrs and clean with pipe cleaner (b); test fit, then apply welding cement to pipe (c), then fitting (d); rotate fitting ¼ turn as you put it on (e).

PVC and ABS plastic pipe are joined with fittings "welded" with a special solvent cement. The cement and a cleaning solvent are available in individual containers wherever plumbing supplies are sold.

Plastic pipe should be cut with a hacksaw or fine-toothed cross-cut saw. When cutting make sure the cut is square so the end of the pipe fits evenly into the fitting. Use fine sandpaper to smooth off the burrs on the end of the pipe after cutting. Clean off grease or dirt on the pipe and fitting with pipe cleaner and a clean cloth.

Before you apply the welding cement, test-fit the pipe and fitting together to make sure the fitting goes on smoothly. With a pencil, mark the orientation of the fitting on the pipe so they will line up properly when joined. The solvent will weld the plastic together in about 5 seconds, so once it is applied you don't have much time to get the fitting and pipe aligned. The orientation of the fitting doesn't matter for couplings, but for elbows and tees it is critical.

With the fitting and pipe marked, apply the solvent to the outside of the pipe and the inside of the fitting. Fit them together one-quarter of a turn out of alignment and as you push the fitting on, turn it to line up the pencil marks. This will evenly spread the PVC cement and make a good seal. Let the joint stand for a few seconds to cure, then wipe off any excess cement.

WASTE WATER PIPES

Barns that have sinks and floor drains require a waste-water system. This is commonly called the DWV (drain, waste, and vent) system. Cast-iron, copper, and plastic are the three types of pipes used.

Because it is lightweight and easy to work with, plastic drain pipe is the best material for the homeowner to install. There are two types of plastic drain piping; PVC (polyvinyl chloride) and ABS (acrylonitrile-butadiene-styrene). Some codes permit only the use of ABS plastic. Both types are available in 10-foot lengths and range in size from 1½ to 4 inches in diameter.

Plastic pipe is cut with a regular saw or hacksaw and the fittings are welded onto the pipe with a solvent that fuses the plastic together. It is an extremely quick and easy procedure to assemble drainage pipe.

Plastic pipe can be run from the building to the septic tank. Special perforated pipe is made for the distribution lines in leach fields. The only places that many codes don't permit the use of plastic pipe are under concrete slabs and where the pipe penetrates foundation walls. Here, cast-iron pipe must be used, and a professional plumber must do the installation if the joints are to be leaded. For the homeowner, no-hub, cast-iron pipe is available that joins together with neoprene gaskets and clamps.

THE WATER-SUPPLY SYSTEM

If the main house is already served by a well that can meet the additional water requirements of your outbuildings, it is best to run an underground line from the house pressure tank to the barn. If, however, your water requirements are greater than the well's capacity, then

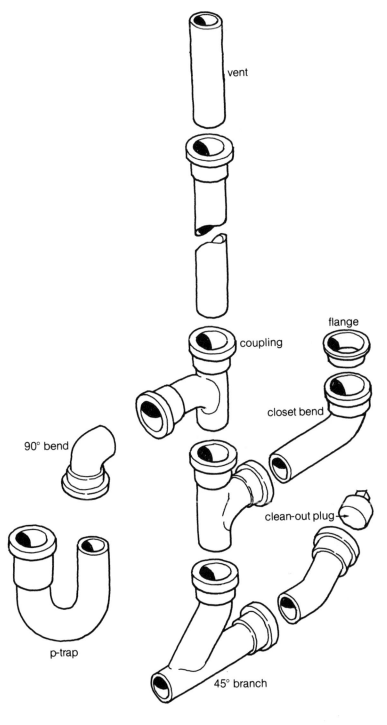

11-8. Drain, waste, and vent systems include various pieces, made of cast-iron, copper and plastic.

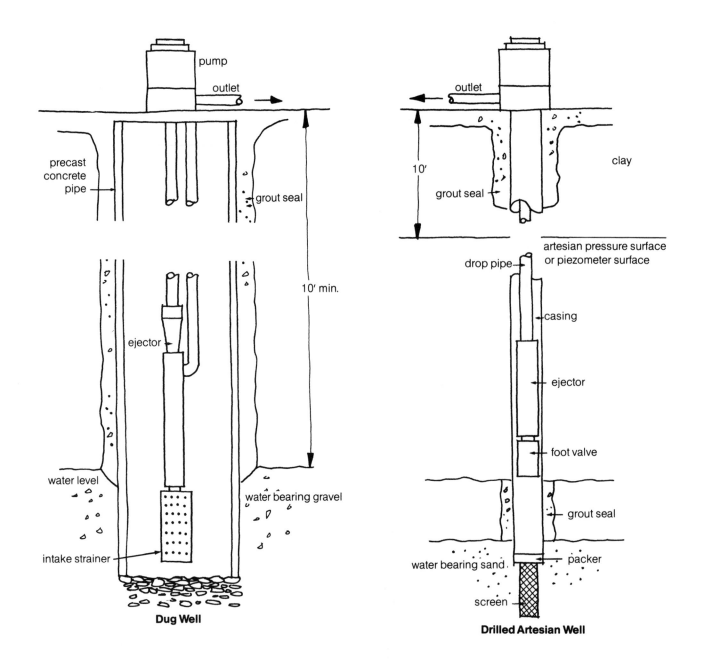

Dug Well

Drilled Artesian Well

11-9. At left is a simple dug well, with a minimum depth of 10 feet and a pre-cast concrete liner; at right is a drilled artesian well, which is dug with heavy equipment.

you must find a new source. Developing a new water supply is almost always expensive, regardless of whether it is a spring, well, or reservoir.

Finding water is not always easy. Even if a neighbor has a well 30 feet deep that supplies 100 gallons a minute, you might drill to 120 feet and find nothing. Three sources of help are available to the rural homeowner looking for water. One option is to have a professional geologist look over your land and suggest the most likely place to find water. Another is to have a dowser with his divining rod walk the property and point to the location of underground water. (Don't laugh: my father is a dowser, and I've repeatedly seen him locate underground water and describe approximately how deep the well will have to be drilled to reach it.) Lastly, you can also consult neighbors about their success with wells. States often keep records of drilled wells that indicate the quality and quantity of water found in your area. These days it is also necessary to consult the natural resources or environmental agencies in your area about the possibility of contaminated groundwater. In some instances a reservoir and water-purifying system may be the only answer.

The simplest type of well is a surface spring that can be dug out and lined with concrete, well casing, or stone. If the spring is on an elevation above your barn, water can be siphoned off by gravity flow, eliminating the need for an expensive pump. Water from this type of well, however, often won't meet health regulations for human consumption because of the possible surface water contamination, and it should be tested before providing it to livestock.

Three things to keep in mind when developing a spring well are: the spring box should be located above any possible source of contamination; the box should be fully encased and covered to keep out animals and dirt; and the well must flow year-round, not just in the spring.

Normally, deep wells are drilled with heavy equipment that can bore through the earth and rock and drive the well casing into place. When the casing hits an underground water fissure, artesian pressure forces the water into the casing and up to the surface. A submersible or tank-mounted pump then lifts the water to a pressure tank. Usually, 1½-inch polyethylene pipe connects the well and pressure tank. This must be run well below the frost line to avoid freezing.

When digging a new well, always ask a state or private laboratory to test the water to determine if it is fit for consumption. Shallow wells can be contaminated easily by sewer lines or spilled chemicals and even deep groundwater can be contaminated by nearby chemical dumps or natural radioactivity in the bedrock. Water also contains many different minerals that may not pose a health hazard but may cause pipes to scale or deteriorate.

All but gravity-feed systems must have a pump and pressure tank to store water drawn from the well. The size and type of pump depends on the depth of the well and how many gallons per minute (gpm) you need to supply. All pumps should have an automatic pressure switch that turns on the pump when the pressure falls below a certain point and turns it off at a pre-set high limit. Most systems are designed to operate between 30 and 40 pounds per square inch (psi) of pressure.

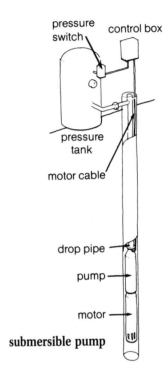

submersible pump

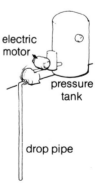

shallow well pump

11-10. Two common types of water pumps.

173

From the tank, the water supply lines or risers lead to individual fixtures. Whether you use plastic or copper tubing, keep in mind the following points when installing the lines:

- **The entire system should be drainable, to protect against freezing and to make necessary repairs. Always slope the pipe at ¼ inch per foot of run and put drain valves at all low points in the system.**
- **Make your pipe runs as short and straight as possible. The longer the pipe run, the smaller its diameter, and the more elbows and turns it takes, the greater the loss of pressure. To compensate for pressure drop, ¾-inch pipe is commonly used for the main supply lines and ½-inch for lines to individual fixtures.**
- **Don't run water lines in outside walls if you can avoid it. Even pipes in insulated walls may eventually freeze, causing water damage and making major repairs necessary. Install pipes where they will be warm and where they can be inspected and serviced easily.**

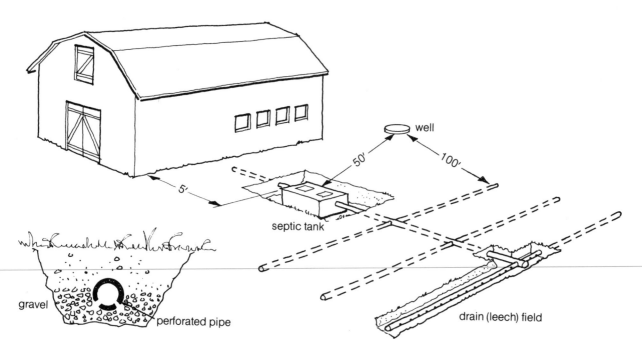

11-11. When planning a septic system, check local codes to ensure that the tank and leach field are adequate distances from the outbuilding.

- Always support overhead pipes with hangers or clamps set a maximum of 10 feet apart. Copper clamps that fit over different sizes of tubing can be nailed into the joints or studs. Galvanized strapping is also available in rolls to hang pipe from the ceiling at a desired height.
- Install shut-off valves on all pressure tanks, hot water heaters, and at points before fixtures such as sink faucets. These components can then be isolated for repair without having to shut off and drain the entire plumbing system.

WASTE SYSTEMS

An outbuilding that has simple outdoor hose bibs for livestock water does not need a waste-water drainage system. The water simply seeps into the ground. If you have sinks and floor drains, however, regulations probably require you to dispose of waste water in an approved septic system or lagoon, not just a drain that runs outside to daylight or to a drywell. This requirement is especially likely for livestock buildings where manure might mix with the waste water.

Check with your local Extension Service agent about the waste-management requirements for your specific type of outbuilding and its uses.

CHAPTER 12

Finishing Outbuilding Interiors

12-1. Proper finishing of barn interiors is as important as choosing the proper barn style and size for your needs. Shown is a typical stable arrangement with stalls on both sides and a walkway between. Note the sliding steel stable doors and steel stall guard windows.

Many barns, sheds, and outbuildings require little or no additional work in finishing off their interiors. For instance, a machine shed is basically finished when you put on the roofing, siding, and a concrete floor and doors if desired. Of course, a few shelves will help, and you may wish to add in a workbench for on-the-spot repair work.

Even a general barn can be left as is, simply wiring up gates or panels to partition it off as needed. This is one of the real advantages of pole barns, especially those with truss roofs, where the entire inside of such a barn remains open and free of supports. The inside of the barn can be adapted to suit almost any purpose. A good example is one of our barns. It started off as a storage barn for hay. We were one of the first in our neighborhood to grow alfalfa, and for several years our main cash crop was alfalfa hay. We regularly put up 10,000 to 12,000 bales of hay in the barn yearly to be sold when the market prices were right.

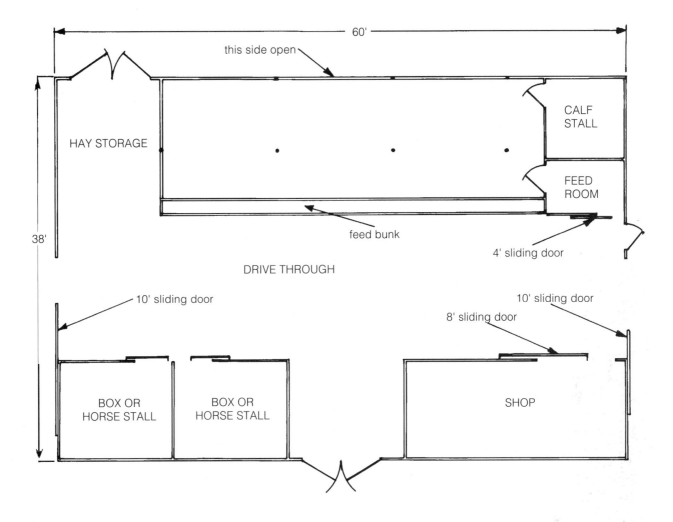

Text inside the figure:

60'

this side open

HAY STORAGE

CALF STALL

FEED ROOM

38'

feed bunk

4' sliding door

DRIVE THROUGH

10' sliding door

10' sliding door

8' sliding door

BOX OR HORSE STALL

BOX OR HORSE STALL

SHOP

Actually, the barn was designed as a horse barn, but when the hay situation came along we just left the interior of the barn open. Since then it has been adapted several times to suit different farming needs, including raising hogs, and has even served as an automobile shop for a teenage boy's automobiles. The floor plan shown is for all-around use, but is just one suggestion. It includes hay storage and a cattle loafing/feeding shed on one side, as well as a calf stall and feed and equipment room. Also included are two box or horse stalls and a small workshop. A driveway through the center of the barn can be used for tractors, trucks, and other equipment. This arrangement leaves the south or loafing side open and the feed bunks open to the center of the barn for feeding small hay bales. Like most folks, however, we've gone to big bales and in reality the barn shown has the cattle loafing shed portion partitioned off with a solid wall from the rest of the barn. Hay in big bales is fed just outside the loafing shed. This allows us to close off the entire barn and use it for equipment storage and as a workshop.

12-2. Floor plan of typical 60' by 38' general barn.

177

Any livestock barn covered with metal siding, plywood, or hardboard should have the lower parts of the interior stall or loafing shed portions covered with solid 2x lumber. In the past this has often been yellow pine tongue-and-groove planking, but a better choice these days is treated 2x6s. This planking should be run up at least 5 feet high and the top edge finished off with blocking between the planking and outside wall so nothing can fall down between the two.

Interior dividers can be permanent, fastened securely between posts, or they can be movable, tying them in place with wire. One alternative I've found extremely useful in creating a more versatile building is to hinge stall and interior dividers to an outside wall. They're basically nothing but gates used to divide an interior into several stalls. It works well for hogs, sheep, and cattle. For cattle use 2x6 material; for hogs, 1x6s will suffice; and for sheep, 1x4s can be used.

Feed bunks are also often a major part of a barn interior. The sheep feeder shown is used both as a feeder and as a divider for the sheep and goat shelter (see Project No. 8, page 217). One end can be left open and you can actually walk through the feeder to place hay or feed as necessary.

On-wall cattle feed bunks are also used when feeding small hay bales. If the barn has a loft for storage, a loft ladder is also needed. Shown is an arrangement that my grandfather had in his barn.

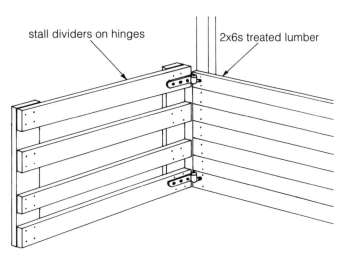

12-3. Gates can act as stall dividers.

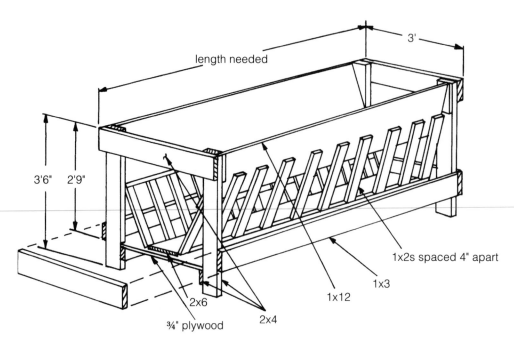

12-4. Sheep hay and grain feeder.

An elevated walkway was placed between two sides of the barn. The height of the walkway is the same as the bottom of the feed bunk. This means you don't have to lift hay bales quite so high to get them into the bunks. A loft ladder and opening directly over the walkway allows hay to be thrown down into the walkway, then over into the bunks. As a teenager I found I could actually throw the bales directly into the bunks — sometimes. Notice that the loft ladder is made of 2x4s with the steps notched into the uprights. Use 16d nails to anchor the steps in place.

Another barn interior feature often needed is a milk-

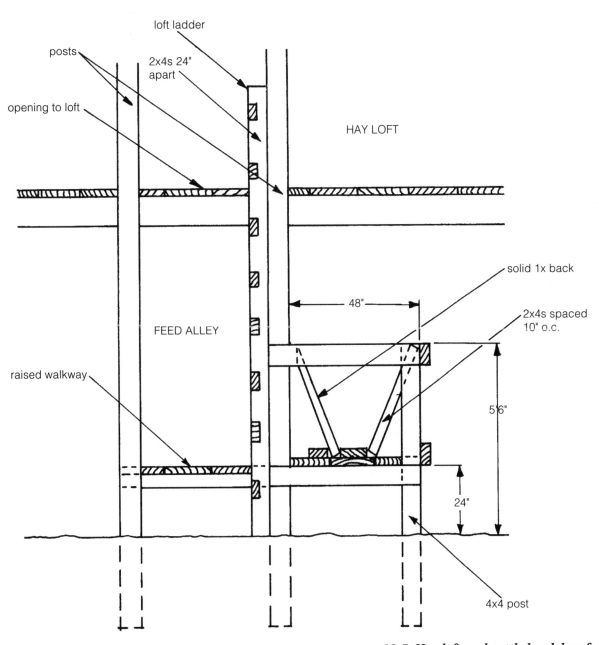

16-5. Hay loft and cattle bunk hay feeder.

ing stanchion. The stanchion shown is for a single cow. First a concrete floor is poured, including a slope off to a drain. Then the stanchion is assembled from 2x4s and anchored to the barn wall or stall divider. Note that a feed box is placed inside the stanchion. The rough dimensions are shown, but should be adapted to suit the breed and size of your cow.

Interiors of horse stalls require complete coverage. In most instances treated 2x6s are installed up to about 5 feet above the floor. Then ¾-inch tongue-and-groove planks are installed horizontally above that. Commercial stall guards (steel barred windows) are used to finish off the stall dividers on the inside.

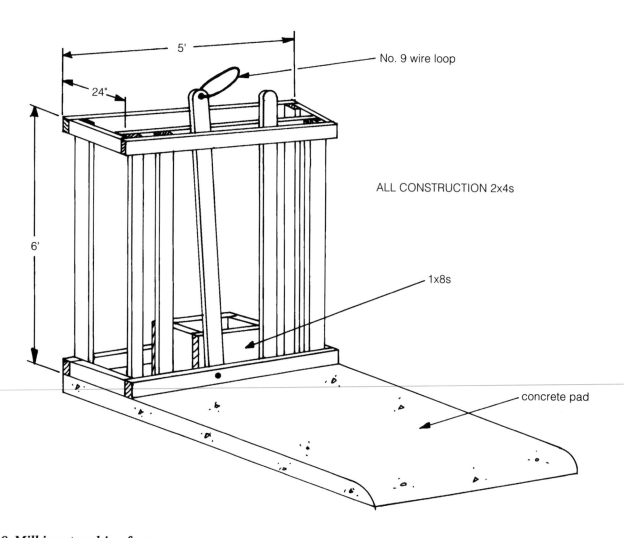

12-6. Milking stanchion for one cow.

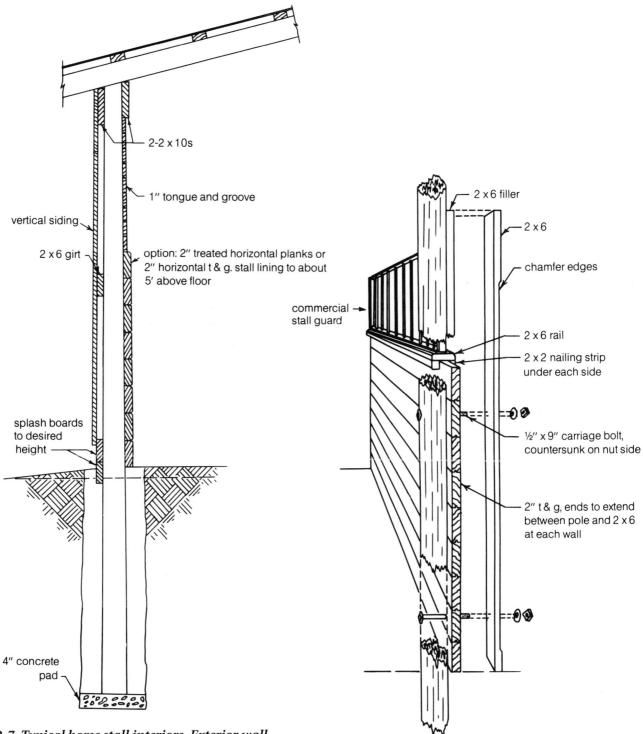

2-2 x 10s

1" tongue and groove

vertical siding

2 x 6 girt

option: 2" treated horizontal planks or 2" horizontal t & g. stall lining to about 5' above floor

splash boards to desired height

4" concrete pad

2 x 6 filler

2 x 6

chamfer edges

commercial stall guard

2 x 6 rail

2 x 2 nailing strip under each side

½" x 9" carriage bolt, countersunk on nut side

2" t & g, ends to extend between pole and 2 x 6 at each wall

12-7. Typical horse stall interiors. Exterior wall detail for the horse barn shows splash boards on poles at or on the earth surface.

12-8. A cutaway section showing details of stall partition, suitable for pole barns.

Project Plans
for Barns, Sheds, Garages, Studios
& other Outbuildings

CHAPTER 13

BARNS

The five barns included in this chapter present a wide variety of building styles and methods. Three barns are general-purpose, with two traditional barn building styles, and one pole barn illustrating this popular and economical method of construction. The three general-purpose barns can be used for a wide variety of purposes by dividing them in a number of different ways — adding stalls, grain and feed storage rooms, concrete floors, and feed alleys for cattle, hogs, etc. The sizes can also be readily adapted to smaller or even larger barns as needed. The other two barns in this chapter, the milk parlor and horse barn, are specific barn projects. Again these designs can be adapted to your particular needs and size requirements.

PROJECT 1

All-Purpose Barn # 1

Typical of many older barns in the Ozarks, this general-purpose barn was constructed around 1900 and made entirely of native materials. Oak was used for the framing as well as the siding. Metal roofing covered the original barn. Atypically, however, the barn was constructed on a fairly large lot in town. When Tom and Ruthelen Heitman decided to renovate the old place into a Bed and Breakfast they redid the barn as well, adding flakeboard siding over the front, then a coat of decorative paint.

The interior of the barn is well thought out with a feed

stall area and feeding alley along the left side. This left the center of the barn (with the highest roof peak and most storage space) for the storage of feed such as hay or oats. The right alley provides more room for animal stalls or equipment storage. A small feed room at the back of the barn on one side as well as a small storage or tack room on the opposite side completes the layout. Note the top door on the front. This allowed hay bales to be passed through the upper part of the barn after the stack has advanced above the open bottom door. A similar door is on the back.

Construction is a modified post-and-beam style, which was popular at the time. It utilizes 4x6 posts, resting on 2x6 sill plates with 2x6 top plates and nailing girts. The solid oak siding boards are then nailed over the nailing girts. The entire structure is supported on a concrete footing and perimeter foundation.

The dividing walls between the areas also consist of 4x6 posts on concrete foundations with 2x6 beams across their tops to help support the rafters. Divider walls are short, or just enough to be able to reach across to get feed and pass from one area to the other. The rafters are 2x6s and are 22½ feet long. It's extremely hard to find good structural materials that long these days, so they would

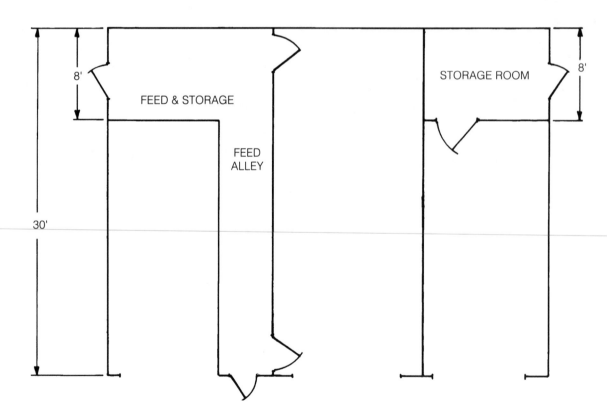

GENERAL BARN #1 FLOOR PLAN

186

probably have to be joined with a wood splice. This should be done over the center support poles and beams. The rafters are held together with 2x4 collar ties.

Good structural materials can, of course, be substituted for the native materials used in the original barn. Treated 4x6 poles would be a good choice for the posts. The 2x6 materials should be selected for strength and be free of serious structural blemishes. Construction, of course, begins with the footing and foundation. The perimeter, as well as the dividers are poured as one pour. Make sure anchor bolts are properly spaced and installed.

The walls are constructed by nailing the bottom sill plates, the outside nailing girts, and the diagonal braces to the posts while the posts are lying on the inside of the building perimeter. The walls can be constructed in sections, joining the girts and plates in the center of a post. Build the two side walls first. Then comes the hard part. It will take several helpers to stand the walls up, then lift them up and over the anchor bolts and install them in place. Once a wall is in position, brace it solidly and erect the opposite side wall in the same manner. The back wall is constructed in the same manner, then erected and joined

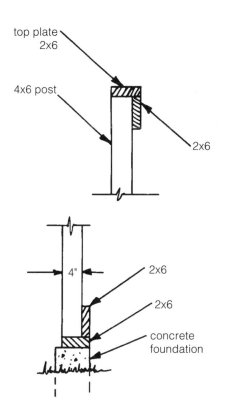

GENERAL BARN #1 DETAIL A

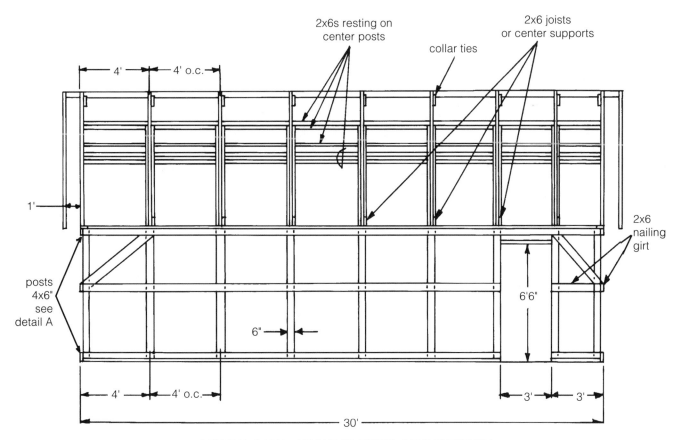

GENERAL BARN #1 RIGHT SIDE VIEW (LEFT REVERSED)

187

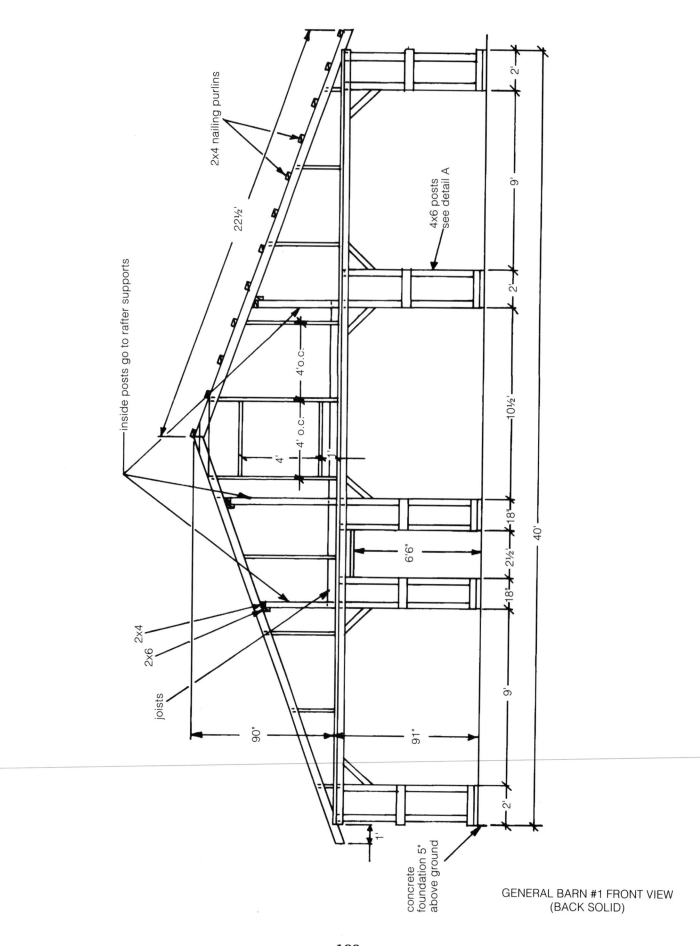

2x4 nailing purlins

inside posts go to rafter supports

22½'

4'o.c.

4' o.c.

4'

2x4

2x6

joists

90"

1'

concrete
foundation 5"
above ground

4x6 posts
see detail A

2'

9'

2'

10½'

18"

6'6"

2½'

18"

40'

9'

91"

2'

GENERAL BARN #1 FRONT VIEW
(BACK SOLID)

188

to the side wall end posts by nailing through the nailing girts. Corner diagonals further strengthen the joints of the walls. The front wall is constructed as pieces between each opening. The final step in locking the walls together is to install a top plate, nailing down over the posts and into the top girt. Make sure the top of the walls are straight before fastening the top plate in place. The top plates should also overlap the corners to tie the walls together.

The next step is erecting the rafters. It's a good idea to construct the outer rafters, then erect and brace them to determine the exact height of the center support posts and beams. The rafters can simply rest on these beams, or a bird's-mouth can be cut to place the rafters exactly on the support beams. This requires precise cutting of the rafters and positioning of the center posts and beams. Once the exact height is determined, install the center divider posts and support beams. Then follow with the remaining rafters. Nailing purlins are fastened over the rafters. These not only provide nailing for the metal roofing, but bracing for the rafters. Once the rafters are erected, install the gable end blocking or studs. A good safety precaution, especially in windy conditions, is to nail diagonal bracing to the underside of the rafters until the roofing has been completed. Two "joist" tie beams can be used to prevent the barn walls from spreading out, although the pole and beam supports in the center alleviate most of the problem.

This particular design also lends itself well to placing a hay loft above the center section. Use 2x10 floor joists spaced 24 inches apart, anchored to the center poles and to joist headers fastened to the poles.

With basic framing completed, roofing and siding is applied.

All-Purpose Barn #2

This barn was built by my dad about 40 years ago. Although we no longer own the property and the barn needs a paint job and a bit of repair, it's still a fine example of a barn for storing hay, with a few stalls for cows, sheep, or swine.

The barn was a joke in our neighborhood because Dad decided to build the barn around a standing stack of hay. But when all the neighbors came in for the barn raising, they stopped laughing. It was an easy way of installing rafters. All you had to do was stand on top of the hay stack. I still remember the fast and furious work that was done to complete the barn and get the hay covered before a summer storm blew in.

The barn utilizes a concrete foundation and a modified simple box framing design. The framework of the walls was constructed of native red oak lumber. One unusual part of the framing is that the nailing girts are notched into the studs.

The basic 24x30 foot barn provides plenty of hay storage for a small herd of up to 25 cows.

The basic barn is constructed first, then the shed added, as is often the case. Or you can construct the entire barn at one time. The first step in construction is to pour the

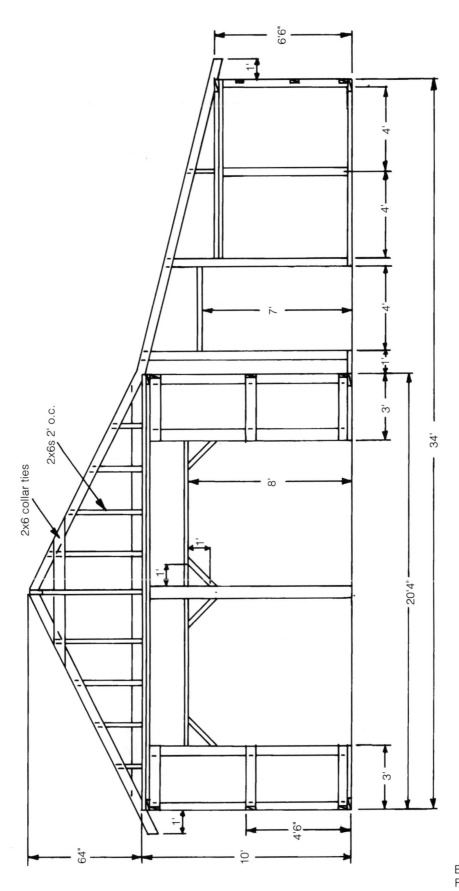

BARN #2 GENERAL
FRONT ELEVATION

191

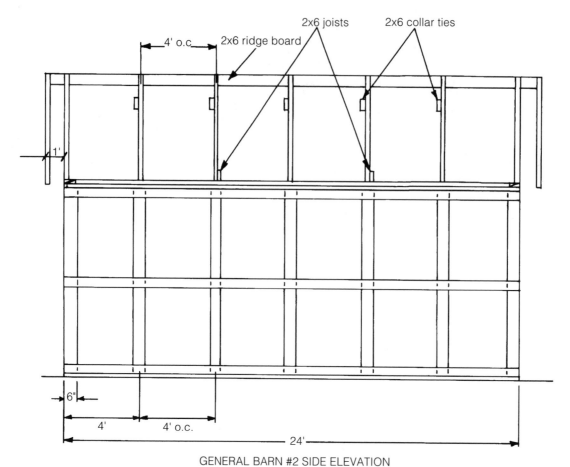

2x6 joists

2x6 collar ties

4' o.c.

2x6 ridge board

1'

6"

4'

4' o.c.

24'

GENERAL BARN #2 SIDE ELEVATION

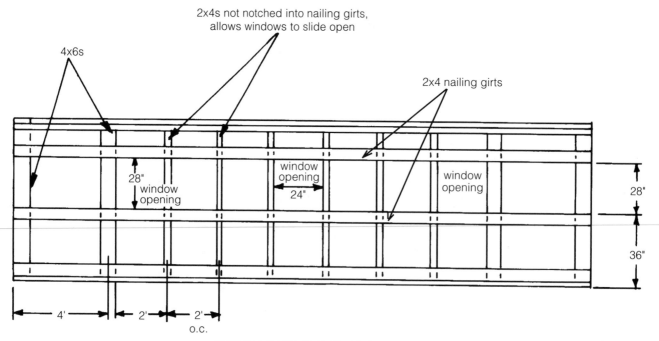

2x4s not notched into nailing girts,
allows windows to slide open

4x6s

2x4 nailing girts

28"
window
opening

window
opening
24"

window
opening

28"

36"

4'

2'

2'
o.c.

GENERAL BARN #2 SHED SIDE ELEVATION

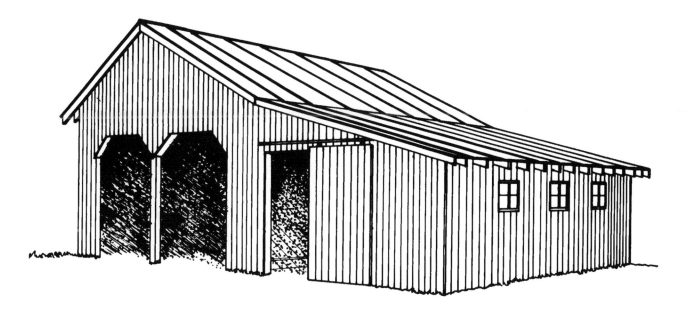

foundation and place the anchor bolts in the foundation while the concrete is still soft. After the foundation has set and the form boards are removed, the framework is assembled. Each wall is assembled in one piece while lying on the ground outside the foundation, using 16d common nails. It's easiest to assemble the walls with the outside face up; this makes it simple to fasten the nailing girts in place. Then, with a couple of friends, you can turn the side over before standing it in place. Mark the location of the anchor bolts and bore the bolt holes in the sill plate, then assemble a side. Then, again with the help of several friends, tip the side up over the anchor bolts and stand it in place. Make sure that it is braced securely, then do the other sides in the same manner. Dad assembled all sides of the basic barn, then called in a few friends for a "barn raising." It didn't take long to lift the sides in place and anchor them solidly.

Regardless of how you erect the walls, once the two sides have been raised, plumbed, and braced securely in place, the back wall is raised and plumbed, then fastened in place to the two side walls, anchoring the back and side walls together with the overlapping upper plates and at the corners.

The front wall with the two open barn doors, however, is assembled by first placing the middle post in position and propping it securely. Then the two front sections are assembled and fastened in place and the top plates nailed across to tie the front together.

After the wall framing has been completed the rafters are laid out and cut. One end pair of rafters is assembled and fastened in place with the ridge pole between them, then the opposite end rafter pair is erected and fastened to the ridge pole. Once these have been positioned, a 2x4 purlin is used to brace them securely. Then the rest of the rafters between the starting rafters are nailed in place.

After all the rafters have been installed, all the nailing purlins are nailed in place and the rafter supports and cross-bracing installed.

Corrugated metal roofing was installed, then car siding or boxing (also called shiplap) installed to cover the outside of the barn.

The shed is constructed in the same manner, except that the side framing is somewhat different to allow for the three 2x2 windows. A sliding barn door is also installed to close off the shed portion.

Pole Barn

Pole barns are simple to construct, economical to build, and, most importantly, extremely versatile. This is especially so if the roof is trussed, as this provides an interior space without supports that can be adapted to almost any need. The author's barn shown was initially constructed to house small square alfalfa hay bales. It has also housed pigs, served as a boat storage building, and even played auto garage as one son worked on restoring a Mustang. The barn began as a basic pole barn, but now has a lean-to shed attached. A divider between the barn and shed allows livestock to be sheltered in the shed side, and farm machinery and a woodworking shop to be kept in the enclosed portion. A sliding door on each end provides for drive-through convenience with large equipment.

Constructed of treated poles, treated lumber at ground level and covered with metal, a pole barn of this nature can be extremely long-lasting and an attractive addition to your homestead or farm. The barn shown is 30x60; and the shed adds another 8x60 space—all in all a fairly large barn. The size of a pole barn, however, can be almost any size from one much smaller to one the same width and whatever length you wish to build it.

As with any construction, the first step is to lay out the

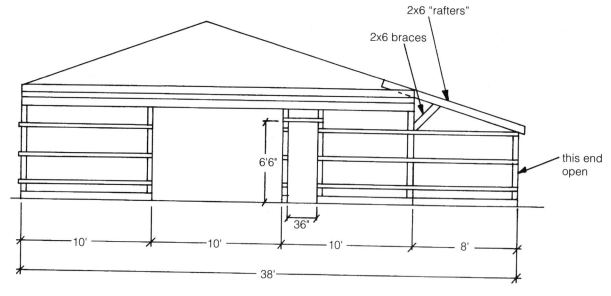

2x6 "rafters"

2x6 braces

this end open

6'6"

36"

10'   10'   10'   8'

38'

POLE BARN FRONT VIEW

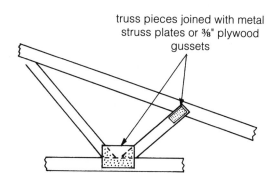

truss pieces joined with metal struss plates or ⅜" plywood gussets

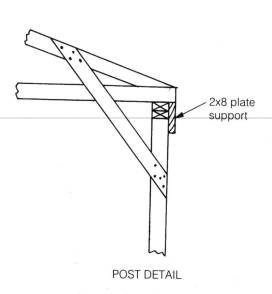

2x8 plate support

POST DETAIL

barn and make sure it is square. Actually, this is more important on pole barns than with other types of construction because you don't have a square slab to build on or square sides to erect. Once the outline of the barn has been established square, mark locations for the poles. You should have the outline of the barn established with string lines on batter boards.

In many instances, the hardest job of the barn construction is digging the holes for the poles. These can be done by hand with a post-hole digger, but are much easier to do with a power post-hole auger. These augers may be powered by either a tractor power take-off or a gasoline engine. The latter can be rented in many towns and cities.

Once the holes have been dug, position treated poles in the holes and reestablish the string line outline of the barn to ensure that the poles are positioned correctly. Using a 4-foot level, plumb the poles in both directions and temporarily anchor them in place with 2x4 braces nailed to the poles and to stakes in the ground. This is an awkward stage of construction and often the most frustrating. When possible, nail the braces to the posts at a position where they won't be in the way when you attach the purlins.

Although the ground may be quite unlevel, you will need to *establish a level line* at the top of the poles for the plates. Start at the pole on the lowest portion of the slope and tack-nail the 2x8 plate support board in place, then level the board to the next pole. Do this around the entire top edge of the poles. Check to make sure the boards are

196

level in all directions, then nail them securely in place. Cut off the poles even with the tops of these leveled plate support boards. A water level can also be used to establish the plate line.

The next step is to nail the plate boards down on top of the poles and the support boards. Doubled plates are needed, staggering the joints and positioning them over the poles to create strength.

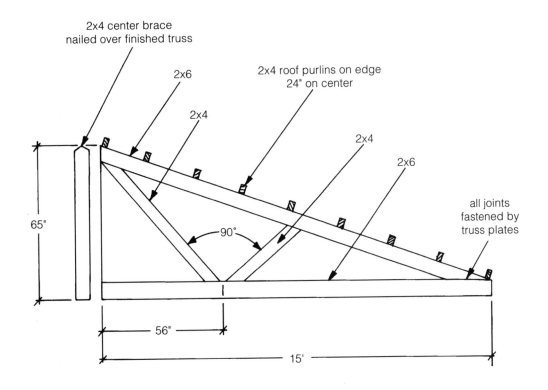

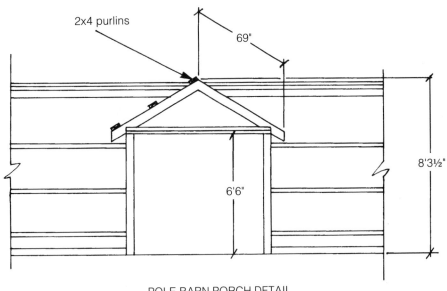

POLE BARN PORCH DETAIL
Note: Extends 6' from barn side

197

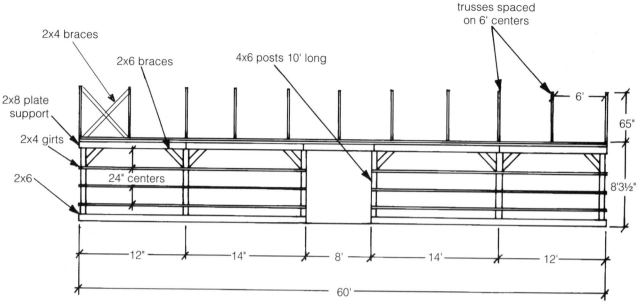

2x4 braces

2x6 braces

4x6 posts 10' long

trusses spaced
on 6' centers

2x8 plate
support

6'

2x4 girts

65"

2x6

24" centers

8'3½"

12"   14'   8'   14'   12'

60'

POLE BARN SIDE VIEW

The bottom skirt boards, made of treated wood for ground contact, are fastened in place, this time starting with the pole on the highest ground. Again make sure all the boards are level before securely fastening them in place. Then follow with the girts between. A string level or water level can be used to establish a level line and speed up this chore. Once all girts have been nailed in place, cut diagonal 2x6 bracing boards, two for each post and nail through the plate support board and girt into the braces. Toenail through the braces into the upper plate and posts.

Truss construction is the next step, and you can often have this done at building supply dealers, or you can make up your own at a greatly reduced price. If constructing your own, make a pattern board for each piece, then cut a "pattern" truss and fasten it together on a flat, smooth surface with metal truss plates or plywood gussets, Resorcinal glue, and nails. Use this as the master truss, placing the pieces for a second truss directly on top of this finished truss and nailing them together. This ensures that all trusses will be assembled the same way.

Now comes the most difficult portion of the job, particularly on a barn of the size shown: erecting the trusses. This is hard work, a bit tricky, and potentially dangerous. Make sure you take all precautions and do not allow anyone below the trusses until they are securely fastened in place. (The author used a tractor bucket to lift the trusses in place, each end resting on the plates.) Slide the trusses carefully into position, starting with the center and working toward each end. Position 2x6 stop blocks at the truss locations to hold the ends in place. Use long push poles chained or tied to the trusses to stand them up. The first two trusses will have to be securely braced in place with support poles to the ground. Once the third truss is installed the 2x4 roofing purlins are started, nailing them into the trusses to begin securing them in place. Before anchoring each truss, make sure it is plumb. With all trusses erected, finish installing all roofing purlins and nail a center stabilizing support across the top of the bottom truss supports to tie the trusses all together at the bottom as well. 2x4 braces nailed diagonally between the trusses and bottom truss supports will further strengthen the roof structure. Don't attempt to erect trusses in a high wind, and make sure they are all strongly anchored and braced before quitting for the day.

The lean-to portion of the barn is also constructed by first setting the poles and following in the same manner for the rest of the construction. Instead of trusses, 2x6 rafters are positioned alongside each truss, and resting on the outside lean-to plate line; then they are anchored in place. Braces from the rafters to the poles add strength.

Installing the metal siding and roofing is the next step, starting with the sides, then the roof. Corner edging is installed last, and the doors constructed and hung.

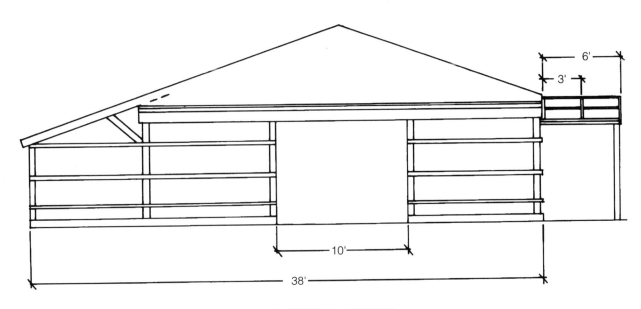

POLE BARN BACK VIEW

199

Horse Barn

Although at 36x84 feet, the horse barn shown can't be considered a small barn, it's a fine example of a stable. Located just outside Humansville, Missouri, at Oak Ridge Stables, the building was constructed by Koehler Construction Company for owner Bob Allen, who raises thoroughbreds. The barn features four standard 12x12 stalls (actually 11'4" inside), two on each end, with an office and large foaling stall on the front side and 7 stalls on the back side. Actually, the building could be constructed half the length for a smaller stable.

Construction is standard pole-barn style with trusses forming the roof. The entire building, including the roof and sides, is covered with metal and finished off with standard metal finishing pieces.

The first step in construction is to lay out the location of the poles, remembering that all outside poles should be set back 1½ inches from the outer perimeter of the building line to allow for the 1½-inch thickness of the nailing girts. Once all poles have been erected and braced in place, the nailing girts are then nailed in place, including those over the doors, etc. The upper and bottom 2x6 girts are nailed in position, then the 2x6 upper plates, and finally the bracing.

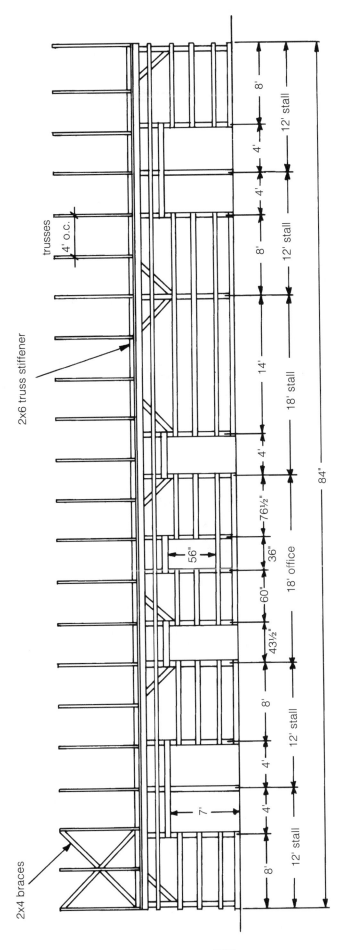

2x6 truss stiffener

trusses

4' o.c.

2x4 braces

8'

12' stall

4'

4'

8'

12' stall

8'

14'

18' stall

4'

76½"

56"

36"

18' office

60"

43½"

8'

12' stall

4'

7'

4'

8'

12' stall

84"

HORSE BARN
FRONT ELEVATION

201

The trusses are assembled using metal truss plates or glue and screwed wooden plywood gussets. Then the trusses are erected and braced together until the 2x4 nailing purlins are installed to tie the trusses together. Diagonal wind bracing and lower stiffeners are added to the trusses to tie them all together.

Metal siding and roofing is applied and finish trim added.

The sliding stall doors were welded up and installed by the builder, and the sliding end doors and pre-hung steel office door were assembled and installed as well.

Finishing for the stables is shown in Chapter 12 on finishing barn interiors.

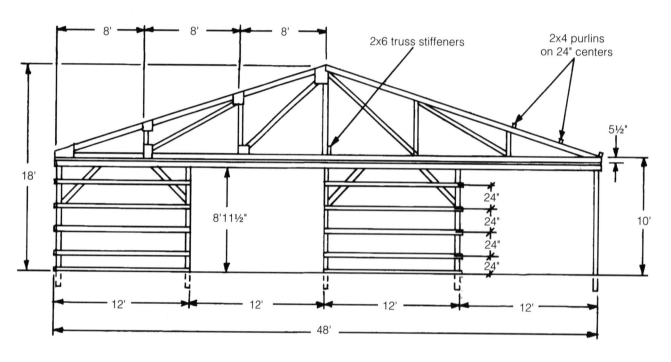

HORSE BARN SIDE ELEVATION

Milk Barn

The milk barn shown is typical of many small milk barns scattered across the country. Many of these smaller dairies have been put out of business these days by the big conglomerates. Before constructing a dairy barn, check with your local health officials as to the details and specifications regarding a milking parlor construction.

The barn shown is a traditional concrete block on perimeter foundation with poured concrete floors. The first step is to lay out the footings, pour them, then pour the foundations and, finally, lay the block walls. Be sure to install anchor bolts for the top plates by filling the top row of blocks with mortar and inserting the bolts while it's wet. The windows and doors are normally purchased steel pre-hung units. Steel lintel bars set in the blocks provide support for above doors and windows.

Once the block walls have been assembled the roof construction is standard 2x4 rafters on 16-inch centers. The first step is to fasten the 2x6 plates to the top of the concrete walls with anchor bolts set into mortar placed in the top row of blocks and extending up above the plates. Both plates must be installed before tightening down the bolts. Make sure to stagger the plate joints for added strength.

Cut the rafters and ridgeboard and erect the roof structure. Place 2x4 ceiling joists parallel to the rafters and resting on the top plate adjacent to each rafter. Nail through the joist into both the rafter and the plate. Finish the gable end blocking, add sheathing, hanging rafters, and fascia boards, and then you're ready to shingle.

The interior concrete walls are erected at the same time as the exterior, locking them in place with metal tie bars or girder reinforcements. These are placed between blocks in the mortar, overlapping interior and exterior walls. The interior arrangement should first be determined by your equipment needs.

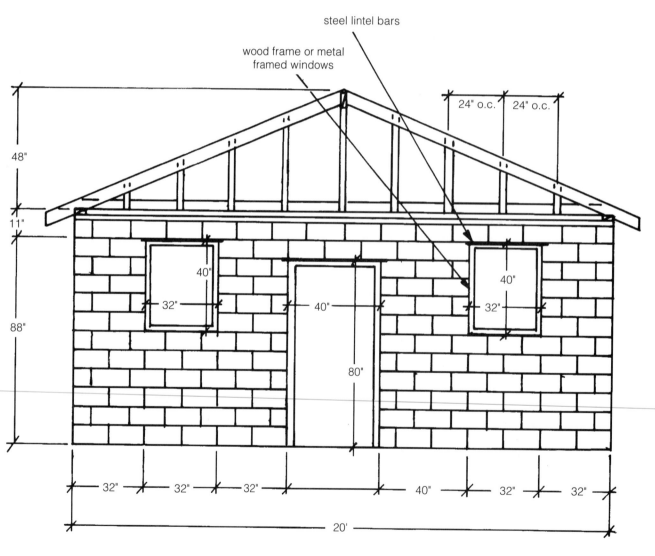

MILK BARN
FRONT ELEVATION

204

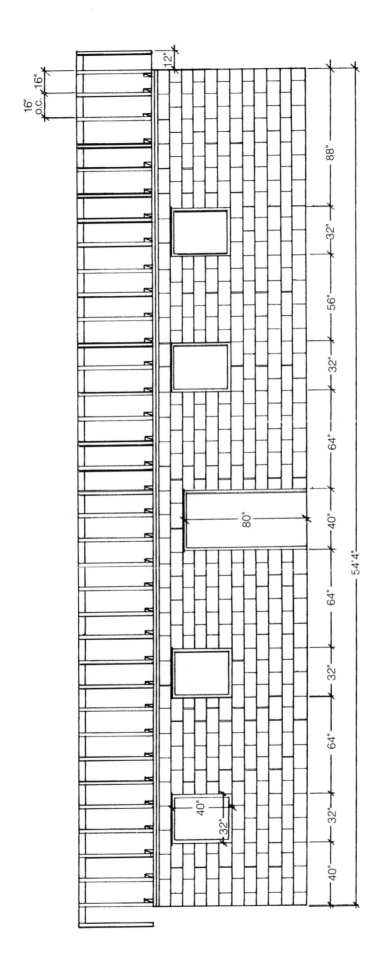

12"

16" O.C. 16"

88"

32"

56"

32"

64"

80"

40"

64"

32"

64"

32"

40"

40"

32"

54'4"

MILK BARN
SIDE ELEVATION

205

CHAPTER 14

SHEDS

It seems as if we can never get enough sheds and small outbuildings to answer for all of our needs around the homestead. The greenhouse/woodshed, however, can serve many purposes in one small building. The equipment or machine shed is a standard and a necessity on the farmstead, while the sheep or goat shed, food preparation house, and the two garden sheds round out the projects and provide a variety of sizes, shapes, and building styles.

PROJECT 6

Machine Shed

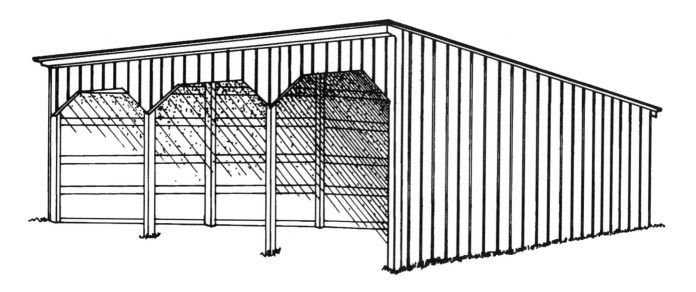

A **strong, weatherproof storage building** for housing farm equipment is a must. Equipment will last longer and work better with less hassle, making farm life much easier. An equipment or machine shed doesn't have to be fancy, and it usually isn't. It can be completely enclosed, but in many cases one side is left open, usually facing to the south or away from the direction of drifting snow in snow belt country. The shed shown is typical of many across the country and is a simple, easily constructed, yet effective pole structure covered with galvanized metal. This is the most economical and trouble-free structure you can put together for this purpose. Three open stalls, not quite 10 feet wide provide plenty of space for a tractor and other small equipment. Actually, the shed can be constructed in almost any length desired in the 10-foot pole spacings shown.

The first step, of course, is to lay out and bore the holes for the poles. Note that each 10-foot section has a front, back, and center pole. Make sure the building is square and the poles in line with each other or you'll have lots of construction problems. Erect the poles, brace them in place plumb, and fasten them securely in place with concrete around them. Once the concrete has set, nail the wall nailing girts in place, making sure they are level. Cut and nail the diagonal braces in place.

Then cut and install the rafters. *Note there are two sets for each set of poles.* Each rafter is nailed in position on the sides of the poles, then the pole tops are cut flush with the

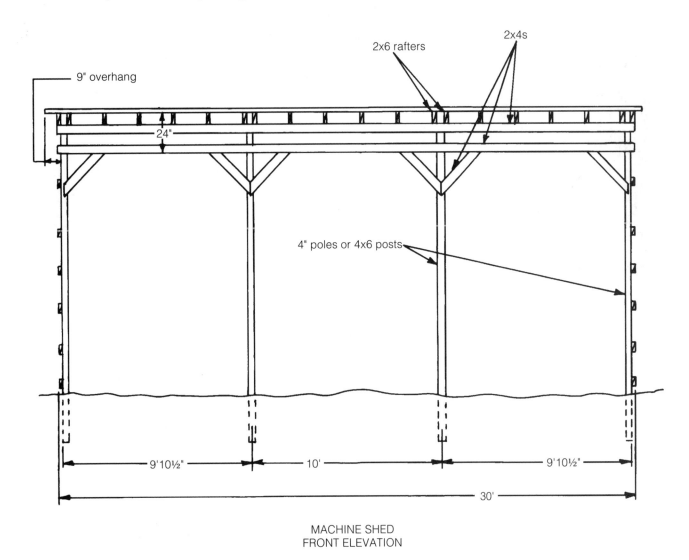

MACHINE SHED
FRONT ELEVATION

207

tops of the rafter edges. Rafters between poles resting on girts are fastened in place with metal straps or blocks of wood. Nailing girts are nailed down on the rafters, spacing them 24 inches on center.

The final step is to install the metal siding and roofing. A rock or gravel floor will prevent dust and mud problems and makes working on equipment much easier, although the ultimate in convenience is a concrete floor.

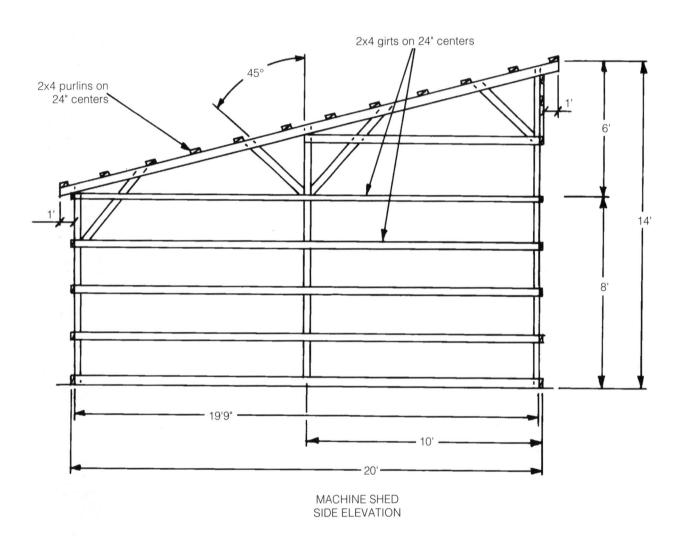

MACHINE SHED
SIDE ELEVATION

Garden Tool Shed/ Greenhouse/ Woodshed

When a storm damaged the full-size greenhouse connected to our house and an old chicken house used for
garden tool storage finally rotted away, we decided to
build one shed combining the two, plus adding a firewood
storage area to one end. Ideal for the small farm or suburban homestead, this shed provides plenty of tool storage,
space for garden and lawn tractors and mowers, and a
work and potting bench. The wood storage area will hold
a generous supply of firewood, keeping it dry and neat.
The greenhouse is basically a "window" style reached
from inside the shed. This solves the problem of having an
exterior door into the greenhouse with the resulting heat
loss and construction problems. Actually, more than half
the greenhouse is inside the building. The front portion of
the greenhouse utilizes the ground as a planting bed,
offering space to grow winter vegetables much like a giant
cold frame. The back of the greenhouse provides plenty of
space for starting seeds, etc. The greenhouse is also solar

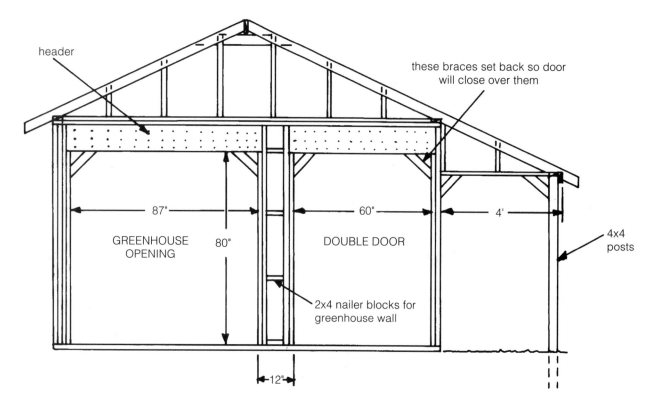

header

these braces set back so door will close over them

87"

GREENHOUSE OPENING

80"

60"

DOUBLE DOOR

4'

4x4 posts

2x4 nailer blocks for greenhouse wall

12"

GREENHOUSE/TOOL SHED FRONT ELEVATION

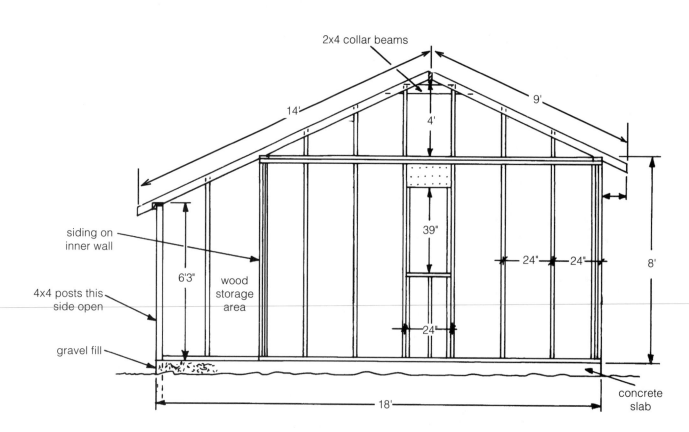

2x4 collar beams

14'

4'

9'

siding on inner wall

6'3"

wood storage area

39"

24"   24"

8'

4x4 posts this side open

gravel fill

24"

18'

concrete slab

GREENHOUSE/TOOL SHED BACK ELEVATION

210

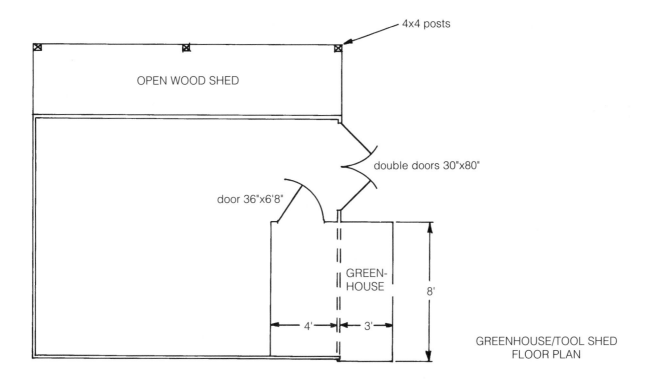

4x4 posts

OPEN WOOD SHED

double doors 30"x80"

door 36"x6'8"

GREEN-
HOUSE

8'

4'   3'

GREENHOUSE/TOOL SHED
FLOOR PLAN

designed, utilizing a heat mass on the concrete floor and insulated wall, and a door on the inside which opens to the shed for access to the greenhouse but which, when closed, reflects the heat back into the house. Two 50-gallon steel drums of water complete the heat gathering mass. The greenhouse glazing is composed of Filon Solar Grow Home Greenhouse Panels. They feature an exclusive solar prism surface that breaks up the sun's rays, providing a diffuse light that promotes more uniform plant growth and superior yields. The panels are made of fiberglass and are shatter-resistant. Automatic vent fans provide heat control.

The shed actually combines and provides an example of two different types of construction. It is also the main project used to photograph the various construction methods used throughout this book. The main building is constructed on a concrete slab, while the greenhouse and the outer portion of the woodshed feature pole construction. The greenhouse utilizes treated materials. Because it functions as a greenhouse, the shed was constructed with the greenhouse side facing the south. This placed the woodshed side on the east, which is relatively storm-free in our part of the country and also handy to our fireplace.

The first step in construction is to lay out the building including the main building slab, the greenhouse, and the piers or hole locations for the posts. Because of the different types of supports and the complex design, careful layout is important.

After the building has been laid out, the batter boards erected, and the excavation chores completed, the slab and footing/foundation are poured. Due to the hardness of ground on our old Ozark hillside farm, the posts were treated poles set on concrete punch pads rather than piers and posts. This meant merely boring the pole holes and pouring a concrete pad in the bottom.

Before the concrete sets up, the anchor bolts are installed in the slab and foundation, making sure they are not located in a stud position.

Standard platform stud framing of 2x4s on 24-inch centers is used for the walls, with the front wall utilizing 2x4 headers for the door and greenhouse openings. All four walls of the main building are constructed first on the slab, stood upright in place, and anchored together. Then the poles are set in place. The poles are tied together with a top and side plate, which also provides support for the rafter ends. Note that the rafters on the shed side have a bird's-mouth cut for both the pole and inside wall building sides. In addition diagonal bracing was installed on both the front and side openings of the shed.

Roof construction is standard 2x4 rafters on 16-inch

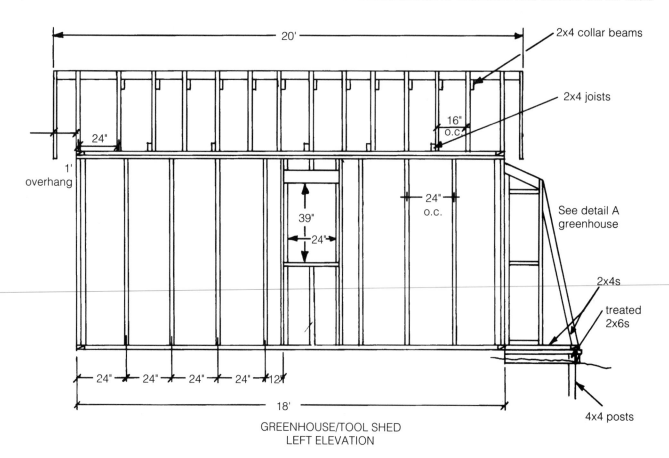

GREENHOUSE/TOOL SHED
LEFT ELEVATION

centers covered with sheathing. The rafters are joined at the top of the building with a 2x4 ridgeboard. There is a 1-foot overhang on each end for hanging rafters. Note that the rafters on the shed shown extend from the roof peak to complete the shed in one rafter. This provides a fairly low side on the woodshed. If you wish to build a higher shed roof, use one set of rafters for the main building and then shed roof rafters for the shed, raising the end of the shed roof. This will, of course, change the appearance of the building quite a bit. If you elect this course, it's a good idea to use a scale ruler to draw the building to get an idea of how it will appear. End blocks are installed to support the end rafter and provide nailing for the siding. Notch them to fit flush with the outside face of the end rafters.

Roof sheathing is applied. Extend it 1 foot past each of the building ends, then nail the hanging rafters underneath the sheathing boards.

The two sides (including the inside side), back, and upper front are covered with textured, grooved plywood siding. The roof is covered with Masonite Woodruf, a fast and extremely durable substitute for wood shingles. The soffit is installed. Then the doors, windows, and corners are trimmed with rough-sawn western white cedar. Incidentally, it's much easier when doing new construction such as this to paint the trim boards first, then install them in place.

The windows are standard aluminum storms. The doors are constructed by hand utilizing 2x2 back bracing, siding to match the building, and rough-sawn white cedar trim. The door hinges and hardware are installed with bolts through the 2x4s.

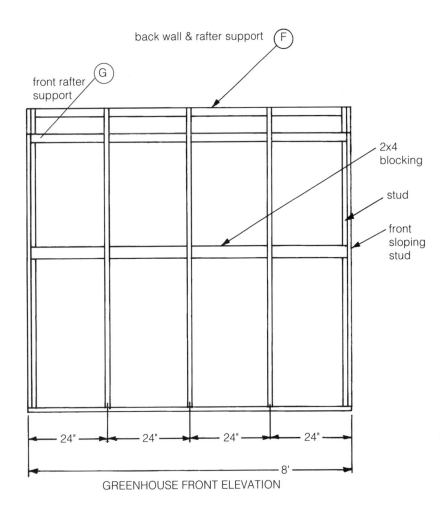

GREENHOUSE FRONT ELEVATION

Greenhouse

The final step is construction of the greenhouse. It is framed entirely of treated 2x4s and 2x6s. The bottom utilizes a treated wood frame fastened to treated wooden

13-1. *Greenhouse is constructed entirely of treated wood. Note the greenhouse shown is on down slope and treated 2x6s are used to create a "foundation," then framework erected on foundation.*

13-2. *Greenhouse fiberglass panels are installed over framework.*

13-3. *Note corrugated wood strips used to seal and fasten panels in place.*

posts sunk in the ground. Make sure the wood used has a stamp or tag noting that it is for use in ground contact. Treated wooden blocks are first anchored to the concrete slab at the inside corners of the greenhouse with concrete "screws." Then the front posts are set in concrete and the greenhouse 2x6 "skirting" is nailed to it, making sure the skirting boards are all level. The posts are cut off flush with the tops of the skirting and a 2x4 "sill plate" is nailed over the skirting edges and the tops of the posts. The greenhouse will be fastened down to this plate. The greenhouse sides are constructed first, creating a short "wall" of the back (A) and center (B) studs as well as the bottom "upper sill plate" (C), cross braces (D), and the end rafter (E). Framing is also included in the upper section for wall vents, fans, etc. Note that the end rafter is set in notches cut in the outside edges of the wall studs. Once these sections have been assembled, nail them in place to the building front and down on top of the greenhouse sill plates. Cut the back rafter support (F) and the remaining rafters. Nail the back rafter to the back wall support, then nail the wall support with rafters attached in place between the side walls and to the building front. Cut the front rafter support (G) and nail it in place between the side walls and nail through it into each rafter. Cut the front sloping wall studs starting at one end and fastening them in place spaced on 24-inch centers. Then install the center blocks between the sloping uprights. The first block is nailed in from the

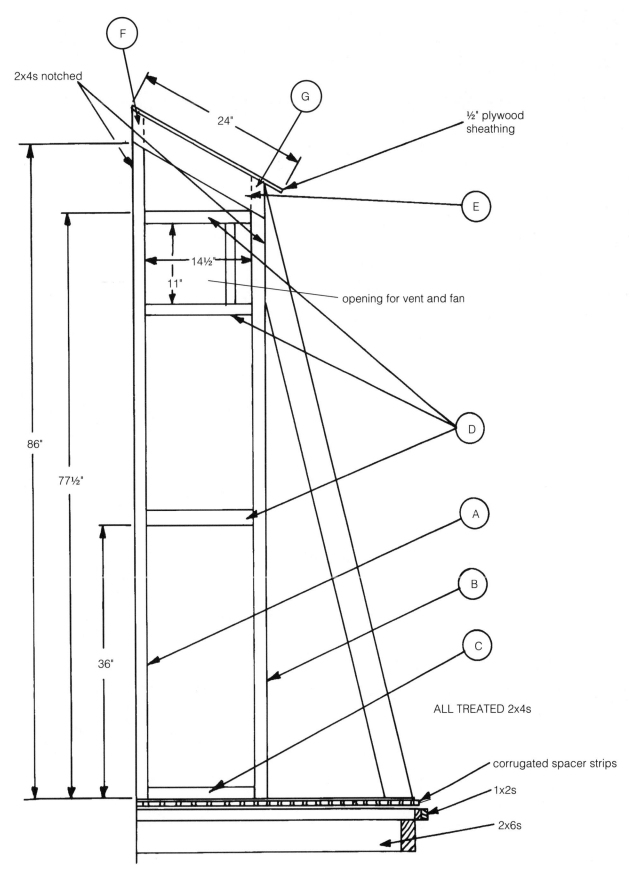

2x4s notched

F

G

24"

½" plywood
sheathing

E

14½"

11"

opening for vent and fan

D

86"

77½"

A

B

C

ALL TREATED 2x4s

36"

corrugated spacer strips

1x2s

2x6s

GREENHOUSE DETAIL A SIDE ELEVATION

215

13-4. Wooden trim strips are applied after panels are in place. It's much easier to rip and cut trim strips to shape, paint and then install.

ends, but the next few will have to be toenailed at one end.

With the framework completed, the next step is to install the fiberglass covering. A 1x2 support strip is first nailed around the top edge of the skirting. Then special cedar "spacing" strips (available with panels) cut to the exact shape of the corrugated glass panels are nailed in place above the support strips. This provides an easy means of fastening the panels in place, as well as a superior insulating factor. Panels are then installed with corrugated metal fasteners and neoprene washers. Once panels are installed, clear caulking is applied to any air leaks or openings. 1x2 trim strips are used to finish the outside edges of the greenhouse.

One-half-inch plywood sheathing is applied over the rafters and roofing applied over that.

The final step in completing the greenhouse is the inside finishing. First insulate between the rafters and fasten a piece of flakeboard or ¼-inch plywood "ceiling" in place. The inside walls dividing the greenhouse from the rest of the building interior are framed, stood upright in place (including) the door opening, and fastened to the concrete slab with concrete screws and to the top joists and walls of the building. Ceiling joists are nailed in place and a flakeboard ceiling installed. Because the greenhouse must hold heat, it is totally insulated both on the ceiling and inside building side walls, covering the outside of the inside walls with flakeboard or plywood. The complete interior of the greenhouse is painted white to reflect light, then the black-painted barrels of water are placed inside and the door is hung in place. Black plastic is placed on the ground inside the greenhouse, then compost is added to come flush with top of the concrete slab.

Black plastic and a 4-inch layer of crushed rock completes the floor in the wood storage area, and is used to create a "ramp" into the building shown to compensate for the steep slope of the ground.

Sheep or Goat Shelter

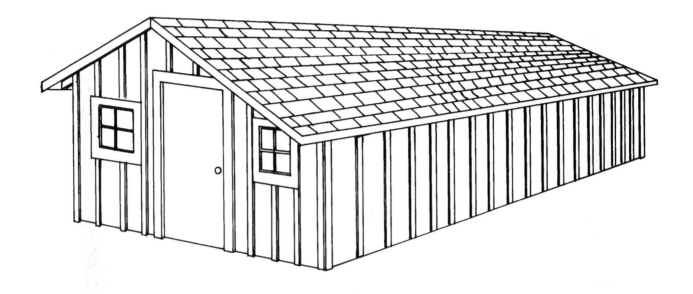

An economical, enclosed barn especially designed for sheep or goats, this structure follows a fairly traditional design. It utilizes the old-time post-and-beam wall construction, with the walls sitting on concrete footings and a perimeter foundation. This prevents drafts as well as keeps predators from digging into the barn. One section of the barn is a feed room, and it should have a solid concrete floor to prevent rodent problems. One unusual feature is that the nailing girts are notched into the sides of the posts.

The first step is to construct the two end walls. Cut the posts to the correct size, assemble an end wall complete with rafters, then stand it in position and brace it. Fasten the wall to the foundation with anchor bolts. Complete the opposite end wall in the same manner. Then construct the two side walls and stand them up between the two end walls and fasten together.

Scissor trusses are used for the remaining roof supports. This provides more headroom space at the center of the building. These trusses are constructed as shown, bolting them together and using plywood gussets and glue and nails for the top rafter joints. Erect these and brace

217

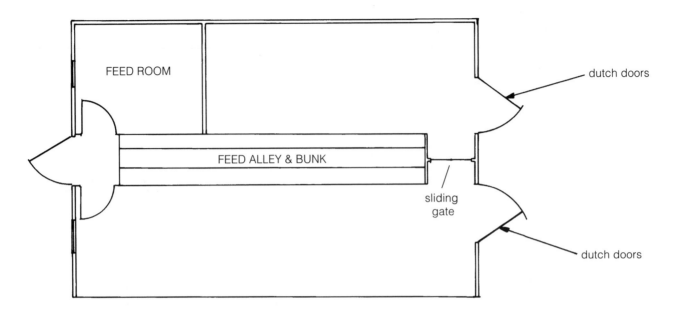

FEED ROOM

FEED ALLEY & BUNK

dutch doors

sliding
gate

dutch doors

SHEEP OR GOAT SHELTER FLOOR PLAN

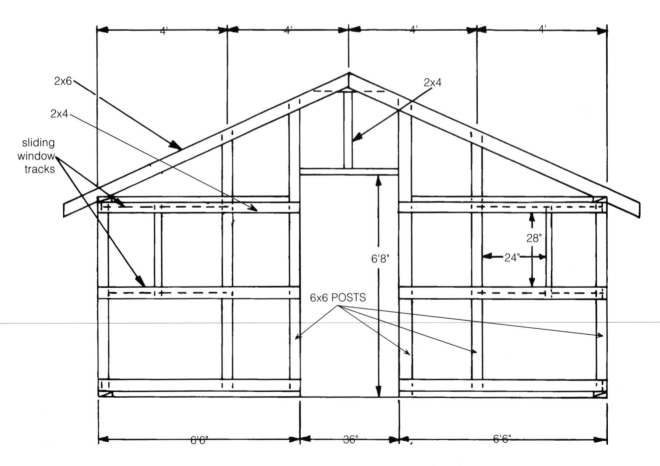

4'   4'   4'   4'

2x6

2x4

2x4

sliding
window
tracks

6'8"

28"

24"

6x6 POSTS

6'6"   36"   6'6"

SHEEP OR GOAT SHELTER FRONT ELEVATION

them in place. To make the building more snug you can use blocking between the trusses, fastening them down on the top plate. Then add the plywood sheathing, extending it far enough past the building edge for the hanging rafters. Add hanging rafters and fascia boards, then cover the roof with asphalt or fiberglass shingles.

The exterior of the building can be covered with metal, plywood, or hardboard siding, board and batten, or shiplap. The interior of the building should be insulated in cold climates. The door is a shop-made door while the windows are fixed sash windows that slide on the center girt between the wall covering and 1x4 tracks nailed in place on the upper and lower girts.

Construct the interior dividers for the feed room and construct and hang the interior door.

A feed rack and alley run the length of the building, except for a sliding gate at the end opposite the feed room. Instructions on building the rack are shown in Chapter 12 on finishing barns.

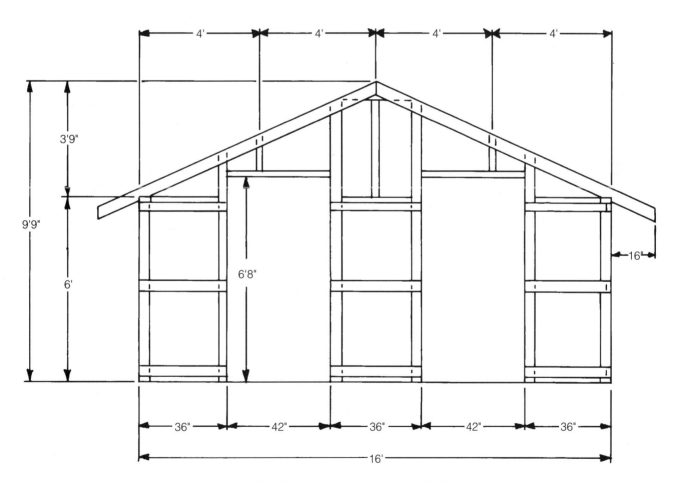

SHEEP OR GOAT SHELTER BACK ELEVATION

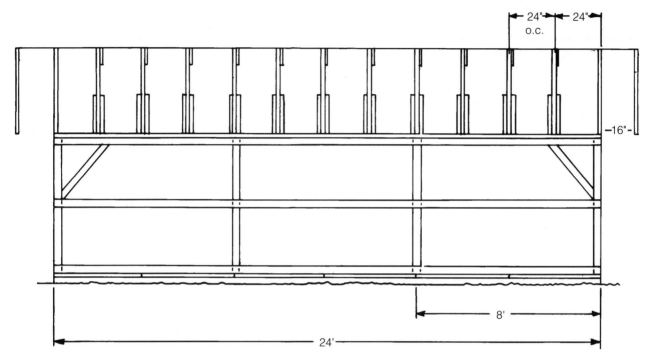

SHEEP OR GOAT SHELTER SIDE ELEVATION

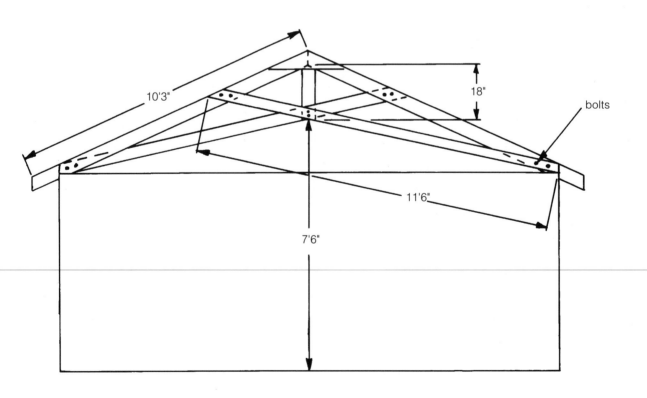

SHEEP OR GOAT SHELTER SCISSOR TRUSS

220

Canning and Food Preparation House

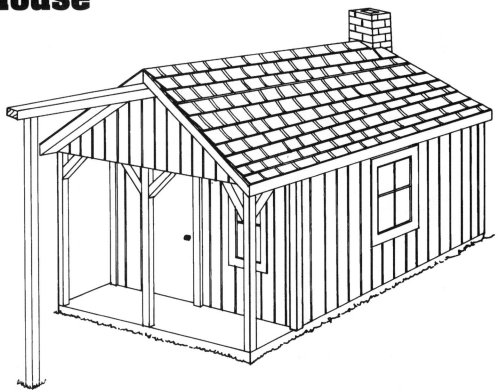

The building shown is actually an old-fashioned "summer kitchen" or food preparation house but could also be used as a guest house, hunting lodge, etc. Canning during the heat of the summer months cannot only make the main house unbearably hot, but create a mess as well. Equipped with a sink, stove, and plenty of cabinet space, this handy little outbuilding can make old-fashioned food preparation easy. There's also room to hang bags of onions, peppers, or other items that need to be dried for the winter months. A flue in the back end of the house accommodates a small stove or heater to heat the building during the colder months.

The house is a simple building on a slab with a front "porch" overhang. A heavy beam runs from the inside of the porch to a post set in front of the building. This provides a handy "hanging" pole for dressing wild game, etc. outside the building.

221

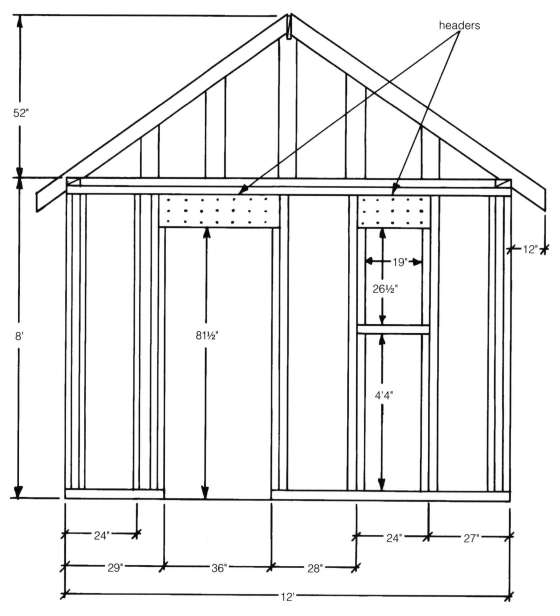

CANNING HOUSE FRONT ELEVATION

The inside of the house can be finished off with fiberglass panels, which are frequently used for finishing the inside of hog finishing houses or dairy barns. It's easy to wash these panels clean.

You may not wish to include all the details in the house as shown, but you should be able to get some ideas on how to build a similar type of building to suit your own needs. The construction of the house is quite simple and a standard procedure.

It utilizes plywood siding and solid plywood sheathing plus asphalt shingles. The soffits and gable ends are also finished off with soffit material and fascia boards to pro-

222

vide a snug, tight, insect- and rodent-proof building. Ventilation vents should be installed in the soffitts. The windows are standard double-hung sash windows and the doors are solid-core, pre-hung units.

The first step is to form the concrete pad. Note that there must be a pitch to the floor and an interior drain installed. In addition, a frost-proof hydrant installed inside the house as well as one outside provides a means of hosing down and cleaning. Anchor bolts must be installed in the pad for all walls, as well as for the three posts on the outer end of the open shed portion.

After the concrete pad has set up, the end walls are assembled and stood in place for the house proper. Then the side walls are assembled and erected. All windows and doors are headed over with 2x12s fastened together with a

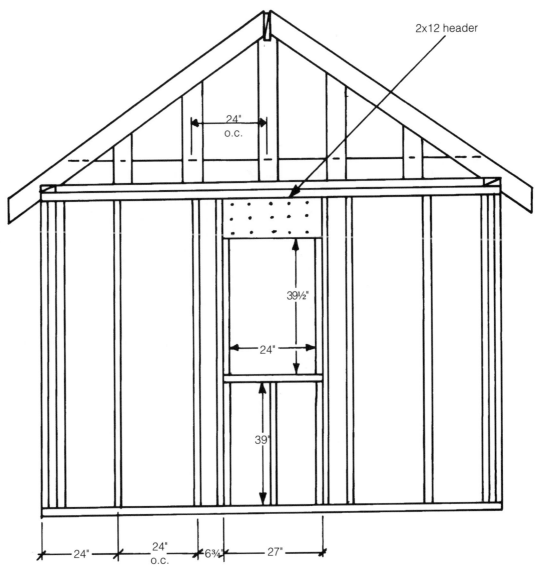

CANNING HOUSE FBACK ELEVATION

223

½" plywood spacer between at top and bottom to make the assemblies a full wall thickness.

The side and end walls are fastened together, then the top plates are installed and the end posts and top plate portion of the porch erected.

The next step is to cut the rafters and assemble them in place to the ridgeboard. Then nail the gable-end supports in place.

If a chimney is desired, construct it, making sure you follow local building codes, then cover the outside of the building with the appropriate siding material and install the windows, doors, and trim on the outside. Then install the plywood roof sheathing and shingle the roof.

The interior is then finished off to suit and the cabinets and any other features installed.

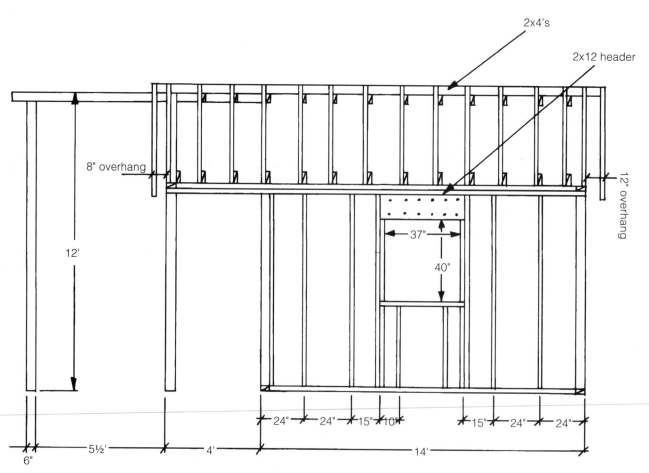

CANNING HOUSE SIDE ELEVATION

Two Garden Sheds

Michael Pruitt is a nut about outbuildings. In fact, his back yard in Bronough, Missouri, is filled with them. A high-school vocational arts (shop) teacher by vocation, Michael enjoys constructing small buildings as a hobby. The two shown are extremely handy small storage buildings. Both are quite simple to construct and are assembled in the same manner. They are portable buildings, constructed on skids of treated 4x4s. The ends of the skids are first cut to the angles shown so they can be pulled quite easily. Then holes are bored in the ends for fastening heavy wire, rope, etc. A box-sill and joists frame is created on top of the skids.

The first step is to nail the two end joists to the outer sills. Then cut the remaining inside joists and nail them in place to the outer sills. Make sure the entire frame is square, then toenail the joists to the skids. Finally, nail a ¾-inch plywood floor down over the joists and sills. Actually, you've created a simple platform framework, just as with permanent building construction. Now it's a matter of constructing and erecting the walls, adding the rafters, covering with siding and roof sheathing, and finishing

225

doors and windows. Both buildings are sided with standard wood lap-siding.

The rafters are installed and ⅝-inch solid wood sheathing nailed over the rafters, then asphalt shingles applied over the sheathing. Because there is no soffit, wooden blocking must be installed between the rafters to prevent mice, birds, and insects from entering the houses, and to provide a nailing surface for the top siding pieces.

The windows are simple aluminum storm windows installed over a roughed-in opening with trim installed on the outside. Doors are shop-made of plywood, 2x2 back framing, and 2x4 exterior trim and framing. All are nailed together securely from the front through the plywood and into the wood framework.

Building Details for Shed #1

Because the salt box roof creates two separate wall heights and roof lines, the first step is to create the front wall. Erect it and fasten in place, bracing until it is fastened to the

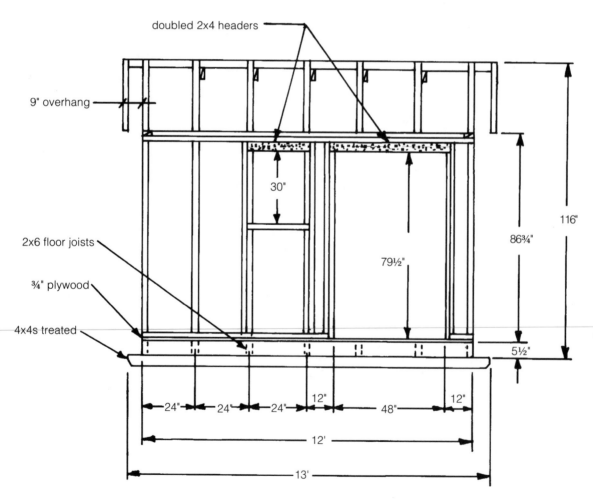

GARDEN SHED #1 FRONT ELEVATION

226

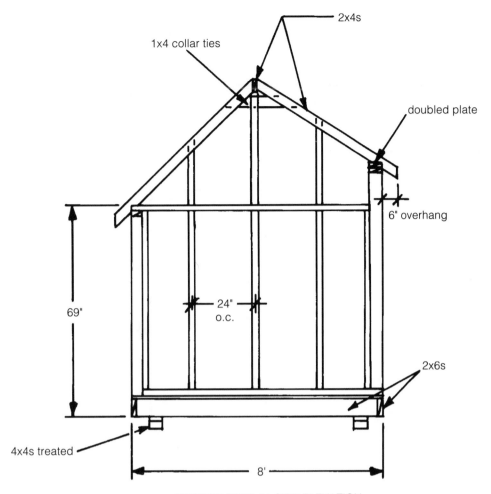

GARDEN SHED #1 SIDE ELEVATION

other walls. Then construct the back wall and brace it as well. Finally, construct the two side walls and erect them. Note that the side walls only have one top plate. The side-wall plates extend over the top of the top plate on the back wall, so studs are 1½ inches longer on the side walls. Rafters are braced with collar ties.

Building Details for Shed #2

A gambrel-roofed "barn," this small building even has a "loft," a handy space for storing long items such as ladders, etc. Once the platform-style floor is created on the skids, the side walls are constructed and erected. Then the end walls are constructed and raised between the two side walls. Notice that the back end wall is the same as the side walls. Standard end blocking is used above the wall plates between the rafters. The gambrel rafters are actually "mini-trusses," using plywood gussets glued and nailed on both sides of the joints. In addition the "ceiling joists" creating the loft are also nailed in place on these trusses before they

227

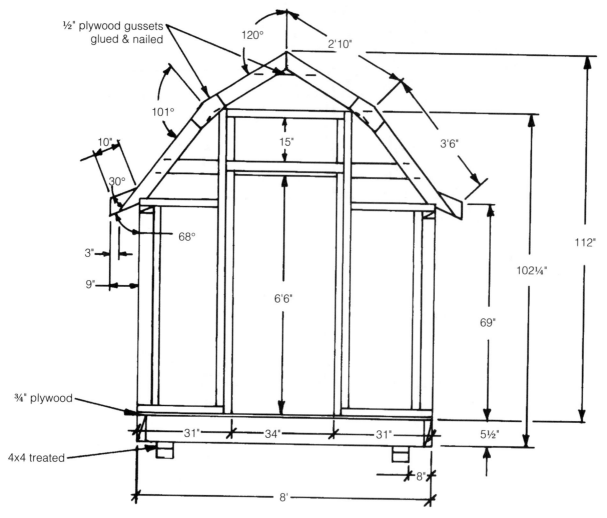

½" plywood gussets glued & nailed

120°

2'10"

101°

3'6"

10"

30°

15"

68°

3"

9"

6'6"

112"

102¼"

69"

¾" plywood

5½"

31"   34"   31"

4x4 treated

8"

8'

GARDEN SHED #2 FRONT ELEVATION

are fastened to the walls. Note that the outer upturned ends of the trusses are simply extensions nailed on top of the bottom rafter sections. Although the angles and rafter lengths are shown for this building, each building will vary somewhat. You will have to cut and fit these pieces for your particular building.

Position the back end truss in place, brace it, make sure it is plumb, then toenail it into the wall plates. Position the next truss in place and nail a solid sheathing board to both trusses on the bottom edge, making sure the sheathing extends at least 8 inches past the back truss, and over 24 inches past the front truss. Erect the rest of the trusses and anchor them in place with the sheathing board. Install the rest of the sheathing boards, then snap a chalk line on the top of the sheathing boards 8" from the outside face of the back end truss. Cut along this line with a circular saw to establish the back roof overhang. Then create the back truss, eliminating the outside gussets and the collar tie.

This is then nailed to the sheathing as a "hanging rafter." The front of the barn roof has an extended peak. It is created in the same manner as the back, except you will have to snap a second chalk line on each side of the roof to define the extended peak. Then cut the front overhang, including the extended peak, along the chalk line. In this case, the rafters are individually cut and nailed to the underside of the sheathing boards. Cut the lower pieces first, then hold the upper pieces in place, or use a bevel and a tape measure, to determine the angle of the cuts and the length of the pieces. Cut the rafters and nail them in place under the sheathing overhang.

The loft can be floored with solid sheathing boards, then the siding and roofing installed and the doors created and hung.

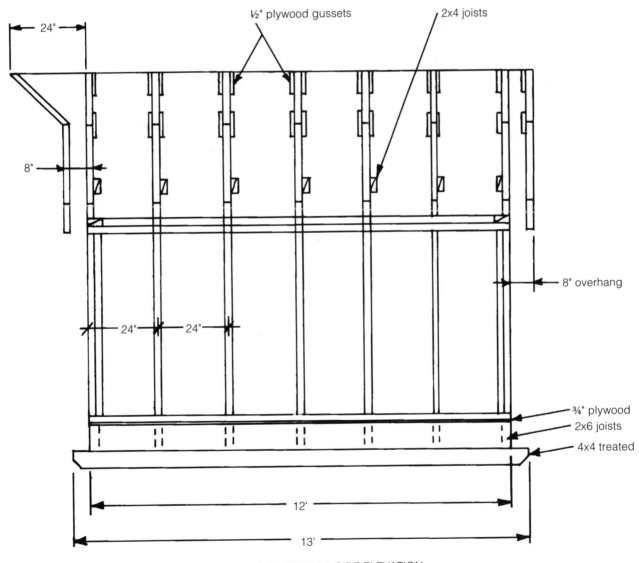

GARDEN SHED #2 SIDE ELEVATION

CHAPTER 15

GARAGES, STUDIOS, ETC.

Workshops and garages are often farm necessities. Two garages, one attached and one free-standing, plus a compact, but highly efficient, concrete workshop are featured in this chapter. There's also a beautiful studio for artists. It has an entire wall of glass for the best in north light. Also shown is a two-story home office with space below for a shop or garage. A small guest house rounds out this chapter of small building construction projects.

Workshop

Retired building contractor Dorman Coppage constructed this small shop for his part-time gunsmithing hobby, which quickly turned into a full-time operation. The building could just as easily hold any kind of small woodworking shop, or a craft or potter's shop. The design is such that the interior could be left completely open if desired. Actually, the same building design could be used for a small dairy barn by changing door locations, or even as a small auto shop by adding a garage door to one end.

Although rafters were used in the design shown trusses can also be used. The building shown is quite simple and

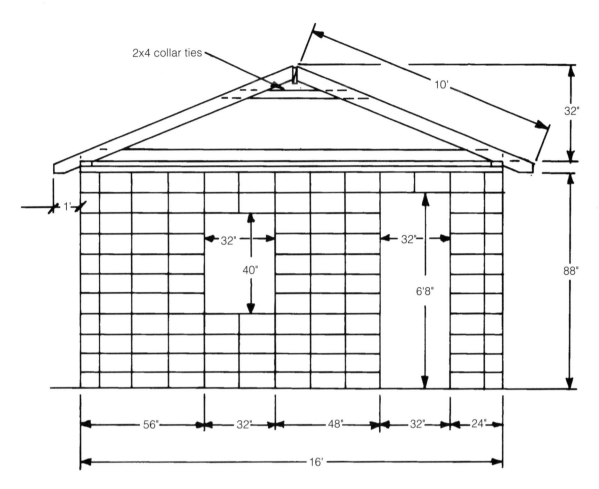

2x4 collar ties

10'

32"

1'

32"

32"

40"

6'8"

88"

56"   32"   48"   32"   24"

16'

WORKSHOP FRONT ELEVATION

has a small front "display" or sales room, a bathroom and shower (a safety factor for gunsmithing, in case the body comes in contact with spilled caustic materials). Steel windows and a steel door enclose the building.

The building is constructed of 8x8x16 concrete blocks laid block-on-block fashion with reinforcing inserted in the mortar between the blocks. The block walls are laid directly on a poured concrete slab, and of course the first step is to lay out and pour the slab, making sure it is level in all directions and that appropriate drainage and plumbing lines are installed before the concrete is poured.

Two options can be used for the block walls. First, you can subcontract the block work if you don't feel up to the task. Laying concrete block is not particularly hard to do mentally, but it can be exhausting physically if you're not in good shape. It is fairly easy to learn how to do, though, and one advantage is that you can take your time and lay only what you have the time or inclination for from one day to the next. Since the job doesn't have to be finished all at once, you can simply lay a few blocks each night after work or on weekends, and you'll be surprised at how easy

231

the task becomes when tackled in this manner.

It is extremely important to make sure the walls are not only square but plumb in all directions, and that the window and door openings are of the correct size and are perfectly square, especially if you're using steel windows and doors as was done on the building shown. Once the last row of blocks has been completed, install the anchor bolts in the mortar in the top row of blocks.

Allow the mortar around the bolts to set overnight, then position the bottom plate board in place next to the bolts, mark the bolt locations, and bore the holes. Anchor the bottom plate board in place, then nail the upper plate board in place, staggering the joints. Cut the ridge pole and a pair of pattern rafters and test-fit them in place. Make any necessary modifications in the pattern rafters, then cut the remainder of the rafters and nail them to the ridge board and down on the plate board, spacing 24 inches on center. Make sure the ends of the rafters are squared on the bottom to create a 4-inch projection for the soffit board and fascia board. With the rafters nailed in place, cut and nail the 2x4 collar ties to the rafters. Fasten the ceiling joists.

The final roofing step is to install sheathing and outer hanging rafters and then apply roofing of your choice.

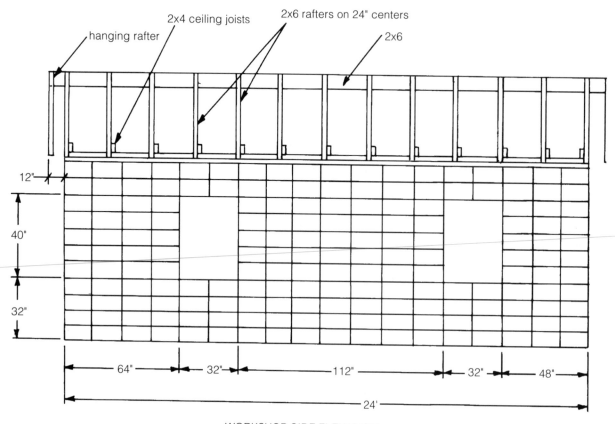

WORKSHOP SIDE ELEVATION

232

Studio

Regardless of whether you're a painter, sculptor, woodcarver, or potter, good north light is essential in the studio. The small studio shown is ideal for a one-person working area and has an entire north wall of windows. The windows are Andersen Feature Windows, available from building supply dealers selling Andersen windows and doors. There's an almost infinite variety of combinations available in many different sizes and shapes. Although the rough opening is shown for a typical window, it's best to order the specific windows and then follow manufacturer's specifications exactly when roughing in the wall containing these windows. The studio also has space for a small fireplace or flue that can be used with a free-standing fireplace or wood stove, providing auxiliary heat when needed and a comfortable place to sit by the fire and relax. The studio is one large room except for a bathroom and storage room in the end opposite the window wall.

Construction is fairly standard except for the window wall and the use of collar ties as ceiling joists. This allows for a high semi-peaked ceiling just above the windows. A 6x8 beam (solid or laminated from 2x6s) can be placed across the center of the building to provide more protection against horizontal wall stress.

233

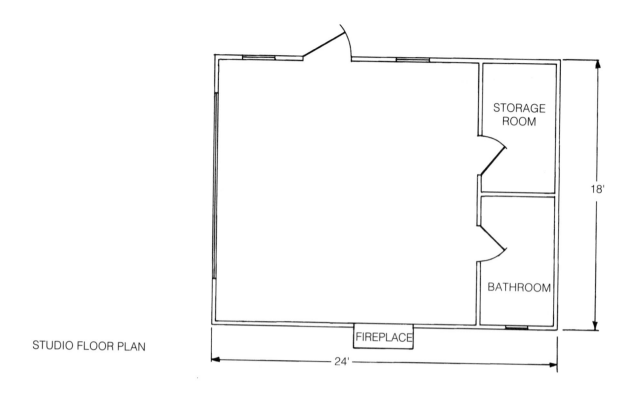

STORAGE ROOM

BATHROOM

STUDIO FLOOR PLAN

FIREPLACE

18'

24'

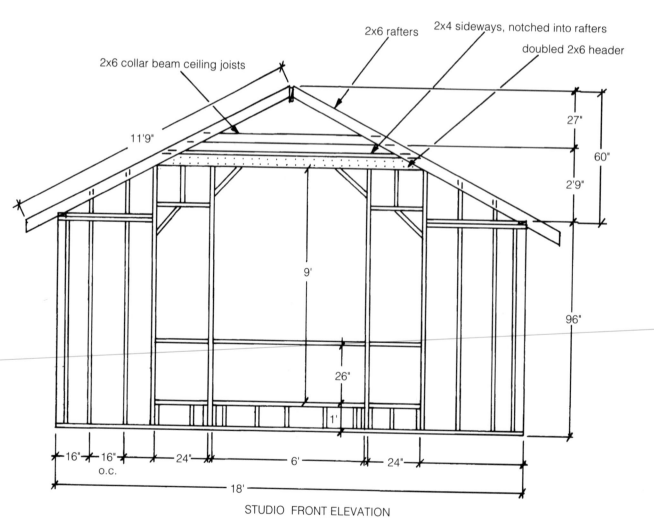

2x6 collar beam ceiling joists

2x6 rafters

2x4 sideways, notched into rafters

doubled 2x6 header

11'9"

27"

60"

2'9"

9'

96"

26"

1'

16"   16"   24"   6'   24"
      o.c.

18'

STUDIO FRONT ELEVATION

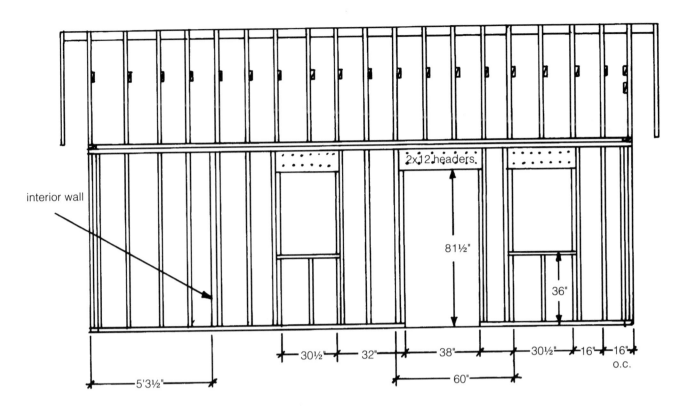

interior wall

2x12 headers.

81½"

36"

30½" · 32" · 38" · 30½" · 16" · 16" o.c.

60"

5'3½"

STUDIO SIDE ELEVATION

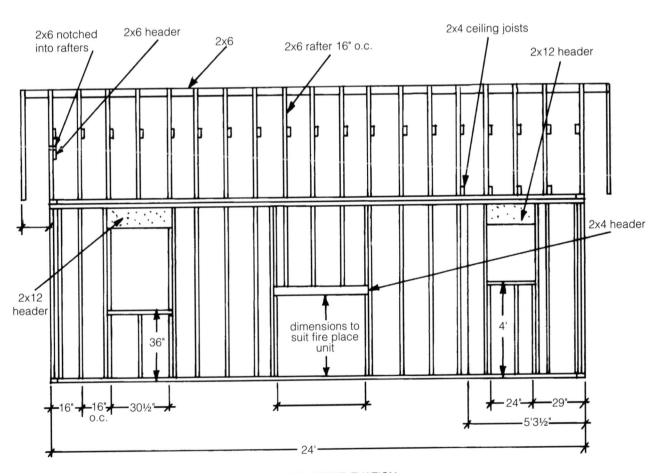

2x6 notched into rafters

2x6 header

2x6

2x6 rafter 16" o.c.

2x4 ceiling joists

2x12 header

2x4 header

2x12 header

36"

dimensions to suit fire place unit

4'

16" · 16" o.c. · 30½"

24" · 29"

5'3½"

24'

STUDIO SIDE ELEVATION

235

The building can be constructed on a concrete slab, although a footing, foundation, and wooden floor will make the days spent standing in the studio much easier on the body and feet. In the latter case, the box sill frame is first created, then the plywood subflooring is put down and the walls are constructed on the finished "platform" and erected.

The window end wall is constructed first on the platform or slab, and an unusual construction technique is used for this. The two rafters on that wall are also cut and assembled to the wall. A short 2x6 block between their top ends is used to space them properly. This is removed after the wall is up and the ridgeboard placed between them instead. Note that the 2x6 header over the windows is cut at an angle to fit flush with the underside of the rafters, while the top 2x4 placed edgewise is notched to fit over the rafters from the back side. This wall is stood up and braced well, the two side walls erected, and, finally, the back end wall is assembled and stood in position to complete the wall framing.

The back end rafters are cut and raised into position and the ridgeboard placed between them and braced in

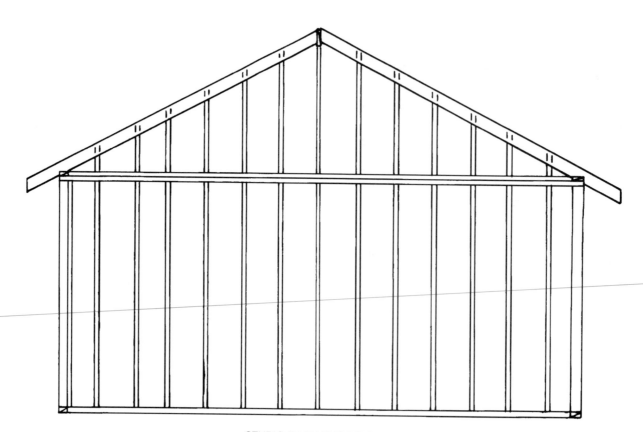

STUDIO BACK ELEVATION

place. Then the remaining rafters are cut and installed, including the collar tie joists. Make sure these are assembled to each rafter straight and level, as this is the "ceiling" of the interior except for the bathroom and closet portion. Location for collar ties is measured on end rafters, then chalk line maped on under sides of rafters so all collar ties will line up. The end blocking is installed on the back end.

The exterior is finished off as desired using plywood or hardboard siding or horizontal shiplap. Then the plywood roof sheathing is installed, making sure it extends past the building edge for the hanging "fly" rafters. These are then installed and the fascia board installed around the entire roof edge. The roof is shingled using shingles of your choice. One excellent choice is the Masonite Woodruf that simulate wood shingles. These are applied in 4-foot sections for a fast and extremely durable roof. Finally the windows and door are installed, including the large shaped windows. The soffit is constructed and all the trim needed to finish off the building is fastened in place.

The interior divider wall is then assembled and erected, and the interior walls covered with paneling or drywall.

Office/Garage/ Guest House

The two-story building shown is the author's combined office/garage, although it could just as easily be adapted to serve as a small guest house or living quarters for an elderly relative. The bottom portion provides space for two cars, or more than enough space for a small woodworking shop, as well as a wash room and space for a laundry room or off-season storage. The upper floor consists of two office rooms, a storage room, bathroom, and the author's darkroom.

The building shown was also constructed partially underground to take advantage of a typical Ozark hillside location. For this reason, concrete block walls make up the lower portion of the construction, with standard wood-frame balloon construction utilized on top of the block walls. A concrete floor completes the lower garage portion. A 2x8 tripled girder runs the full length of the center of the building, supported by a 6x8 center post which also helps support the upper floor.

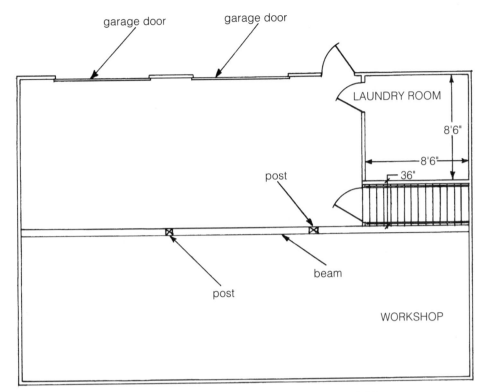

OFFICE/GARAGE/GUEST HOUSE LOWER FLOOR

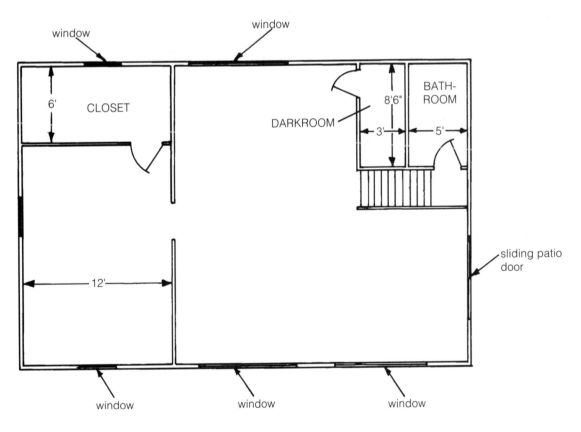

OFFICE/GARAGE/GUEST HOUSE UPPER FLOOR

The office/garage building sits close to the main house with a deck connecting the two, providing convenient access to the house, yet offering private office space. If constructing the building on a level surface above ground an exterior stairway should be added, although there are interior stairs.

The upper floor has a cathedral ceiling design to utilize the most of what would normally be "attic" space and cut down on the height of the building. One disadvantage of this design is that the ceiling must be thoroughly insulated with 6-inch batts because there is no attic space. Ventilation vents are placed in each end.

If building on a flat area the concrete-block bottom portion can be eliminated and standard wood-frame construction used instead.

The first step in construction, as with any building, is to lay out the corners of the building and make sure they are square. Batter boards and string lines are fastened in place to mark these dimensions and corner locations during

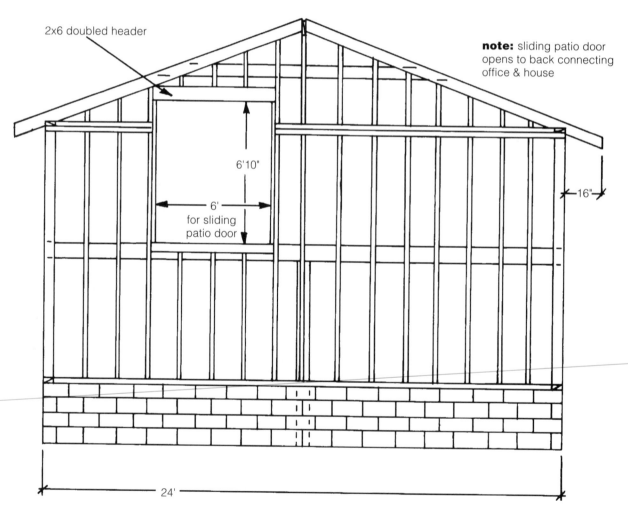

OFFICE/GARAGE/GUEST HOUSE FRONT ELEVATION

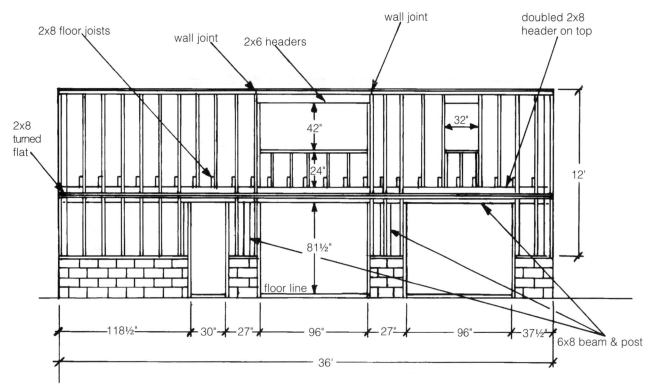

OFFICE/GARAGE/GUEST HOUSE SIDE ELEVATION

construction. Then, in case the building is to be constructed on a sloping lot, the back portion is excavated to the proper depth. Footings are then excavated and poured, making sure they're level and smooth.

The concrete block lower wall portion is then laid and anchor bolts are set in position in the top of the concrete block wall by placing mortar in the block openings and setting the anchor bolts in place. These should be allowed to set up overnight. The bottom plate boards cut, the stud locations marked on them, and the plate boards laid in position on the top of the block wall. Anchor bolt locations are marked on the plates, and holes bored for the anchor bolts.

Once the concrete block walls have set up, coat them thoroughly with asphalt to make sure they don't leak, then place gravel around the bottom of the excavation (and drainage pipes if needed for your area) and complete the filling, making sure you don't apply too much pressure on the walls at this time.

The stud walls are constructed in sections, with a bottom plate, studs, door or window headers, and lower top plate, staggering the joints of the lower top plates and upper top plates to assemble the wall sections together. Even at that, it takes a bit of effort to lift the wall sections up in place. They can be constructed on the inside of the

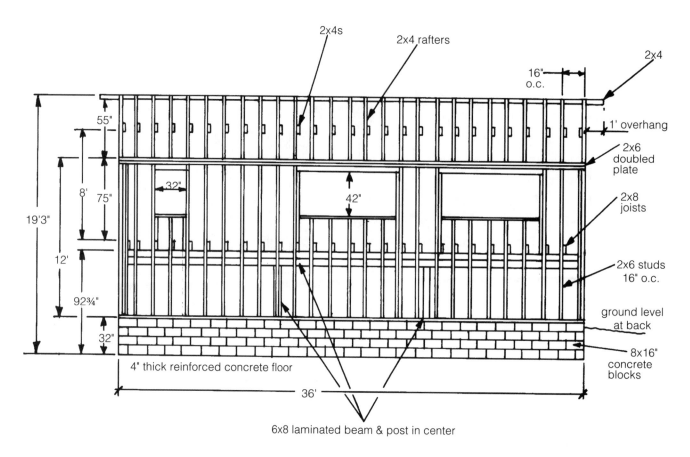

OFFICE/GARAGE/GUEST HOUSE SIDE ELEVATION

walls, if the floor area is level and smooth, and then carried outside and erected in place. Make sure they are firmly braced in place until all are anchored together securely, especially when quitting for the night.

Once all the walls have been erected, the rafters and ridge boards are erected. Then the collar ties (which are actually the top ceiling joists as well) are anchored firmly in place.

The walls of the building shown were first covered with sheathing, a layer of tar paper, then vertical 1x12s of western white cedar spaced ¾ inch apart. 1x2 battens nailed over the joints complete the outer skin.

Roof sheathing is fastened in place and fiberglass shingles are used to cover the roof.

The concrete floor is poured and allowed to cure, then the center support post anchored in place. The lengthwise support beam is constructed and nailed in place to the outer wall studs and resting on the support post. The interior stairs are constructed, the upper floor joists nailed in place to the outer stud walls, and the inner ends resting on the support beam fastened together with cleats on both

242

sides as shown in the detail.

The floor sheathing is installed and, finally, the building is enclosed by installing the windows, doors, etc.

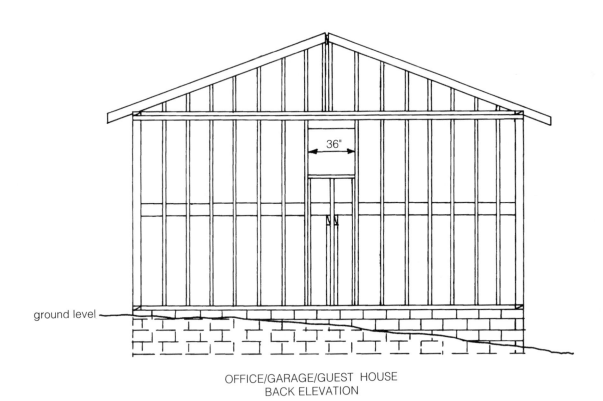

36"

ground level

OFFICE/GARAGE/GUEST HOUSE
BACK ELEVATION

joint

floor joists

cleats or scab

beam

post

2x12

2x12 stringer

8"

10"

Add-On Garage

As our small publishing company grew, our two-story garage/office eventually began to overflow. Our vehicles sat out in the weather while the garage filled with books. The answer was to construct an add-on garage on the back of the building. Due to the steep slope of the site, the width of the building, and a second-floor window, the add-on provided several challenges. We also wished to construct it large enough so that additional storage space could be utilized along the outside wall. The garage ended up measuring 28x28, whereas most standard two-car garages these days are 24x24. The garage was constructed on a monolithic footing/slab pour, which due to the steep slope required a great deal of fill rock on the downhill side, then quite a lot of gravel fill for the front and earth for the finish grade.

Once the footing/slab was created most of the hard work was done. The construction is standard platform-frame style with 2x6 studs on 24-inch centers for the walls. The roof framing uses trusses. This is due to the long span and the desire to leave the garage without interior supports.

The first step in construction is to create the slab floor. Anchor bolts are installed before the concrete sets up, making sure they are spaced so they don't fall where a stud will be located. The walls are nailed together on the concrete floor, then stood up and fastened together, anchoring their top corners together with the second upper plate board. In our case, the old siding on the side of the structure joining the garage was removed to be replaced. The walls were anchored to the side of the old building, nailing through the new wall stud into the top and bottom plates of the old wall. Then new siding was installed as the garage was sided.

Our family of five, with Michael, Jodi, and Mark (aged 10, 12, and 19 respectively), at the time, did all the construction work, including building the trusses. The long header over the garage door opening was constructed and lifted in place, proving to be one of the hardest jobs for our "crew." In retrospect the entire wall should have been constructed at one time and then "lifted" in place with push poles.

An amusing anecdote occurred during the construction of the trusses. Running out of nails, I drove into town and left the kids nailing a truss together on top of my "pattern" truss. Metal mending plates had been used to fasten the trusses together, and when I returned I discovered that Michael, the youngest, had filled every hole in the plates with nails. At least I know that truss will stay together!

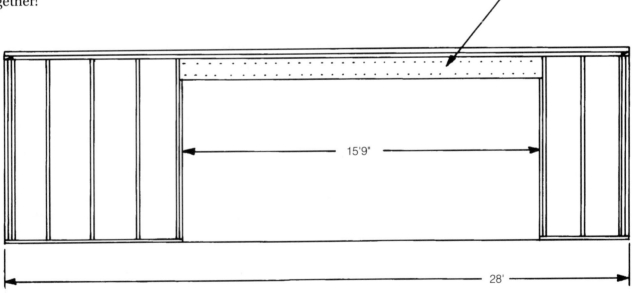

2x12 header

15'9"

28'

ADD-ON GARAGE
FRONT ELEVATION

245

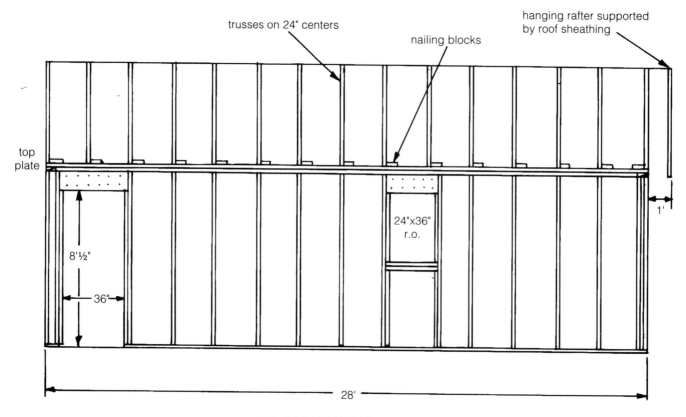

trusses on 24" centers

nailing blocks

hanging rafter supported
by roof sheathing

top
plate

8'½"

36"

24"x36"
r.o.

1'

28'

ADD-ON GARAGE BACK ELEVATION

As each truss was assembled it was erected and temporarily anchored in place with 2x4 bracing. One of the first investments I made when we moved to the farm was to purchase a tractor with a front-end loader. I've used it almost continually, and one excellent use has been lifting trusses into place. In this case, I placed the tractor in the garage. We assembled a truss, placed it on the tractor bucket, chained it in position by looping a chain around the truss top and over the bucket to a cross arm on the tractor bucket. The truss was slowly hoisted in place, then slid back and forth on the bucket until each end was pushed over the top of the outside walls and the truss was in the correct position with a 1 foot overhang on each side. Then I stood in the tractor bucket and lifted the truss erect. One family member on each side wall nailed the truss to a block previously nailed to the upper plate. These blocks are installed so the trusses are lifted up and laid across the wall tops, then pushed up against the nailing blocks to be in the correct position.

Note: Make sure that the walls are thoroughly braced diagonally and from the outside as well with 2x4s running to stakes driven in the ground, since the weight of the trusses can push the walls outward until all are anchored in position.

246

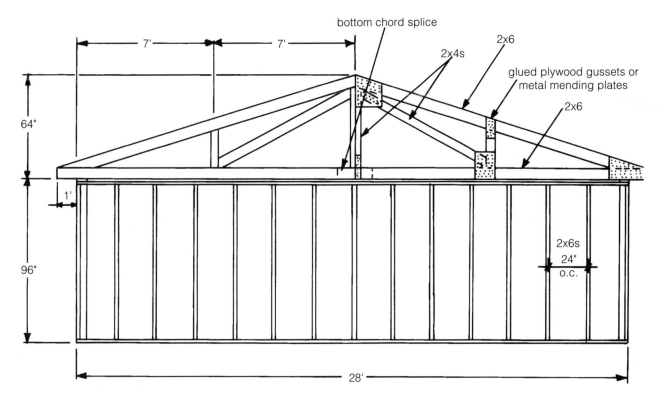

ADD-ON GARAGE END ELEVATION

Rafters are installed against the side of the existing building the same size and length of the truss, then the first truss is positioned in place and 2x4 bracing nailed between the two. Once the third truss is in place the plywood roof sheathing begins, further strengthening the building as the construction progresses.

In our structure, the walls were covered with exterior plywood grooved siding, and western rough sawn white cedar was used for the trimming to match the existing building. Once the roofing and siding had been installed on the garage, new siding to match was installed on the existing building, bringing it down over the shingles on the roof for further water protection.

Once the slab floor was complete, it took our family of five a little less than a week to complete the garage, shingles, siding, and all.

Two-Car Garage

An unattached garage is a popular kind of outbuilding on many farmsteads, as well as in many small towns. The two-car garage shown was constructed to match the prevailing house style in the small town of Humansville, Missouri. Featuring lap siding, it blends in perfectly with any number of homes across the country that were built during the 1950s and '60s.

Also typical of the time period is the practice of constructing on a concrete block foundation, with a poured floor between. The foundation sits on a wide footing and extends approximately 6 inches above the poured floor. This makes the wooden wall height shorter than 8 feet, actually 90 inches added to the foundation height to make up the full 8-foot ceiling height inside the garage. Measuring 24x24 with a single 16-foot-wide overhead garage door, the garage provides plenty of space for storage along both side walls and the back. Two windows in the back and a window and door in the house side provide air, light, and entry into the garage.

The first step is to lay out, excavate, and form the

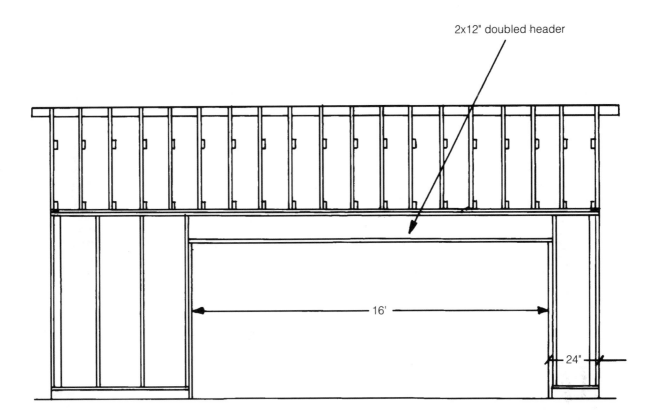

2x12" doubled header

16'

24"

TWO-CAR GARAGE FRONT ELEVATION

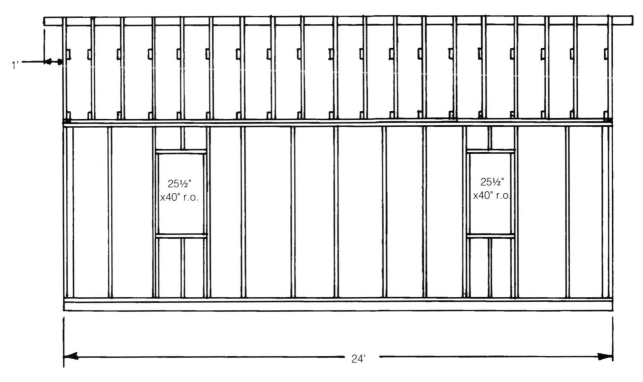

1'

25½"
x40" r.o.

25½"
x40" r.o

24'

TWO-CAR GARAGE BACK ELEVATION

249

footings. Then the concrete block foundation is laid, adding anchor bolts to the concrete filling the top holes of the blocks. The walls are constructed with a bottom sill plate and single upper plate in the usual fashion, stood up on the foundation, and over the anchor bolts. Bolts are tightened, and wall braced in place. The upper top plate is then installed to anchor the walls together. The wall studs are spaced 24 inches apart on center.

The rafters are erected, spaced on 16-inch centers. 2x4 ceiling joists are nailed across the top plates and to the rafters and an additional 1x4 collar tie is fastened to tie the rafters together.

In most instances of this type of construction solid 1x12 sheathing was then applied diagonally before the siding was installed. Plywood or flakeboard sheathing would be a good alternate choice these days. Then the lap siding is installed starting at the bottom and nailing to the wall studs and finishing corners with metal corner pieces.

During that time period the roof was also usually sheathed with solid 1x12s, then tar paper and shingles were applied. Again plywood sheathing offers a faster, cheaper alternative. After it is installed, tar paper is fastened in place with staples and the shingles of choice applied.

Finishing involves adding the soffit and fascia boards

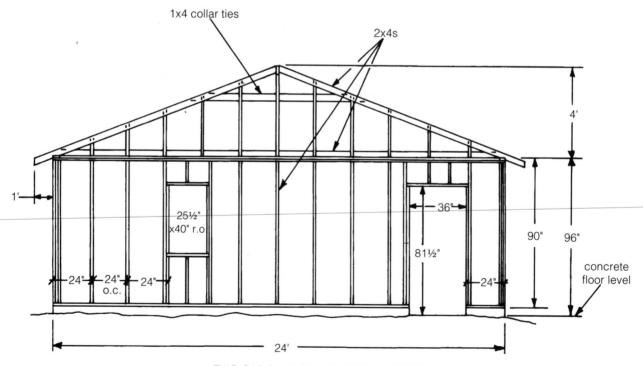

TWO-CAR GARAGE LEFT SIDE ELEVATION

as well as window and door "trim." The windows and doors are simply aluminum storm windows and doors, and are installed over the exterior trim pieces. This provides an economical means of closing off the building, but a more secure method would be to install regular doors and windows first, then the storms.

The inside is finished off with shelves running the length of the closed side and back, providing lots of garage storage.

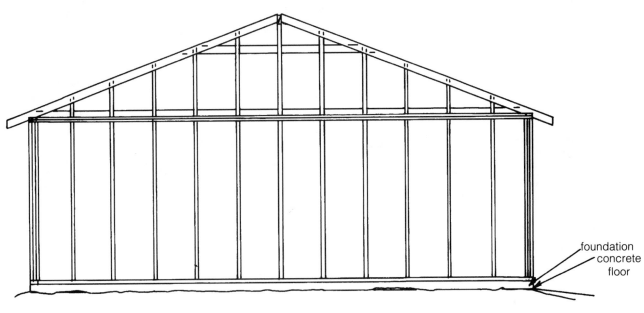

TWO-CAR GARAGE RIGHT SIDE ELEVATION

251

Guest House

When John Mangum constructed Bucksaw Point Marina and Lodge on Truman Lake in central Missouri, he added several small housekeeping cottages along the bluff overlooking the lake. Although John's place is a sprawling complex, these cottages are also ideal for a small resort, or for someone wishing to construct a guest house out back. They consist of a simple platform-frame construction on a foundation with the front roof line extending past the building to form a comfortable porch. The porch roof is supported by treated 4x4 poles and the porch is constructed entirely of treated 2x6s.

The interior is simple and efficient. Divided down the center, the front half is a combination cooking/dining and living area. There's more than enough space for a small work counter, sink, stove, refrigerator, table, and chairs in one end. The opposite end accommodates a couch, a couple of chairs, and a television set.

The back half consists of two bedrooms with a bathroom in between. There's more than enough space for two couples or a large family.

Because the houses shown were constructed on a

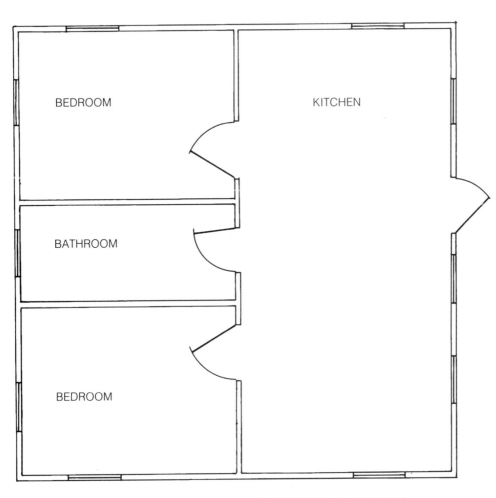

GUEST HOUSE FLOOR PLAN

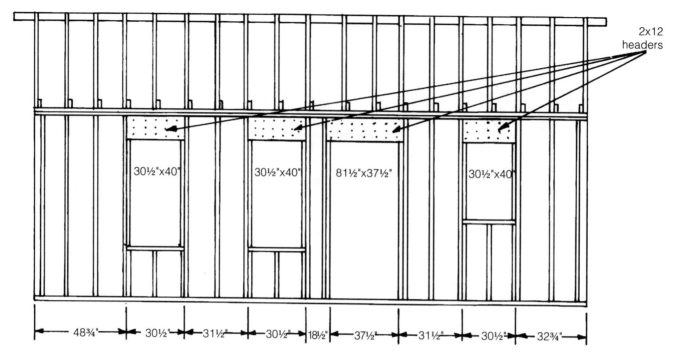

2x12 headers

30½"x40" 30½"x40" 81½"x37½" 30½"x40"

48¾" 30½" 31½" 30½" 18½" 37½" 31½" 30½" 32¾"

GUEST HOUSE FRONT ELEVATION

2x4 studs, rafters, & ceiling joists   studs, rafters on 16" centers

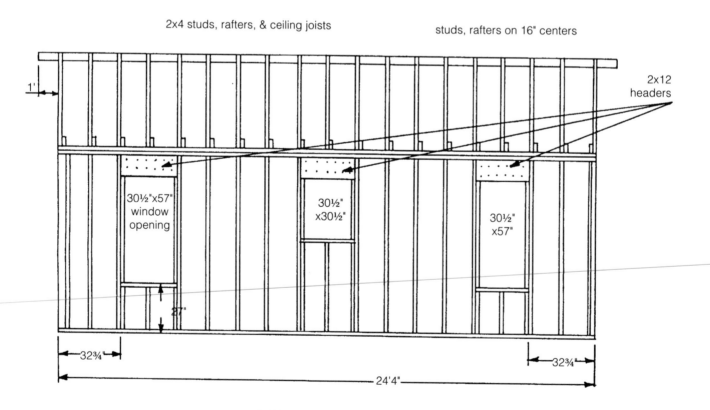

1'

2x12 headers

30½"x57" window opening

30½" x30½"

30½" x57"

27"

32¾" 32¾"

24'4"

GUEST HOUSE BACK ELEVATION

254

steeply sloping lot, the porch was lowered and a step-up provided. The low, slanting roof is no problem. If the house is to be constructed on a flatter surface, the porch floor will be raised to the interior floor level, and the porch roof will have a shallower pitch than the rest of the roof to provide the necessary headroom.

Construction is basic platform on foundation style. The building is quite simple, with little "fancy" details that can quickly become complicated. It's a good "first house" for a beginning builder.

The first step is to excavate, form, and pour the footing and foundation. Anchor bolts are installed in the foundation before the concrete sets up. Once the concrete has hardened a sill plate is bolted in place, then the sills and floor joists are installed. Floor sheathing is nailed in place on the joists to complete the floor framing.

The walls are constructed on the floor, stood up and nailed in place, then tied together with an upper plate, doubling the top plates. Standard rafter and ridgeboard construction is used for the roof framing. Once the roof line is established with a "trial" rafter on each end, the porch's treated poles are set in place and the long rafter support beam fastened in place on top of the support poles.

The house is covered with prefinished hardwood vertical siding, and the roof is covered with asphalt shingles. The final step is to install doors and windows and the soffit

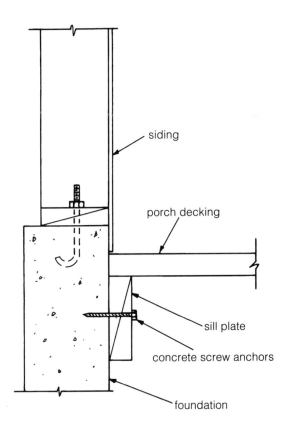

siding

porch decking

sill plate

concrete screw anchors

foundation

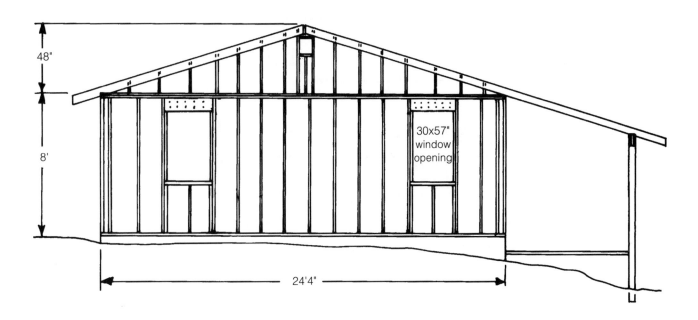

48"

8'

30x57"
window
opening

24'4"

GUEST HOUSE SIDE ELEVATION

255

under the house eaves, as well as the porch roof.

Actually, construction of the deck is more complicated than the house. The first step is to fasten a sill board of treated lumber to the concrete foundation. This can be done by using concrete nails, or, better yet, concrete anchors and bolts or screws. A sill is also nailed to the outside of the support posts. (Note: All nails should be galvanized for long-lasting fastening ability and to prevent rust streaks on the wood.) Floor joists are then nailed between the two, using support hangers or nailing blocks into the back sill board and nailing to the blocks. The floor is then installed and finally the rail and seat assembled. The first step in rail assembly is to cut the rail supports to the correct shape. Then fasten them to the front deck joist with anchor bolts. Fasten the upper rail and then the lower rails to the rail supports. Fasten the seat supports in place and nail down the seat boards. The final step is to install the steps, railing post, and the step-down railing.

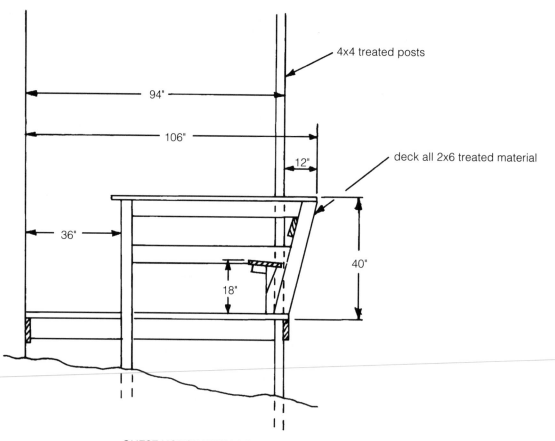

GUEST HOUSE DECK & RAILING DETAIL

CHAPTER 16

OUTDOOR PROJECTS

In addition to the standard outbuildings, barns, and sheds, numerous other small, often "for-fun," but extremely useful buildings are also needed. The roadside stand shown can be used for a number of selling jobs, while a couple of unusual chicken houses and an insulated dog house provide more small projects. A simple but effective bottle calf house provides a means of sheltering these popular small farm animals while the children's playhouse offers a fun house the entire family can enjoy.

PROJECT 18

Roadside Stand

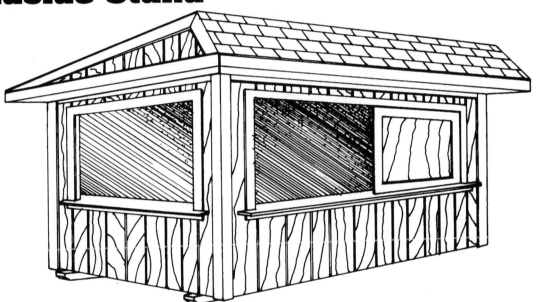

A multipurpose portable building on skids, this handy building can be used as a produce or fireworks stand, or even as a concession stand for ball games and such. It features three open sides with drop down doors over countertops that extend to the outside of the building. The exterior is covered with textured plywood siding with 8-inch on center grooves and is roofed with asphalt shingles.

The first step in construction is to cut the skids from 4x6s to the proper length. Then bevel their ends so they won't dig into the earth when the building is moved. Holes can also be bored through the ends for fastening rope, wire, or chains.

A sill and floor joists assembly is constructed first, over the skids, anchoring the joists to the skids by toenailing them in place. Then ¾-inch exterior plywood is fastened

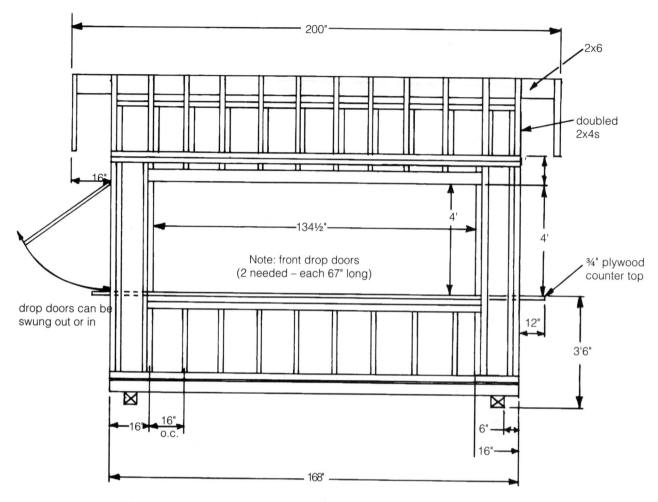

200"

2x6

doubled 2x4s

16"

134½"

4'

4'

Note: front drop doors
(2 needed – each 67" long)

¾" plywood counter top

drop doors can be swung out or in

12"

3'6"

16"
16" o.c.

6"

16"

168"

ROADSIDE STAND FRONT VIEW

down over this to provide flooring. The front and rear walls are constructed first, assembling them on the floor, lifting them up, and bracing them until the sidewalls are completed and joined to them. Note the doubled 2x4s that provide more support under the open countertops.

Assemble the sides, stand them in place between the front and back walls and fasten them in place.

The next step is assembling the roof, and in this case the rafter situation is a bit unusual. The ceiling joists are also the collar beams that hold the rafters. Because there can be a great difference in how each person marks and cuts angles, it's a good idea to make up a sample rafter set, cutting and adjusting as needed. Once the rafter set suits, use the samples to cut the remaining rafters. Then, starting at one end, nail one angle-cut 2x4 joist (A) down on the top plate of the end wall by toenailing it in place. Nail an upright (B) against the inside of the joist. Then nail a second upright 2x4 (C), shorter than the first by the width of the bottom 2x4 joist, on top of the joist and into the inside upright. Repeat this on the opposite end of the building. Then nail the remaining 2x4 joists in place to front and back by toenailing them in place. Nail the remaining uprights against the joists. Cut a top plate (D) to go over the uprights and nail it securely in place. Then cut a second top plate (E) to go over this. Now begin adding

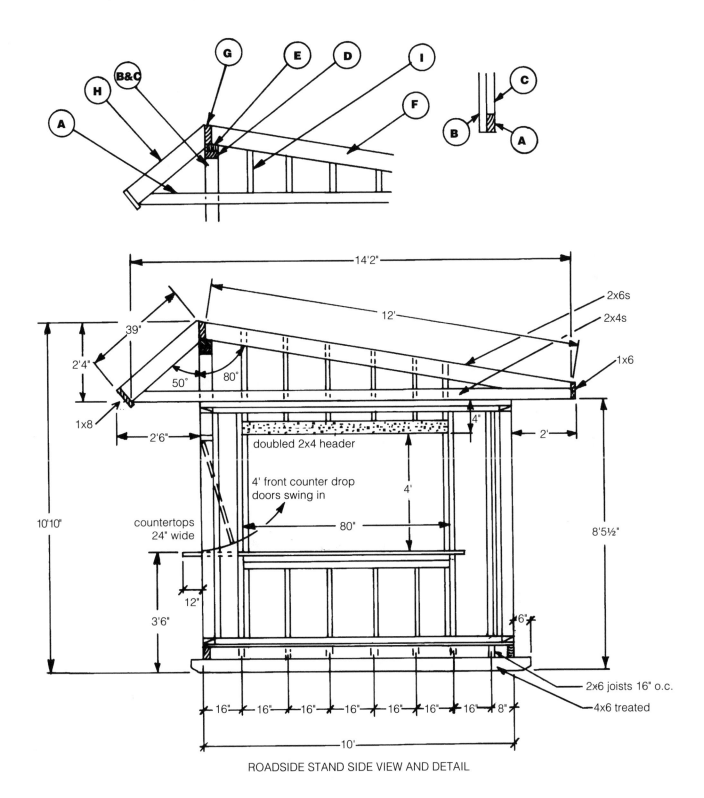

ROADSIDE STAND SIDE VIEW AND DETAIL

259

the rafters (F), starting at one end and toenailing the back rafter into the outer end of the 2x4 joist at the back of the stand and into the doubled plate at the front. Once this is done to all back rafters, nail the ridgeboard (G) to the rafters, making sure the ridgeboard extends past the end rafters 16 inches on each end. Fasten the front rafters (H) in place by toenailing them into the ridgeboard and from the bottom side of the 2x4 joists into the bottom edge of the rafters. Complete the roof construction by adding end blocking (I) between the end rafters and the top plates, and then sheathing boards over the rafters, extending the sheathing at least 16 inches on each end. Chalk-line the extended sheathing and cut it straight. Nail the hanging rafters on each end under the sheathing and to the ridgeboard, then nail a hanging "joist" on each end as well.

Cut the siding to fit and nail to the front, back, and ends. Then cut matching siding to fit over the upper end sections. Cut a front and back fascia board and add a bottom fascia board over the siding to match the front and back fascia boards. Cut upper fascia trim boards to match

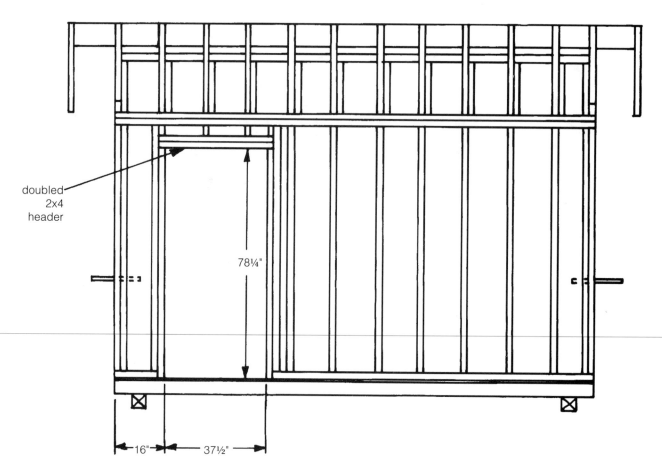

ROADSIDE STAND BACK VIEW

260

the angle cuts of the rafters and nail them over the siding and into the hanging rafters.

Cut the soffit hanging boards (2x4 blocking between the rafter ends) and nail in place on the building ends between the wall and the hanging rafters.

Cut the trim boards and nail on each corner as well as around the windows and under the back edge of the soffit.

Cut the countertops, making sure that they extend around the wall framework both inside and out. Full cabinets or bracing can be used under the countertops on the inside to help support them.

Then construct the drop-down doors from plywood siding and wooden trim strips. Hinge the doors at the top. They can be made to swing in or out as you prefer. Doors that swing in are easier to handle, although they do cut down on headroom. Hooks and eyes are used to hold the doors in position, and the same can be used to lock them shut when the building is not in use.

The final step is to build and hang the back door and paint or stain the structure to suit.

A Chicken House with a View

The **Maples Bed and Breakfast** in Humansville, Missouri, not only offers excellent food and a chance to spend some time in a completely restored Victorian home, but a touch of small-town farm life as well. Directly out the back window is a view of the chicken yard with this attractive chicken house. One unusual feature is the plate glass window on the side, allowing easy observation of the hens on their nests.

The house is a simple platform construction. For a permanent installation, the best tactic for ease in cleaning is a concrete slab floor with sill plates made of treated lumber. The same house can also be constructed as a movable house, built on "skids" of 4x6 logs. In this case the floor rests on treated 2x6 joists fastened to the skids. Treated plywood is used for the floor.

Regardless of which method is used, the size of the

262

building makes it an ideal choice for a beginning project. After building the platform, the next step is to construct the walls. Stand them in place, brace them, and then add the upper top plate to tie them together. Cut the rafters, nail them to the ridge board, and fasten collar ties to each set of rafters except the ends. Add gable end blocking or studs. The roof is sheathed with plywood, then the hanging rafters are installed. Then plywood siding is fastened in place, the fascia boards are nailed on, and the roofing is installed.

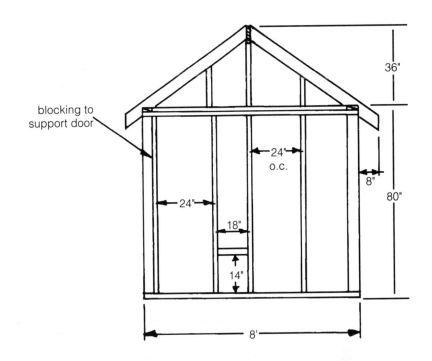

CHICKEN HOUSE FRONT ELEVATION

NOTE: BACK SAME (door on same side, no small door)

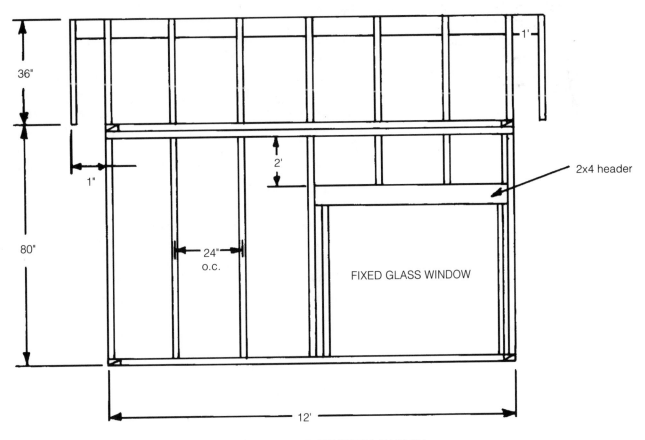

CHICKEN HOUSE SIDE ELEVATION

Portable Laying Hen Shelter

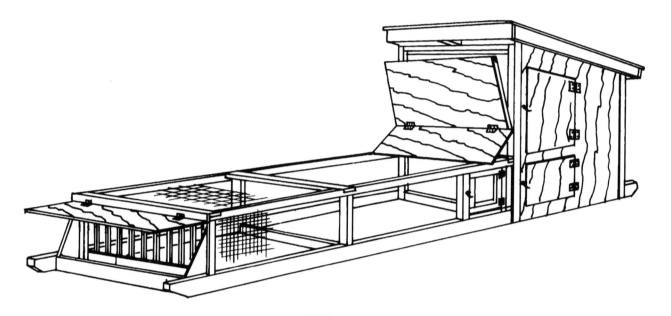

This unusual, but highly useful small, portable hen house provides space for up to a dozen laying hens. It has an enclosed house with roost and four nests and an open poultry-wire–covered front that allows the hens to scratch and feed in the open. Once an area becomes overworked, merely move the house and pen on the runners. The front of the pen has an enclosed feed and grit trough that is reached from the outside. A small door on the back of the open pen provides a means of filling a small waterer. A drop-down door on the bottom of the house allows you to lock the hens in the house while moving. The drop-down door on the front can be opened to provide ventilation, or closed in severe weather. The front of the house should typically face the south or away from the winds and storms. Doors on both sides allow for ease in cleaning, collecting eggs, etc.

The entire front open section is constructed of treated lumber, the skids of 2x6s and the remaining framing of 2x4s. Notice that the uprights of the pen are notched to fit over the 2x6 bottom sides and upper 2x4 pieces. Then the top cross braces are notched into the side pieces. The feeder is constructed on the front by nailing the inside to the front uprights between the two skids. Nail the two

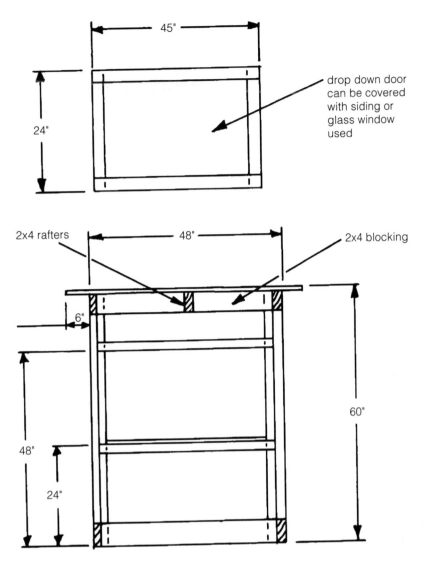

drop down door can be covered with siding or glass window used

45"

24"

2x4 rafters

48"

2x4 blocking

6"

60"

48"

24"

PORTABLE LAYING HEN SHELTER FRONT ELEVATION

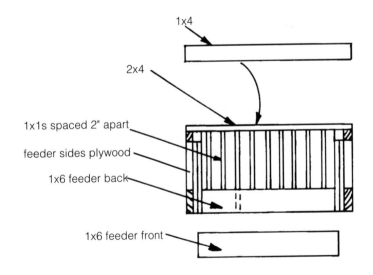

1x4

2x4

1x1s spaced 2" apart

feeder sides plywood

1x6 feeder back

1x6 feeder front

PORTABLE LAYING HEN SHELTER FEEDER DETAIL

feeder sides in place, the bottom and the feeder front. Then cut the 1x1 slats and fasten them in place. Finally cut the top 1x4 and fasten it to the front of the top and hinge the feeder lid to it. The door on the back side of the open pen is constructed of half-lapped 2x2 door frame and hinged in place.

The house is constructed by cutting notches in the upright side studs to fit over the skids and the outside rafters. Then nail these in place and cut the cross pieces for the sides and nail them in position. Cut the cross pieces for the back and front bottom and tops and nail them in position. Then nail the center rafter on top of the upper cross piece.

Cut the inside floor, leaving a 12"x36" open space between the roost boards for the hens to fly up into the house. Assemble the roost from three 2x8 boards cut 14 inches long. Nail 2x2 roost boards across their tops, and a couple of 1x6 brace boards between them at their bottom outside edges. Construct the nest sides, back and center dividers of 1x4s with a 1x8 for the front. Construct the doors of 2x2s cross lapped and cover with siding. Cover the sides, upper and lower front portions (down to the upper side of the open pen), and the back with plywood siding. Nail the 1x12 drip stop over the lower portion of the front and then hinge the front drop-down door to the top and

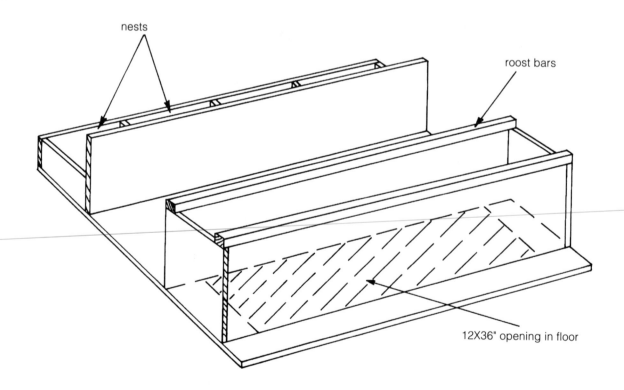

nests

roost bars

12X36" opening in floor

the lower raise-up door to the bottom sections. Install hardware cloth or poultry netting to the inside of the top front opening.

Hinge the doors in place and add catches. The final step in construction of the open run is to cover it entirely with poultry netting.

Cut fascia boards for the front and back and then cover the top with ⅝-inch plywood. Asphalt or fiberglass shingles or roll roofing can be applied to finish off the roof.

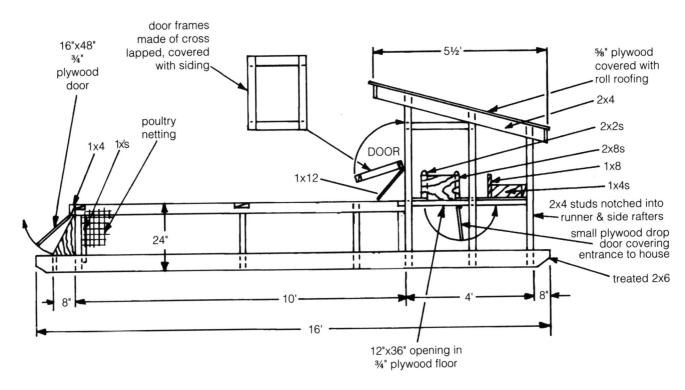

PORTABLE LAYING HEN SHELTER SIDE ELEVATION

Insulated Dog House

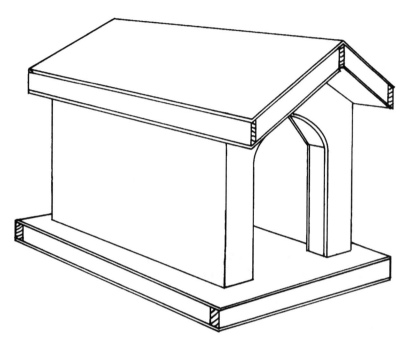

The best home for a farm, homestead, or hunting dog is a good solid house built specifically for the animal. One of the ways of combatting cold weather is to build an insulated dog house. If the house is properly built and sized for your particular dog, the dog's body heat will keep it warm during the coldest weather. The house illustrated is about right for one medium to large dog. A smaller house makes it easier for a smaller dog to keep warm. Incidentally, the insulation will also help keep the dog house cool in the summer.

The dog house is built and insulated from the inside out, the reverse of how a regular house is constructed. The bottom is assembled first from 2x4s. Use standard 3½ (4") fiberglass batt insulation in the bottom. A great deal of cold comes up from the ground, and adding this insulation helps alleviate that problem. If the house is set on a concrete pad you won't have to cover the bottom. If it is to be placed on the ground, however, a plywood bottom cover should be installed.

Build the sides and ends from 2x4s, gluing with Resorcinal glue and screwing them together. After the basic framework has been assembled, cover the inside with ¼-inch prefinished hardboard or flakeboard. Place insulation between the studs from the outside, stapling it in place from the back side. This can prove a bit tricky. The vapor barrier (paper surface) must face the inside of the house on all sides and on the top and bottom. After the insulation has been installed, cover the outside of the house with siding to match your home or other outbuildings. The roof is then made by first cutting the rafters, the ridgeboard, and "fascia" boards. Cut the two roof undersides from flakeboard and nail them in place on top of the

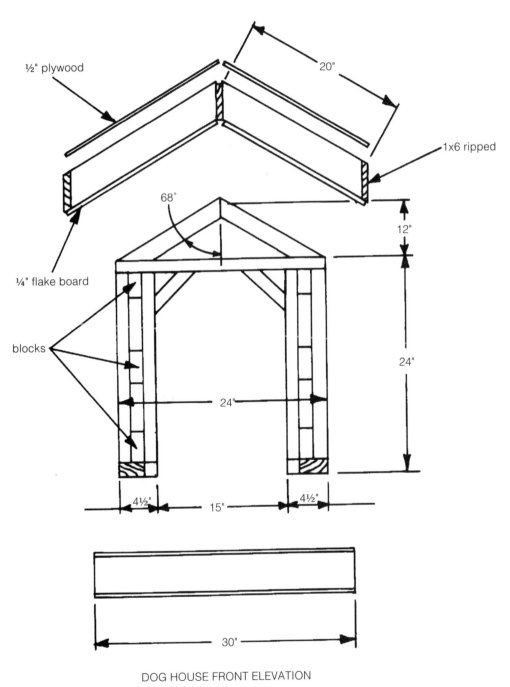

DOG HOUSE FRONT ELEVATION

269

sides and ends. Nail the rafters to the ridgeboard, and the fascia boards to the outside faces of the rafters. The entire rafter assembly is then placed down on the roof undersides and fastened in place by toenailing into the wall plates and the angled 2x4 end rafter supports. The insulation is put in place and the top pieces of ½-inch exterior plywood are nailed in place. All joints should be sealed with latex caulking compound. The roof can be painted with a good light-reflecting paint. Several coats should be applied to insure that the roof is well sealed. You can shingle the roof if you like, but the house will then be almost twice as heavy to raise for cleaning.

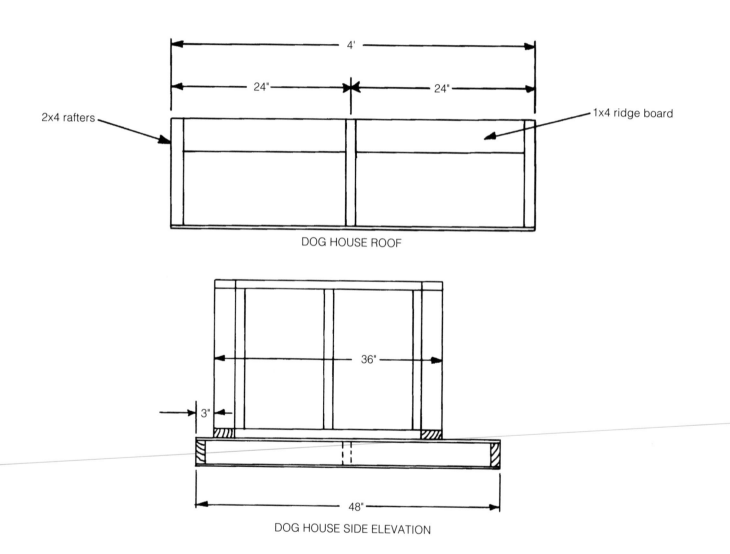

DOG HOUSE ROOF

DOG HOUSE SIDE ELEVATION

Individual Calf Hutch

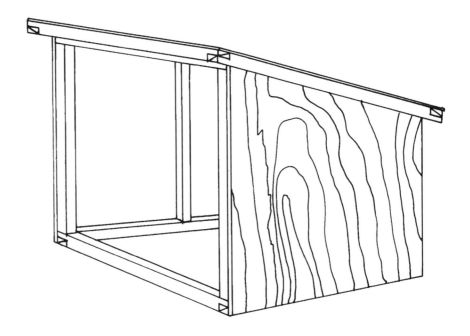

Raising bottle dairy calves in individual hutches or pens is considered the best method of preventing disease. You can build your own hutches quite easily using two sheets of ½-inch exterior plywood and 2x4s of treated lumber ripped into 2x2s.

The first step is to rip the 2x4s into two equal pieces. Then cut the framing pieces for the back wall and assemble them together using two 3-inch No. 10 wood screws in each joint. Lay the frame on a smooth flat surface, use a carpenter's square to make sure it is square, and then nail the plywood securely in place with galvanized nails.

Assemble the front frame, then cut the bottom side pieces and fasten them in place to the front and back frames with wood screws. Cut a plywood side piece and nail it to the back frame, front frame, and bottom side frame piece. Hold a piece of stock over the inside of the top of the side frame to determine the angle of the two end cuts and then cut it and install with screws from the upper back and front frame members. Use nails from the plywood side. Hold the center frame support in place on the top of the bottom side frame piece, mark its length and the angle

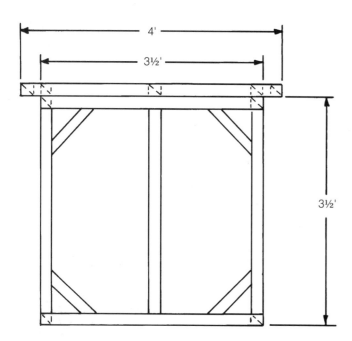

CALF HUTCH BACK ELEVATION
(note: front is same, but 48" high and
without center support)

of cut from the bottom edge of the top side frame piece, and cut to length. Install with screws and nails through from the plywood side. Repeat this process for the other side.

Cut the roof framing pieces and fasten them together with screws. Then lay the frame on a flat, smooth surface and fasten the plywood on top with nails. Finally, position the roof in place and fasten securely with 3-inch No. 10 screws.

A rack for holding bottles and also one for hay can be nailed to the inside of the side wall at the front.

Give the exterior a couple of coats of good paint for protection.

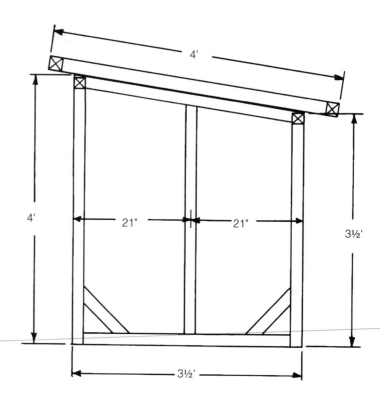

CALF HUTCH SIDE ELEVATION

Child's Playhouse

Another of **Michael Pruitt's** backyard projects, this attractive little playhouse was built for daughter Sarah. The playhouse is also constructed on skids to be portable like Michael's other projects (see Projects 10 and 11). Construction is fairly simple, and quite similar to the construction of the gambrel shed, Project No. 11. Refer to Projects 10 and 11 for information on constructing the skid and platform.

Once this has been constructed, the sides and back are assembled and erected on the platform floor framing. Then the front is assembled by first making the door frame then cutting and installing the pieces between the door frame and the sides. Finally the rafters are cut and nailed to the ridgepole and sides, and 1x6 collar ties are nailed in place to secure the rafters.

Siding and roofing are installed, then the aluminum storm windows and shop-assembled doors installed.

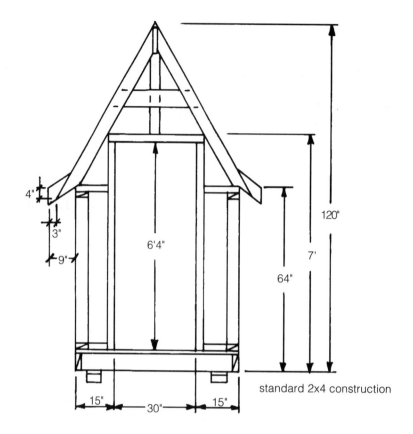

4"

3"

9"

6'4"

64"

7'

120"

standard 2x4 construction

15" 30" 15"

PLAYHOUSE FRONT ELEVATION

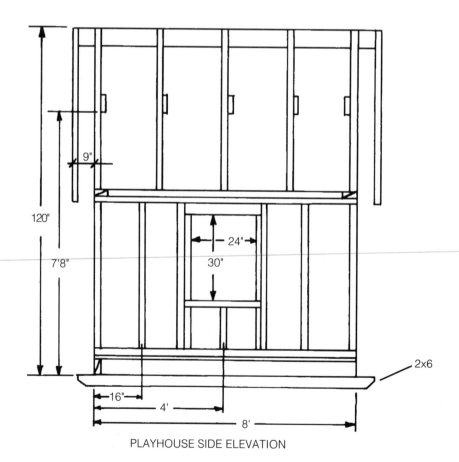

9"

120"

7'8"

24"

30"

2x6

16"

4'

8'

PLAYHOUSE SIDE ELEVATION

274

Resource & Suppliers List

American Plywood Association
P.O. Box 11700
Tacoma, WA 98411

American Wood Preservers Institute
1651 Old Meadow Road
McLean, VA 22102

Andersen Windows, Inc.
100 Fourth Ave. N.
Bayport, MN 55003-1096

Filon Division of BP America
P.O. Box 5006
Hawthorne, CA 90250-5006

Foley-Belsaw Company
6301 Equitable RD. Box 419593
Kansas City, MO 64141

Goldblatt Tool Company
P.O. Box 2334, 511 Osage
Kansas City, KS 66110

Hickson Corporation
1100 Johnson Ferry Rd., Suite 680
Atlanta, GA 30342

Masonite Corporation
One South Wacker Dr.
Chicago, IL 60606

Midwest Plan Service
University of Missouri
Columbia, MO 65211
(or check with your local Extension Office)

National Steel Corporation
20th and State Streets
Granite City, IL 62040

Nattinger Materials Company
P.O. Box 4007, 1650 E. Atlantic
Springfield, MO 65808

Owens-Corning Fiberglas Corporation
Fiberglas Tower
Toledo, OH 43659

Shell Rock Industries, Dek-Block
7301 Ohms Lane, Suite 445
Minneapolis, MN 55439

Skil Corporation
4300 West Peterson Ave.
Chicago, IL 60646-5999

Stanley Tools
Division of the Stanley Works
New Britain, CT 06050

United Steel Products Company, Kant-Sag
Hickory St. at 4th, N.E.
Montgomery, MN 56069

Wheeling Corrugating Company
1134 Market Street
Wheeling, WV 26003

Wood-Mizer Products
8180 West 10th Street
Indianapolis, IN 46214-2430

INDEX

Figures in *italic* indicate drawings, photographs and tables.

The Wheel of Time

SAND MANDALA

My Dearest Brel and Elizabeth,

This has been a year of travel in so many forms for each and all of us. Our shared connection with the Tibetan culture feels centering and significant as we continue to journey through the Wheel of Time. Each moment I am aware of your presence in my life, I feel such gratitude, so eternally blessed. I love each of you so deeply and celebrate our unique family this Christmas...

Always
from my heart
Mama Lillian

1998

el *of* Time
ND MANDALA

Visual Scripture of Tibetan Buddhism

FOREWORD BY THE XIV DALAI LAMA

by

BARRY BRYANT

IN COOPERATION WITH
NAMGYAL MONASTERY

HarperSanFrancisco
A Division of HarperCollinsPublishers

FIRST HarperCollins PAPERBACK EDITION PUBLISHED IN 1995
ISBN 0–06–250088–0 (pbk.)

An Earlier Edition of this Book Was Cataloged as Follows:
Bryant, Barry
    The wheel of time sand mandala : visual scripture of Tibetan Buddhism /
    Barry Bryant – lst ed.
        p.   cm.
        ISBN 0–06–250089–9 (cloth)
        1. Kalachakra (Tantric rite)—Tibet.   2. Buddhism—Tibet.   I. Title.
    BQ7699.K34B73   1992
    294.5'37—dc20   91–59044
                                                                         CIP

95   96   97   98   99   ❖   RRD(C)   10   9   8   7   6   5   4   3   2   1

This edition is printed on acid-free paper that meets the American National Standards Institute Z39.48 Standard.

In Appreciation

For their valued wisdom and time, my deepest gratitude to
Ven. Tenzin Yignyen,
Ven. Lobsang Samten, Ven. Pema Lobsang Chogyen,
Ven. Dhondup Lobsang Gyaltsen, Ven. Jamphel Lhundup,
Ven. Tenzin Migyur, Ven. Tenzin Legdan, Ven. Lobsang Gyaltsen,

Very Ven. Kirti Tsenshab Rinpoche,

and

Deborah Moldow and Gregory Durgin

contents

previous page: The Dalai Lama applies the first sand
to the Kalachakra Sand Mandala on the third day of
the Kalachakra Initiation. New York City, 1991.

Acknowledgments

THE WHEEL OF TIME SAND MANDALA: Visual Scripture of Tibetan Buddhism owes its inspiration to His Holiness the Dalai Lama, who has devoted great energy and commitment to the continuity of these splendid and perfect teachings of the Buddha.

I am deeply indebted to the monks of Namgyal Monastery who participated—in particular Lobsang Samten, the first monk to work on the project, who, with Dhondup Lobsang Gyaltsen, Jamphel Lhundup and Pema Lobsang Chogyen, answered countless questions and generously gave of themselves; and to Tenzin Yignyen, who spent one year with me translating sacred Kalachakra literature and explaining the philosophical meanings of Kalachakra, for his steadfast discipline and warmth. Thanks are also owed to Tenzin Migyur, Tenzin Legdan, and Lobsang Gyaltsen.

I would like to pay particular acknowledgment to Sakya Trizen Rinpoche, the 98th Ganden Tripa, Jamgon Kongtrul Rinpoche, Kalu Rinpoche, and Chutgye Trichen Rinpoche for their inspiration and devotion to Kalachakra; and to Kirti Tsenshab Rinpoche, whose hours of explanation of the Kalachakra Tantra on beautiful spring days in Dharamsala added substantially to the text. I also want to thank the many translators, including Kalsang Dhondup, Tenzin Sherab, Karma Lekshe Tsomo, and Tenzin Gyatso from the Dialectic School, Philippe Goldin, Ken McLeod, and Namgyal Korkhor. Thanks also to Tenzin Geyche Tethong, Lodi Gyari, Rinchen Dharlo and Michele Bohana.

I would like to express my heart-felt appreciation and thanks to Deborah Moldow, whose long hours and devotion to this book have been invaluable during the lengthy process of writing, rewriting, and editing. My thanks also to William Meyers for his generous support in completing the editorial process. I am also grateful for the research and editorial work done by Norvie Bullock, Karma Lekshe Tsomo, Daia Gerson, Helen Tworkov, Elizabeth Selandia, Ellen Pearlman, Vivian Kurz, and especially Peter Sheene and Maria Dolores Hajosy Benedetti.

Without the able input of the indomitable Gregory Durgin and my brother Stan Bryant, this project simply would not have been possible. I also greatly appreciate the guidance I received from Valrae Reynolds of the Newark Museum concerning the visual materials in the book.

Other valued advisors included Alexander Berzin and Professor Robert A.F. Thurman, holder of the Jey Tsong Khapa Chair of Columbia University. My thanks also to Phuntsok Dorje, Moke Mokotoff, Barbara Lipton of the Jacques Marchais Center of Tibetan Art, Jean and Francis Paone, Christian Lischewski, Perkins + Will, Daniel Maciejczyk, Somi Roy, Dr. Mikhail Khusidman, John Bigelow Taylor, and Dianne Dubler.

I also want to express my special thanks to my parents Carolyn and Frank Bryant, and to Glenn Bryant, Nancy Scheyer, Warner Scheyer, Joe Cobuccio, Kathryn Feldman, Scott Friedman, Elena Konovalova, Brian Adams, Soho Black and White, Ann Wilson (whose introduction to Malcolm Arth at the American Museum of Natural History opened the door), Dr. Peter Keller of the Natural History Museum of Los Angeles County, Dr. Pratapaditya Pal of the Los Angeles County Museum of Art, Beverly Walton and John Denver of Windstar Foundation, and Jean and Francis Paone.

Thanks are due to Helene Silverman with whom long hours of design esthetics were shared to illuminate this book, with the help of Scott Frommer and the inspiration of Tony Smith and Marcel Duchamp.

Further, I would like to express my gratitude to Fred Segal, Joan Brady, LeVar Burton, Barry and Connie Hershey, Chris Sarazen, Maggie Kress, Barbara and Bruce Bordeau, Dennis Konner, Fred Koenig, Marvin Ostroff, Stanley and Elyse Grinstein, Jane Smith, Sally Sachs, David LaFaille, Bruce Bryant, Madelaine Brostrom, Carol Moss, Szajna Kellman, Tom and Margot Pritzker, Bonnie Lundgren, Rita Narang, Francesca Kress, Jack Mayberry, and the New York State Council on the Arts for their enthusiasm, support, and encouragement.

And for their support in bringing this book from concept to publication, my sincere thanks to Mia Grosjean, Barbara Ascher, Jacqueline Onassis, Judy Sandman, Kara Laverte, Patrick Jordan, Kate Coleman, Ani Chamichian, Dick Carter, Dennis Dalrymple, Karen Lotz, Julia Moore, and especially John Loudon and HarperSanFrancisco.

Finally, my special thanks for their diligence and clarity to Tashi Tsering, Ngawang Topgyal, and Thupten Choephel of the Namgyal Monastery Committee, Michael Cohen, Esq. and Ada Clapp, Esq. of Davis, Polk & Wardwell, Sidney Piburn of Snow Lion Publications, Richard Weingarten of the Tibet Fund, Frank Douglas, Esq., Barbara Hoffman, Esq., Reed Wasson, Esq., the staff of Volunteer Lawyers for the Arts, and last, but certainly not least, Robyn Brentano.

The XIV Dalai Lama

FOREWORD

FROM THE VERY BEGINNING, the teachings of the Buddha have been concerned with fulfilling the wish shared by every living being to find peace and happiness and avoid suffering. Moreover, because all beings have an equal right to pursue these goals and because, even among human beings, there are wide differences of interest, character, and ability, the Buddha taught a vast array of methods.

These mostly deal with the need to develop clarity of mind, purity of behavior, and a correct view of reality. They involve meditation and a code of ethics summarized as seeking to avoid harming others, actively helping them if you can, and a realization of the interdependence, and hence the lack of an inherent identity, of all phenomena.

Of all the different kinds of teachings the Buddha gave, Tibetan tradition regards the tantras as the highest. Outwardly, the practitioner maintains a life-style that accords with pure ethics; internally, he or she cultivates an altruistic intention to attain the state of a Buddha. Then, secretly, through the practice of deity yoga, concentrating on the inner channels, essential drops, and energy winds, the practitioner enhances his or her progress to enlightenment.

The Kalachakra system was one of the last and most complex tantric systems to be brought to Tibet from India. In recent years many Westerners have become acquainted with this tradition as various lamas have given the Kalachakra Initiation to large groups of people. I myself have given it several times in Western countries, as well as in India and Tibet. Such initiations are given on the basis of a mandala, the sacred residence with its resident deities, usually depicted in graphic form. The tradition I follow employs a mandala constructed of colored sand, which is carefully assembled prior to each initiation and dismantled once more at the end.

Due to their colorful and intricate nature, mandalas have attracted a great deal of interest. Although some can be openly explained, most are related to tantric doctrines that are normally supposed to be kept secret. Consequently, many speculative and mistaken interpretations have circulated among people who viewed them simply as works of art or had no access to reliable explanations. Because the severe misunderstandings that can arise are more harmful than a partial lifting of secrecy, I have encouraged a greater openness in the display and accurate description of mandalas.

Consequently, Barry Bryant and Samaya Foundation, assisted by monks from Namgyal Monastery here in Dharamsala, have done a great deal of good work to increase appreciation of the Kalachakra Mandala and its larger significance in the context of Buddhist practice. This book, containing material drawn from authentic sources and presented clearly to be easily understood, is a further welcome fruit of their efforts.

Whereas the Buddha initially gave most of the tantras to individual disciples, the Kalachakra was from the outset given to an entire community, the citizens of the Kingdom of Shambala. Subsequently, in Tibet it became customary for the initiation to be granted to great gatherings of people. I feel that introducing such a profound means of enlightenment in this way creates a strong positive bond among all those who are present and so plants fertile seeds of peace.

With this in mind I congratulate the author and all who contributed to this book and offer my prayers that their good intentions be fulfilled.

June 2, 1992

The
Wheel
of
Time

SAND MANDALA

Introduction

IN OCTOBER 1973, I watched the Fourteenth Dalai Lama of Tibet walk down the steps of an Alitalia jet at the airport in Copenhagen, and my life was changed forever.

The Dalai Lama is the exiled spiritual and temporal leader of the occupied nation of Tibet, of both its oppressed inhabitants and its thousands of refugees living in exile. In 1973, at the age of thirty-eight, he was on his first European tour and was arriving from Rome, where he had just met the pope. His message in Denmark was unlike anything I had ever heard. He helped me to understand that each of us has the right to be happy and that there is a path to happiness, which we can freely choose to follow. But more than his words, it was the Dalai Lama's compassionate presence that shook my foundation of beliefs and assumptions, and allowed me to see beyond my own arrogance and self-importance.

A World War II baby, I had grown up in material comfort during the 1950s in Washington State and had become a fervent participant in the revolution of values that took place during the 1960s. In 1970 I traveled to Copenhagen to visit my twin brother and found the atmosphere so friendly toward my work as an artist that I stayed there three years to develop an environmental art project, which included making films and videos for Danish television.

Even though this work was very rewarding, it didn't satisfy my urge to express something more meaningful or profound.

Fifty thousand students assemble to receive the Kalachakra Initiation from the Dalai Lama at Bodhgaya, India in 1973.

1

Friends led me to a Tibetan meditation center, where I felt instantly at home. I was inspired by what I heard regarding Tibetan Buddhism and particularly by its emphasis on the perfection of the mind.

Having begun to study and meditate there daily, I heard that the Dalai Lama would be visiting soon. As I looked forward to his visit, the focus of my art shifted from concern for the environment to expression of inner vision, using Tibetan Buddhist-inspired images. But it wasn't until I actually met the Dalai Lama that I began to understand the subtle levels of consciousness where my pain and dissatisfaction could be transformed into inner fulfillment.

His heartfelt concern for others was inspiring. He paid particular attention to personal details and always seemed to be genuinely considerate, asking the right question at the right moment. I continually probed him with my '60s-style antiauthoritarian questions. His objective, good-humored responses were always just what I needed to hear. After years of searching, I sensed that I had finally found something that would lead me beyond the limited thinking imposed by my conditioning. I immediately began to film the Dalai Lama as part of the Danish television workshop. When he was teaching I seemed to receive a personal message, as though the Dalai Lama were silently saying to me, "Put your camera down. Just sit and listen." That is how I became a student of Buddhism.

Relatively little public attention was being paid to the Dalai Lama at that time, so I was able to follow him on his speaking itinerary, attend his teachings, and even bring my camera to his private meetings. Twelve years later, I returned to Copenhagen to see that the crowd for his public talk had increased from 100 to more than 10,000.

But my work with the Dalai Lama had only just begun. I requested a private audience, after which my whole life seemed to pass before me like a movie. I saw the deceit, the confusion, the abusive relationships, the struggle with my family, and most of all, the need for gentleness and clarity in my life. My suffering was a result of my own distorted perceptions and ignorance. It became clear that I needed to take full responsibility for my actions.

The following week I went to Germany and Switzerland to video-tape the Dalai Lama as he performed ritual ceremonies and gave teachings, continuing the ancient oral tradition. In Rikon, Switzerland, more than 2,000 Tibetans living in exile in Europe lined up on either side of the road to greet the man they consider to be the living Buddha. Their faces were jubilant as they anticipated receiving the teachings and his blessing.

Before leaving Switzerland, I was invited by one of the Dalai Lama's secretaries to film the Kalachakra (Wheel of Time) Initiation, which he was to confer on 50,000 Tibetans two months later in Bodhgaya, India, the site of the Buddha's enlightenment.

Upon confessing that I had no idea what this initiation was, I was told that it was an empowerment ritual. At the center of the ritual was a sand *mandala*, a complex and colorful design which symbolized the teachings of the Buddha. Through contemplation of this mandala one could attain an enlightened state of mind, free of all obstacles and filled with compassion and wisdom. Just what I had been seeking, I thought.

I had neither been to India nor been part of a Buddhist initiation before, but the thousands of Tibetans around me at Bodhgaya during that Christmas holiday in 1973 made me feel welcome. In spite of the language barrier, I learned much from their ever-present kindness throughout the ten days I was there.

On the night before the final day, the initiation site seemed strangely still under the full moon. Many Tibetans were doing prostrations—bowing down to the ground and then lowering their entire bodies to the earth in a gesture of prayer and homage to the throne of the Dalai Lama. I had never prostrated myself before, but each time I did, I felt as though I were somehow letting go of the accumulated burden of my past misdeeds.

Then I heard someone reciting the mantra (or chant) *OM MANI PADME HUM.* When I recited it myself, I felt heat being generated inside my body, and I began to understand how reciting a mantra could help one to feel and generate compassion toward others.

Tibetans listening to the Dalai Lama during the 1973 Kalachakra Initiation at Bodhgaya, India. The woman in the center foreground is spinning a prayer wheel.

During this profoundly meaningful night, I sat under the brilliant light of the full moon reciting a Buddhist prayer before the enormous bodhi tree where the Buddha had attained enlightenment 2,600 years ago. At about 6:00 A.M., I felt a growing exhilaration as a crowd of thousands of monks began to gather for their full moon morning prayer. The Dalai Lama's attitude of reverence and patience seemed to pervade the entire community. I felt a sense of wholeness within myself that I had been seeking for a lifetime, and I realized

A Tibetan monk prostrates toward the Great Stupa monument at Bodhgaya. In his right hand he holds a *mala,* or rosary, used for counting mantras, prayers, and prostrations.

how such a ritual could engender what the Tibetans call "bliss consciousness."

The only English I remember hearing during the entire initiation was while the Dalai Lama was describing—in Tibetan—the deities of the mandala to the students, who were to imagine or visualize them in their minds. Suddenly, he said clearly in my own language, "Now you enter the mandala."

Those words had the weight of prophecy for me. At that point, which was near the completion of the twelve-day ceremony, the multitude of initiates were lined up to see the Kalachakra Sand Mandala and to receive the blessings of His Holiness the Dalai Lama.

Years later I would learn that each person has a unique spiritual experience upon seeing the mandala for the first time. Some people experience peace; some feel their own suffering; some feel humanity's suffering; others feel bliss; still others see their spiritual teacher. According to my journal entry that night, when I saw the mandala for the first time, "I felt my insides lighting up like a sunburst . . . as if firecrackers were exploding. When they settled, I saw the mandala before me in a large gallery in a North American museum, with thousands of Westerners experiencing the peace and light I felt at that moment."

From that day on, my work was cut out for me. Fifteen years later Samaya Foundation, under my direction, would present the Kalachakra Sand Mandala at the American Museum of Natural History, fulfilling this vision and marking the first time the mandala would be displayed as a cultural offering outside its ritual context.

More than 50,000 museum visitors saw the Kalachakra Sand Mandala at that exhibition, and millions more throughout the world have since had the opportunity to view it in museums and galleries and through its extensive exposure in the electronic and print media.

At the American Museum, visitors marveled at the monks as they went through their rituals—praying and applying sand—and always moving about the gallery explaining the meaning of the mandala. It was as though their every gesture embodied the very essence of their training.

Many visitors had questions: "Are the monks making up the design of the mandala as they go along? Where does the colored sand come from? What are those tools they're using?" I heard repeatedly, "Is there a book we can get about sand mandalas? Or about the

A Namgyal monk uses the tip of a *chakpu*, a metal funnel, to produce a decorative design on the rim of a dharma wheel in the Kalachakra Sand Mandala. The Kalachakra Initiation, Bodhgaya, India, 1973.

Buddha? Or Tibetan Buddhism?" This book was conceived in response to those questions.

The Dalai Lama has said that the practice of the Kalachakra Tantra, an esoteric but profound teaching of the Buddha, ultimately affects everyone. But nothing can communicate the beauty and the power of a Tibetan sand mandala like seeing it in person. My hope is that this book will bring the general reader as close as possible to that experience and serve as an introduction to the tantric path of Tibetan Buddhism.

I have been a student of Kalachakra since 1973, and I feel that I have only just begun to explore its depths. Yet I have come to understand enough to be aware of its potential benefit for the individual, the community, and for all life on the planet. May this book act as a seed for peace through compassion and wisdom, which are the essence of the Kalachakra teachings.

Barry Bryant, 1993

Opposite: At the Kalachakra Initiation,
Bodhgaya, India, 1973.

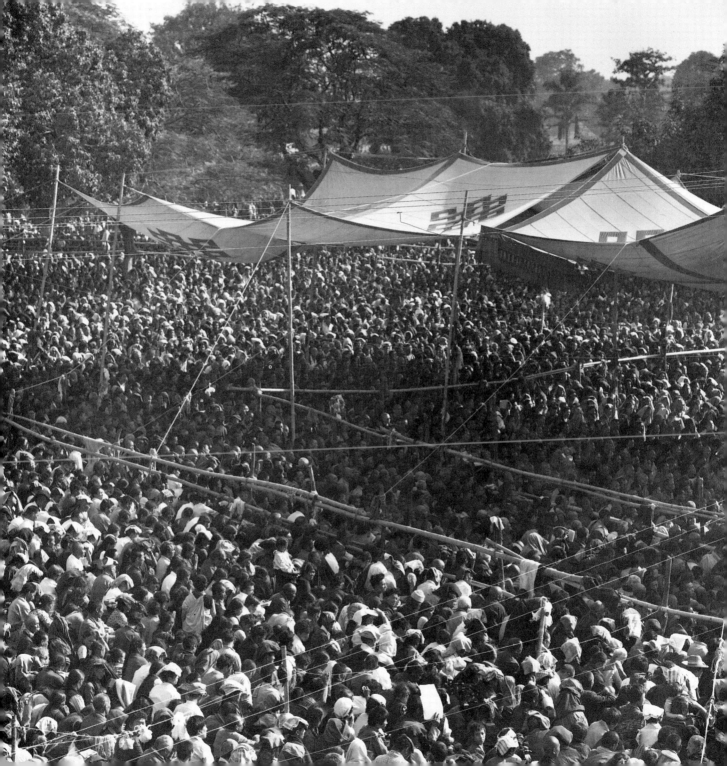

The Path of Kalachakra

Praise glorious Kalachakra,

conqueror of all fears and disillusionment,

at whose feet I bow in great humility.

sharing the secret path

"KALACHAKRA is a vehicle for world peace," the Fourteenth Dalai Lama of Tibet, recipient of the 1989 Nobel Peace Prize, has stated time and again. What does he mean? Can unlocking the mysteries of a secret doctrine taught by the Buddha over 2,600 years ago help to address present-day concerns for our endangered planet and the very survival of our species?

The Buddha and other enlightened beings return to this life to help others when the world is in need. The

This photograph of a Kalachakra Sand Mandala was taken by Basil Crump, who wrote an article entitled "The Meaning of the Great Peace Mandala" for the *Peking Chronicle* on October 30, 1932. The mandala was constructed for the Kalachakra Initiation conferred by the Panchen Lama on October 21 at the Tai-ho-tien, which Crump called "the most important hall in all Peking." The grand scale of the mandala (approximately 32 feet in diameter) may be determined by the figure in the upper left corner. Surrounding the mandala are ten *purbas,* ten vases, and a circle of *torma* offerings.

Crump wrote, "It was interesting to watch them applying the powdered colors through tapered tubes (somewhat like icing a cake) and the skill with which they executed the intricate drawing of the various symbols."

Kalachakra Tantra has been handed down from generation to generation in an unbroken lineage of oral transmission, from teacher to student to the Dalai Lama of our own time. Today it is coming to the West when, for the first time in history, we are beginning to attend to the concerns of our planet as a whole. These include easing international tensions and violence, recognizing and dealing with the threat of global warming, overpopulation, malnutrition, environmental pollution, and epidemic plague—in short, the many physical and psychological challenges of human suffering.

The Kalachakra Tantra is one of the most advanced and complex teachings of the practice of Buddhism. Like all Buddhist teachings, Kalachakra serves as a path to enlightenment.

While he was still living in Tibet, the Fourteenth Dalai Lama conferred the Kalachakra Initiation twice on thousands of followers, many of whom traveled hundreds of miles on foot for this once-in-a-lifetime experience. One of the highlights of the pilgrimage was the opportunity to view the magnificent Kalachakra Mandala, made of colored sand, at the completion of the ritual.

Since he went into exile in 1959, the Dalai Lama has given this twelve-day initiation numerous times throughout the world and has welcomed all who wish to attend. In such empowerment rituals, practitioners must pledge "not to reveal the supreme secrets of the mandala to those who have not entered, nor to those who have no faith." However, the Dalai Lama says, "It's already become an open secret with many misconceptions, and it's my responsibility to correct them. A profound explanation should be available."

In 1988, his commitment to world peace and to sharing the treasures of Kalachakra inspired him to break with tradition. He sent four monks from his Namgyal Monastery in India to construct the Kalachakra Sand Mandala as a cultural offering at the American Museum of Natural History in New York City. The reaction of Westerners to this timeless masterpiece was one of awe, appreciation, and delight.

Like all of Tibetan Buddhism, the precious legacy of Kalachakra is in danger of extinction in Tibet, which has been occupied by the

People's Republic of China since the 1950s. As a result, many Tibetan lamas seeking to preserve these teachings of the Buddha have come to the West, where interest in expanding spiritual awareness has been increasing as faith in material solutions has declined.

the meaning of tantra

The Kalachakra Sand Mandala, like all Tibetan Buddhist mandalas, is part of a tantra imparted by the Buddha. The literal translation of the Sanskrit word *tantra* is "continuity," although an exactly corresponding English word or concept does not exist. "Way of life according to the teaching" is a more comprehensive translation. Tantra is a living expression of the timeless nature of the teaching—its origin, its practice, and its fruits.

The teachings of the Buddha are all classified as *sutras* or *tantras,* both of which exist as written scriptures, recorded centuries after they were taught. Sutras were taught publicly in dialogue form and have always been widely available. There are various types of tantric texts, including medical and astrological tantras, as well as the root texts of Vajrayana meditation practice, such as the Kalachakra Tantra. The Buddha conferred the tantras while assuming the form of various deities. He taught them secretly to individuals or groups because only those who had a certain background and education were prepared to integrate them into their practice. These tantras have until recently been accessible only to scholars and devout practitioners.

Tantra is the core of Vajrayana Buddhism, which has been widely practiced in Tibet and Mongolia for more than 1,300 years. Vajrayana is one of the three *yanas,* or Buddhist paths. The three yanas are *Hinayana,* also known as *Theravada,* which is the path to individual liberation; *Mahayana,* the path of altruism, or the attainment of enlightenment through compassionate dedication to the liberation of all sentient beings; and *Vajrayana* (the "diamond path," also known as *Tantrayana* and *Mantrayana*), which is said to be the most direct path to transformation and enlightenment through tantric practice. "Tibetan Buddhism" refers to the Indo-Tibetan Buddhist tradition which came to Tibet directly from India, and which includes the practice of all three yanas.

the tantric deities

Kalachakra is one of thousands of tantras taught by the Buddha. At the center of each tantra is a deity, which is regarded as a manifestation of a particular aspect of Buddha mind. Rather than gods to be adored, the deities are perceived as personified states of mind, to be attained and mastered as one progresses on the path toward enlightenment. In addition to the central deity Kalachakra (wheel of time), three deities important to the history of Kalachakra are Manjusri (wisdom), Avalokiteshvara (compassion), and Vajrapani (power).

Buddha literally means "awakened one." A Buddha is one who has "gone beyond," beyond the endless cycle of cause and effect, or suf-

Above: Avalokiteshvara, Buddha of Compassion, holds a lotus flower in his left hand. 10$^{7}/_{8}$" high, cast gilt copper with jewel inlay, Nepal, 14th century.

Right: Vajrapani, Buddha of Power, wields the ritual implement called the *vajra*. 8" high, cast gilt copper with painted detail, Tibet, 17th century.

fering, to realize the state of complete awareness of all phenomena—both matter and mind—and their essential emptiness. *Buddha mind,*

Manjusri, Buddha of Wisdom, is recognized by the sword which cuts through ignorance. Front and rear view. 4¹/₂" high, brass with copper inlay, Kashmir, 10th century.

then, is the state of omniscience, the perfected mind that realizes the union of compassion and wisdom, the state referred to as bliss consciousness or pristine awareness.

Compassion and wisdom, which figure prominently in the teachings of the Buddha, refer to those qualities of mind that are the result of practicing the meditative concentration associated with the tantra or deity. Compassion refers to that altruistic state of being where our actions and their consequences are for the benefit of others. It is the sense of unconditional love and kindness that goes beyond ego in the realization of true selflessness.

Wisdom, from the Buddhist perspective, is the realization of emptiness, the direct knowledge of things as they are. This knowledge is universal in nature and arises spontaneously from pure intention, thought, and action. It is a wisdom of discernment, functioning in accord with the fundamental laws of natural phenomena. Its essence is grounded in altruism.

The realm of the deity of wisdom and compassion known as Kalachakra is one of omniscience—or the knowledge and understanding of all events and perceptions, past, present, and future. His omniscience has been obtained through his practice of abandoning all faults. Kalachakra is not only a deity but a teaching, a tantra. It has its own mandala, its own text, its mantras, prayers, meditation practices, ritual practices, ritual dances, and a long and colorful history. It is through the integration of its many components that one is able to realize and experience the mind of Kalachakra. The more deeply one practices it, the more expansive and profound is its result.

Each Buddhist deity has many forms. In its simplest form, it has one head and two arms, while more elaborate forms have multiple heads or arms. Many principal deities, whether male or female, are in a state of perpetual union with a consort known as *yab-yum,* literally "mother-father," symbolizing the union of wisdom and compassion.

The early exposure of tantric art to Western culture without proper explanation often resulted in yab-yum being misinterpreted as a purely sexual embrace. In the practice of the Kalachakra Tantra, the deities are visualized as being in that state of consciousness where the feminine and masculine principles are in union or balance. Realizing the completion stage of the meditation practice, one is able to master the energies of the right (female) and left (male) channels of the body, to realize the totally awakened and clarified state (Buddha mind) of the central channel, which is represented by this embrace.

The Kalachakra deity has a number of forms, but the one associated with its sand mandala has four faces and twenty-four arms and stands in union with his consort Vishvamata. An example of a complex deity form is the thousand-armed Avalokiteshvara, whose numerous arms represent boundless, infinite compassion.

Opposite: The deity Kalachakra and his consort Vishvamata. Scroll painting (*thangka*), Tibet, 18th century.

14

Every deity has a *mantra,* which is a series of Sanskrit syllables. When recited together, the particular combination of syllables or sounds evokes a specific, desired state of mind.

For example, *OM MANI PADME HUM* literally means "the jewel in the lotus." This mantra is intended to evoke the jewel of enlight-

Above: This visual representation of the Kalachakra mantra is known as the "Power of Ten."

Right: The mantra *OM MANI PADME HUM* carved in stone. These *mani stones* are usually placed in heaps by pilgrims at holy sites as offerings.

enment arising in the purified mind, which is symbolized by a beautiful lotus flower emerging from the mud in which it grows. Reciting this mantra repeatedly with a motivation to be of benefit to others increases the potential to attain such an exalted state. A mantra, like a friend, is always accessible and reliable; association with it deepens the bond. *OM MANI PADME HUM,* the mantra of Avalokiteshvara, is the most widely recited mantra in Tibetan Buddhism.

The principal mantra of Kalachakra is *OM AH HUNG HO HANG KHYA MA LA WA RA YA HUNG PHAT.* This mantra cannot be translated because the syllables, which do not form words, encompass all aspects of the Kalachakra Tantra.

The pantheon includes both benign deities, which embody the gentle nature of the Buddha of compassion, and wrathful deities, which are the seemingly fierce manifestations of a fundamentally compassionate nature. Their wrath corresponds to the righteous anger of a parent scolding a child who plays with fire.

The most peaceful deities are often depicted as white, light blue, or green, whereas the wrathful ones are frequently red, dark blue, or black, with flames spewing from their bodies to dissolve the poisons of the mind and other obstacles on the path. They carry implements to cut through ignorance and delusion and are often shown displaying the trophies of their conquests. Wrathful deities are frequently seen as protectors of the gates who prevent those with bad intentions from entering. They also encourage those with good intentions into the mandala or monastery, etc.

There is no tantra without a deity; likewise, each deity embodies a tantra. With each deity thus representing an aspect of Buddha

The Kalachakra mantra, *OM AH HUNG HO HANG KHYA MA LA WA RA YA HUNG PHAT,* in Tibetan characters. This configuration of the syllables represents the cosmic Mt. Meru with the Kalachakra mantra at its top.

mind, each tantra discloses methods for realizing the awakened state of the deity.

the mind in tantra

In tantra, the mind is seen as the central source of being. It is the stream of consciousness which connects the individual to his or her past and future. This can be observed within the actions of this life alone or, according to the principle of reincarnation, in the actions of past and future lives. The mind is a repository of these actions and the impressions they make, known altogether as *karma,* the law of cause and effect which carries over from lifetime to lifetime.

As the Dalai Lama and other lamas have said, the primary difference between human beings and animals is that human beings have the quality of mind that is able to practice the *dharma,* the law of nature as taught by the Buddha. The dharma includes the natural phenomena of the physical plane and of the mind, as well as their interdependence.

At the center of tantric practice is the subtle mind, which from the Buddhist perspective resides at the heart center (not to be confused with the physical heart) and is the source of our vitality, compassion, and wisdom. This is significantly different from the Western viewpoint, which regards the mind as a function of the brain, with an emphasis on deductive reasoning. The tantric mind holds both the relative truth and the absolute or ultimate truth.

a guide to enlightenment

Each of the tantras contains instructions for entering into the practice, in order to realize the perfected quality of mind represented by the particular deity. Thus the tantra is not only the teaching but also a guide to achieving and abiding within the enlightened state of being. It is also the transmission of its teachings, which are passed from generation to generation by lineage holders who are empowered to do so.

To gain some perspective on the ageless quality of the tantra, we need to understand the Buddhist conception of time. We are presently in the age of the fourth Buddha—the historical Shakya-muni Buddha, known before enlightenment as Siddhartha or Gau-tama—who lived approximately 600 years before Jesus Christ. The Buddha of the previous (or prehistoric) age, who was called Osung, lived when the average life span was 20,000 years; and the one be-fore, known as Serthup, when the average life span was 40,000 years.

Shakyamuni Buddha, 70 feet high, carved into a mountainside two miles south of Lhasa, Tibet in the 12th century. His right hand touches the earth, indicating that the earth is witness to his enlightenment. The bowl in his left hand may be the begging bowl of a monk or the bowl of healing herbs of the Healing Buddha. Students regularly paint the face, hands, feet, bowl, and lotus flowers at the base as an act of devotion. A pile of mani stones is in the right foreground.

The first Buddha, Khorvajig, appeared at a time when the life span had just been reduced to 60,000 years. The Dalai Lama reminds us gently that we Westerners are very impatient, due to our limited view of time.

The Buddha's teachings spread throughout India until they were threatened, and ultimately driven out, by the foreign invasions which took place from the 10th to 13th centuries. During that time Vajrayana, or tantric, Buddhism spread northward to Tibet, where it was practiced and passed down to our own time through the teachings of eminent lamas, including the preservation of ritual arts. An example is the Kalachakra Sand Mandala, which contains symbols embodying the entire Kalachakra Tantra.

the transmission

The Buddha taught students according to their individual capacity and disposition. Some only understood his words, while others received his teachings on a more profound level. This type of learning is called "mind transmission," a direct transfer of consciousness that can pass from a deity to a human or from one human to another. It requires a qualified sender and a receiver whose mind has been tuned to the frequency of the signal being transmitted. This attunement is achieved through the generation of *bodhicitta,* the awakened state of mind that arises through the practice of compassion.

When the Buddha conferred a tantra, he manifested as the deity of that tantra and imparted the complete teaching through mind transmission. This empowerment by the deity included authorization to practice the tantra, instruction in the philosophy of the tantra, training for the practice of meditative concentration, and directions for the practice of the tantra's rituals and ritual arts.

The practice of ritual arts is a means of invoking the deity in one's mind, awakening the Buddha nature which we already possess. Ritual arts include chanting, music, dance, making *torma* (barley flour and butter) and butter sculpture, as well as sand mandalas, all of which are offerings to the deities. They are an integral part of the practice of tantric Buddhism.

The ritual arts serve as a "visual scripture" in communicating the tantra. They not only provide another dimension of the teachings but also the opportunity to develop devotion through their practice.

It is said that seeing the image of the Buddha or a Buddhist deity can activate the seed of Buddha nature in the mind of the viewer. In the Tibetan Buddhist canon such visual imagery includes *thangkas* (scroll paintings), statues, and mandalas. When properly made and consecrated, these works are believed to contain the same empowering energy as the text, the deity, or even the Buddha himself. They are considered to embody that which they represent.

the mandala

The Tibetan word for mandala is *kyilkhor,* which means "center and surrounding environment." The mandala of a tantric deity includes the deity and his palace, which is also a representation of the mind of the deity. Based on symbols familiar to the people of India during the Buddha's lifetime, each mandala is a pictorial manifestation of a tantra. It may be "read" and studied as a text, memorized for visualization during meditation, and interpreted.

The most commonly known mandalas are depicted in two-dimensional form, showing a floor plan—like a blueprint—of the three-dimensional palace of the deity, including the architectural design and the many decorative details. Every component is a symbol representing an aspect of the teaching.

The purpose of a mandala is to acquaint the student with the tantra and the deity and to allow the student to "enter into the mandala"; that is, to enter into the state of being in which the deity dwells.

The texts state that a mandala can be drawn, painted, made of particles, or constructed by meditative concentration. A two-dimensional mandala may be made of powdered and colored rice or flowers, particles of stone or jewels, or colored sand. There are also three-dimensional mandalas constructed out of wood, metal, or other solid materials.

There are two types of meditation mandalas, which are constructed in the mind. In one, the practitioner visualizes the entire three-dimensional mandala in minute detail. The other, the most advanced tantric practice, known as "actualizing the mandala," requires the reorientation of the subtle energies of the body to conform to those of the deity.

Raktayamari Mandala, a painted meditation mandala of Sakya master Kunga Lekpa, teacher of Tsong Khapa. Central Tibet, circa 1400.

the practice

The student wishing to penetrate these mysteries begins his or her practice by "taking refuge" in the Three Jewels: the Buddha, the dharma, and the *sangha* (the teacher, the teachings, and the community of practitioners). In order to practice tantra, he or she receives an initiation that grants permission or authorization. There is such an initiation for each of the thousands of tantras. The ceremony, also referred to as an empowerment, is given by a ritual master who has received this empowerment from his own teacher

through oral transmission, in an unbroken lineage traceable to the original exponent of this teaching, the Buddha himself. The essence of a tantra is transmitted through the empowerment ritual, so that the student is not only receiving permission to practice but also the actual blessing of the Buddha as embodied by the ritual master.

Monks making a sand mandala on the floor of Drepung Monastery in Tibet. This photograph was taken in 1935 by C. Suydam Cutting, who led the first American expedition to Tibet.

Buddhists see the ritual master as conferring, bestowing, or giving the initiation rather than performing or conducting a ceremony. Similarly, the student does not "undergo" an initiation, but simply receives it.

Preparation for the initiation, including the making of the sand mandala into which the deities of the tantra are invoked by the ritual master, is considered very important, and it often takes longer than the empowerment itself. In the case of the Kalachakra Initiation, it takes nine days of purification and consecration before the students may enter the mandala site, then one day of student preparation and two days of initiation.

the kalachakra tantra

Kalachakra (in Tibetan *Du kyi khorlo*) is a Sanskrit word that can be translated literally as "wheel of time." *Kala,* or "time," is not linear time but the flow of all events, past, present, and future. This is similar to our concept of space, which does not imply a particular direction or limitation. The Kalachakra deity represents omniscience, for he is one with all time and therefore knows all. *Chakra,* meaning "wheel," refers not only to the cycle of time but also to the way in which the enlightened experience of great bliss radiates like the sun from the self to all sentient beings. The wheel, with no beginning and no end, is also the universal symbol of Buddhism, representing the teachings of the Buddha.

The Kalachakra Tantra comprises three unique and simultaneous cycles of instruction: the external Kalachakra teachings on cosmology; the internal Kalachakra teachings on the nature and functioning of the human body; and the alternate Kalachakra, the meditative path toward enlightenment.

The Kalachakra Sand Mandala is a visual representation of the entire Kalachakra Tantra. It is also a two-dimensional representation of the five-story palace of the Kalachakra deity, in which a total of 722 deities reside, with Kalachakra and his consort Vishvamata united in an embrace of perpetual bliss at the very center.

Those who participate in making the sand mandala place themselves within the realm of the deity. While we tend to see the mandala as a work of art in the process of creation, the monks who do this work say they are not creating, but reconstructing a representation of something that already exists. By doing so, they are perfecting or attaining the awakened state of mind. Rather than the emphasis being placed on creating something of material value, the monks are engaging in a process that benefits them and will also serve to benefit others.

a teaching for our time

One of the important features of the Kalachakra Tantra is that it is given for a community. The Buddha offered it for an entire country, the mythical kingdom of Shambala, so historically the initiation has been given to large groups. The Dalai Lama has remarked that in earlier times communities were separated by valleys, rivers, mountains, or oceans, whereas today, with instant communication and transportation, our community includes the entire planet. This is another reason why he feels Kalachakra is a teaching for our time.

Although it is particularly useful for subduing crisis or warfare (both internal and worldwide), the Dalai Lama makes it clear that the Kalachakra Tantra alone is not enough to achieve world peace. We need world leaders in economics, religion, sociology, politics, and, perhaps most importantly, the physical sciences to work together, sharing their different traditions in order to find innovative solutions to the problems ahead of us. He states further that this cooperation must be motivated by a true sense of universal responsibility to all fellow human beings, which, as an expression of the union of wisdom and compassion, is itself the very essence of Kalachakra.

The Sand Mandala *as* a Cultural Offering

WHAT WE KNOW today as Tibetan ritual art came to Tibet primarily from India between the 8th and 12th centuries A.D., as part of the Buddhist tantra. Stylistic influences also came from Nepal, Kashmir, and China, as well as other neighboring countries.

Over the centuries, Tibetans have developed their ritual arts—including dance, chanting, music, thangka painting, butter sculpture (a uniquely Tibetan art form), and the elaborate and exquisite sand mandalas—to a highly sophisticated degree. But until very recently, the expression of those arts had never been seen outside their intended

The Kalachakra Sand Mandala, the first Tibetan sand mandala in history to be presented as a cultural offering, captivates viewers at the American Museum of Natural History in New York City, 1988.

27

sacred setting. When the Kalachakra Sand Mandala was presented at the American Museum of Natural History, it was extracted from its larger ritual context. The term "cultural offering" evolved as a means to describe this transplant of a sacred art into a cultural and anthropological institution.

In appreciating the Kalachakra Sand Mandala as a work of art, Westerners are challenged to see beyond their own definition of art, which places value on innovation and self-expression. In Tibetan ritual arts collaboration in the execution of the intricate sand mandala is considered to be more valuable than originality, and the ritual artist's single most important objective is to cultivate a pure motivation to benefit others.

In the museum, the art form is demonstrated for a public generally unfamiliar with the larger ritual. In addition to drawing and painting the sand mandala, the monks' very presence in the museum creates a tranquil atmosphere which conveys the essence of the larger ritual. The cultural offering includes the daily recitation of prayers used in the initiation. The monks recite the Kalachakra prayer in which, through a process of visualization, they generate themselves as the deity Kalachakra. The state of mind resulting from this practice remains with them throughout the day while they demonstrate the sand painting and talk with visitors.

the moving sands

In 1988, Samaya Foundation received a grant from the New York State Council on the Arts for a Tibetan monk from Namgyal Monastery to be artist-in-residence at the foundation and to demonstrate the construction of sand mandalas in New York. We discussed an exhibition with Dr. Malcolm Arth of the American Museum of Natural History, who encouraged us with the simple statement, "Just let me know when the monk will arrive."

When Lobsang Samten arrived at JFK Airport with two suitcases filled with colored sand, Samaya Foundation was waiting for his instructions. He thought it best that we make the Guhyasamaja

Namgyal monk Lobsang Samten painting
the Guhyasamaja Sand Mandala at
Samaya Foundation, 1988.

Sand Mandala first, before tackling the much more complex
Kalachakra at the museum. Lobsang explained the belief that if the
Guhyasamaja Tantra were preserved, then all tantras would be preserved, so it seemed a good omen to begin with this sand mandala.
Also, the Guhyasamaja Mandala is of a size and complexity appropriate for one monk working alone, although it is traditionally made
by several monks together. But Lobsang informed us that first of all
we would have to build a *thekpu,* or mandala house.

The methods of building the thekpu, which would have to be
both strong and portable, were translated into Western engineering
concepts by architect Stan Bryant. Our thekpu was designed to

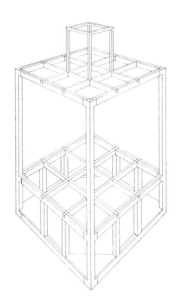

house a video camera directly above the mandala, as well as ceiling lights, all supported by vertical columns made of hollow steel to act as conduits for electrical lines for lighting, audio, and video.

The project became a real community effort, entailing the work of many friends, volunteers, and family members. Samaya Foundation's technical director Greg Durgin designed the base of the thekpu for recorders, circuitry for the lights, and a robotically operated camera, along with other electronics. Since there was no precedent or plan for what we were doing, we had to experiment and improvise. The Samaya thekpu became a beautiful merging of ancient and modern technologies.

Next came the decorative design details. Lotus flowers and other decorations were carved out of wood and painted by Tibetan artist Phuntsok Dorje. Silk brocade valances and banners were sewn and hung.

Lobsang learned English as he taught us philosophical aspects of the Guhyasamaja Tantra. The complex and sometimes obscure

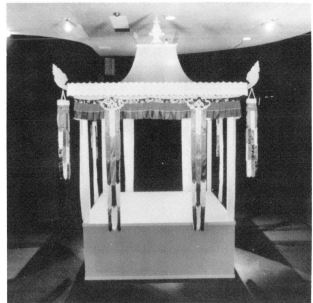

Tibetan explanations of the tantra had to be translated into simple layman's English for the press and the public.

The Guhyasamaja Sand Mandala was constructed at Samaya Foundation during the month of June in 1988, marking the first time a Tibetan sand mandala had ever been presented as a cultural offering. No longer accessible only to Buddhist practitioners, the mandala attracted hundreds of New York artists, as well as eminent lamas and the simply curious.

Lobsang, watching himself create the tiny details of the mandala on a video monitor as he worked, appreciated the capacity of this new tool to help him perfect his skill. And he learned to discuss his work so that we could all understand.

Thus the groundwork was being laid for the American Museum, where the Kalachakra Sand Mandala would be presented during July and August of 1988.

After the Guhyasamaja Sand Mandala was completed, the traditional dismantling ceremony was performed. More than 300 people

The Samaya Foundation *thekpu*, or mandala house, for the American Museum of Natural History exhibition in New York City, 1988. From left to right: artist's sketch; structural drawing; museum installation with silk banners; Phuntsok Dorje painting lotus flowers on the roof detail; the completed thekpu.

came to Samaya Foundation to join Lobsang Samten as he removed the thirty-two deities represented in the mandala and swept up all the delicately applied sand. The sand was then carried to the Hudson River, in a procession that passed by the 110-story World Trade Center, a striking juxtaposition of the ancient and modern worlds.

At the river, Lobsang recited prayers and offered the consecrated sand to the spirits of the water, while cameras recorded the setting sun illuminating New Jersey's industrial landscape across the Hudson. In a shared moment of joy, we poured the colored sand into the water for the benefit of the aquatic life.

John Perrault, an art critic who attended the dismantling ceremony, wrote the first article on the Tibetan sand mandala in the United States, published in the *Village Voice*. He called the demonstration of sand mandalas "performance art of a high order."

living natural history

On July 6, we started work at the Museum of Natural History. We began to learn the Kalachakra Tantra and the museum staff became part of our sand mandala family in this unique, historic event. Lobsang Samten was then joined by three other Namgyal monks: Lobsang Gyaltsen, Pema Lobsang Chogyen, and Jamphal Dhundup.

On the day the sand mandala opened at the museum, *New York Times* reporter Dennis Hevesi wrote, "Amid the clamor and clatter of the city, a pinpoint of pure calm—a 'gateway to bliss'—is being created." Other writers and photographers from newspapers, magazines, television, and radio soon followed.

During our six weeks at the museum, at any time one might find up to 100 people totally involved in the exhibit. People contemplated the monks at work with great respect, as if witnessing a ceremony in a church or temple. Each morning at 10 A.M. as many as fifty people might be waiting to join the monks for their morning prayers, which lasted up to forty-five minutes. Then the special Plexiglas protective cover was lifted so that the monks could begin their work.

The Namgyal monks performing their preparatory prayers at the American Museum of Natural History, as they did each morning before beginning to paint with sand.
New York City, 1988.

More than 50,000 people came to see the exhibition. What they saw in the room was a monk sitting on the waist-high mandala base, applying sand through an elongated funnel to a mandala seven feet in diameter. Viewers walked in a circle around the thekpu, which was surrounded by stanchions because the mandala's fragility made

Above: Live, close-up images of the step-by-step process of constructing the mandala were provided by closed-circuit video feed from the camera located in the ceiling of the thekpu. The detail shown here, actually 3" wide, was magnified 600 per cent on the video monitor.

Left: Museum visitors join the Namgyal monks in contemplating the completed Kalachakra Sand Mandala. New York City, 1988.

it inadvisable for viewers to get too close. What they heard was the sustained rasping of the metal funnel used to apply the tiny stream of colored sand particles.

Painting the Kalachakra Sand Mandala in its ritual context traditionally takes six days, employing as many as sixteen monks. At the museum it was slowed down so that the visitors could experience the entire process in detail over a period of six weeks. Four large TV monitors placed throughout the gallery allowed the visitors to see details of the mandala emerge as they were painted. Accustomed to television, visitors found the enlarged video view provided by the overhead robotic camera both comfortably familiar and educational. This bird's-eye view magnified minuscule details 400 per cent, prompting one woman to say it was "just like being in the monks' shoes."

The exhibition also included a stationary video installation with a twenty-minute film of the Kalachakra Initiation bestowed by the Dalai Lama at Bodhgaya, India, which served to illustrate the sacred context of the mandala.

One of the questions asked most frequently was, "How do the monks feel about doing this sacred work in the museum?" The monks always responded positively: "When we paint a mandala in the monastery, it is usually seen only by our fellow monks. But here

A vase containing the sand from the Kalachakra Sand Mandala is carried in procession through the streets of New York from the American Museum to the Hudson River. 1988.

in the museum, thousands of people are able to see it. Here we are showing the culture of Tibet, and the mandala itself will benefit many people."

The monks learned to describe their work to people who had no knowledge of the culture and its traditions. They were soon able to reduce their detailed explanations of the teachings to succinct phrases, such as one monk's assertion that work on the mandala offered him "peace of mind." Most of New York's television news teams covered the story, interviewing both the monks and the public.

Art writers came from various publications. Kay Larson, in her *New York* magazine column, said, "A sand mandala is an extraordinary thing, collaborative, ephemeral, unsigned, ahistorical—contrary in every way to 'art' as we mean it in the world." This reflected the New York public's fascination with a work of art which, after six weeks of meticulous craftsmanship, would be swept up and offered to the natural spirits of the Hudson River.

We discussed with the museum staff the possibility of preserving the mandala so that more people could benefit from seeing and studying it. We experimented with sprays and other protective coating techniques and received a great deal of advice. Finally, we all

Monks praying at Santa Monica Bay before offering the blessed sand to the marine life of the Pacific Ocean. This followed the dismantling of the Kalachakra Sand Mandala at the Natural History Museum of Los Angeles County in 1989.

agreed that the best way to preserve the mandala would be to follow tradition: sweep it up and pour it into the Hudson.

The dismantling ceremony was witnessed by hundreds of people. The four monks recited prayers requesting the 722 deities which they had invoked during the process of constructing the Kalachakra Mandala to now return to their sacred homes. The monks adroitly removed the colored sands which represented the deities and, within minutes, six weeks of painstaking labor was swept up and put into an urn. The urn was then carried in a ceremonial procession from the museum and the sand was poured into the river.

The dismantling ceremony was as solemn and as important as the construction of the mandala itself. Prayers were said at the

79th Street Boat Basin as television news crews brought the event into millions of homes.

on to the west coast

When Samaya Foundation presented the Kalachakra Sand Mandala at the Natural History Museum of Los Angeles County in 1989, there were several differences in presentation, including a more sophisticated video system and another group of Namgyal monks. Instead of a small, dedicated gallery, the installation was located in the middle of the museum's huge Hall of North American Mammals. Surrounded by dioramas of grizzly bears, buffalo, and other large beasts, the thekpu shone like a radiant jewel. A special area was set apart for children to try their hand at sand painting under the guidance of one of the monks.

Samaya Foundation worked closely with both museum staff and the Tibetan Buddhist meditation center Thupten Dhargye Ling, which organized the Los Angeles Kalachakra Initiation that the Dalai Lama was giving at the same time to 3,000 Buddhist practitioners. Many initiates came to the museum to spend more time studying the mandala; likewise, many museum visitors became interested in learning about Tibet and its spiritual traditions.

The museum staff told us some visitors, attracted by the sand mandala project, had come to the museum for the first time; some had even traveled across the country to see it. Members of the Hollywood film community appeared frequently and treated the monks to tours of Disneyland and the *Star Trek* set at Paramount Studios. Regular visitors became friends of the monks and of Tibetan Buddhism.

The dismantling brought hundreds of spectators to the museum and to the procession (which was by car on the Santa Monica Freeway this time) to the Sand and Sea Club on the Pacific Ocean. There, once more, the blessed sand was offered to the water spirits. It was unique to see the Tibetan monks praying on the beach surrounded by friends, television cameras, roller skaters, beach bunnies, body builders, homeless people, and bikers.

rocky mountain high

From Los Angeles we journeyed to Aspen, Colorado, to be part of Windstar Foundation's CHOICES IV Symposium, attended by 1,700 environmentalists, nutritionists, faith healers, and socially conscious

Namgyal monk Tenzin Yignyen discusses the Kalachakra Fire Offering Sand Mandala with John Denver and the author at Windstar Foundation's CHOICES IV Symposium in Aspen, Colorado, 1989.

promoters of the "New Age." The monks demonstrated the Kalachakra Ritual Fire Offering Sand Mandala, which is traditionally constructed after the completion of a ritual ceremony or a prolonged meditation practice to remove any obstacles on the path.

At the closing ceremonies the monks offered a vase filled with blessed sand from the Fire Offering Sand Mandala to the endangered streams of the Rocky Mountains. Another vase was given to John Denver, a founder of Windstar Foundation, to take to Prince William Sound in Alaska the next day as an offering to the marine life threatened by the Exxon Valdez oil spill. The Wheel of Time was continuing to expand its reach deep into the culture of the 20th century.

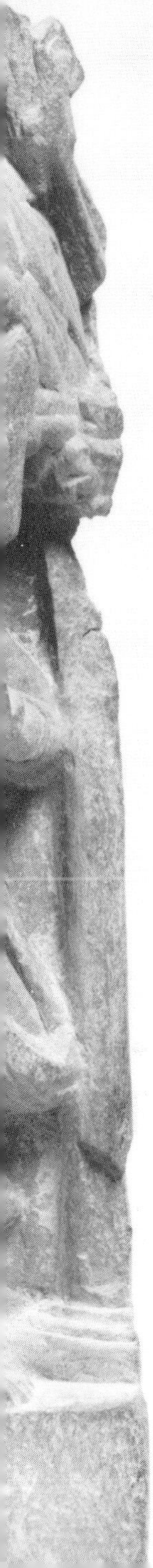

The Life
of
The Buddha

IN TIBET, the literary form of biography, which is otherwise completely foreign to the culture, is reserved for the
highest realized teachers of Tibetan Buddhism. Such a biography is known as *namtar*. Its purpose is to illuminate
the deeds of the master and it is used as a teaching aid.
Usually written by a disciple, the namtar, like other Tibetan Buddhist texts, is considered to retain the enlight-
ening qualities of the teacher himself. Like the life story
of Jesus, the biography of the Buddha is rich with anec-
dotes of his achievements and brings life to his teachings.

The Conception (of the
Buddha). Queen Maya
dreams that a tusk
of a white king elephant
penetrates her rib
cage. Gandhara stone
carving, India, 2nd–3rd
century.

A Buddha is one who has awakened, becoming liberated from the habitual patterns that perpetuate suffering, and has attained enlightenment. Versions of the life of Shakyamuni Buddha vary, from the biography of an exceptional but very human man to legendary deification. His story has inspired innumerable people over the centuries to turn to the dharma as a means to achieve happiness and liberation from suffering.

Shakyamuni, "the sage of the Shakya clan," was born in 527 B.C. The Buddha of "this fortunate era" is regarded as the fourth of one thousand Buddhas, with the others yet to come. After many years of rigorous pursuit of the truth, Shakyamuni Buddha realized his own Buddha nature, or the potential for perfection that is innate within all sentient beings. He spent the last forty-five of his eighty years disseminating his teachings, also known as the *Buddha dharma.*

The Buddha taught in India at a time when Hinduism flourished, with its pantheon of colorful gods and rigid social structure. Defying the strict hierarchical customs that regulated social behavior in the India of his day, the Buddha received outcasts and aristocrats alike, subdued warrior kings and wild beasts, and converted infidels and criminals. His teachings on compassion and his espousal of democratic social views were revolutionary for his day. In defiance of the customs of the times, he allowed women to be ordained.

The Buddha bequeathed to humanity methods for attaining self-realization that could be utilized by individuals according to their differing needs and capacities—by householders and monastics, mystics and pragmatists, scholars and peasants. From this approach, three levels of his teaching and practice evolved: Hinayana (or Theravada), Mahayana, and Vajrayana Buddhism.

The Buddha's basic teaching is that life is suffering, and that this suffering arises from our dualistic view that all things are either desirable or undesirable. This constant state of desiring, or attachment, can never bring us lasting fulfillment. The suffering can only end when we are able to set aside our narrow conceptions of self and gain access to our true, inherently altruistic identity, or pure nature.

The Nativity. To the left of Queen Maya is Indra, king of the gods, holding the infant Siddhartha. To the right is an unidentified goddess. Gray schist, Pala Dynasty, India, 9th–10th century.

the tradition of oral transmissions

Three months after the Buddha's death, 500 devotees gathered to recount his teachings. A cohesive written canon did not appear until several hundred years later. During that time, a vital oral transmission flourished and it continues to this day.

Although a vast number of commentaries have expounded the Buddha's teachings from the time they first appeared as written texts, the authority of the written word has not suppressed or invalidated the oral teachings. The reason for this is that in ancient India oral transmission alone resanctified religious teachings. The written word was considered helpful to merchants but inappropriate for sacred studies.

Consequently this dual tradition, functioning like counterpoint in music, has resulted in an ongoing, energetic dialogue. And the dialogue process has consistently served to revitalize Buddhism throughout its long history, allowing for its adaptation into a great diversity of cultural environments.

the story of the buddha's birth

Shakyamuni, born Siddhartha Gautama, was the son of King Suddhodana and Queen Maya. Prior to the full-moon night on which Shakyamuni's conception took place, the queen dreamed that a tusk of a multi-tusked white king elephant had entered her rib cage. This indicated that a child of strength and majesty would be born.

Toward the end of her confinement, Queen Maya journeyed to her father's house, stopping at the fragrant gardens of Lumbini in what is today southern Nepal. There, according to legend, in a golden palanquin attended by thousands of court ladies, the queen gave birth from her right side at the very moment that she reached out to pluck the flowering branch of an ashoka tree.

The legend holds that the deities Indra and Brahma immediately descended from the heavens to take the infant in their arms, that the other gods blinked in wonder, the galaxies quivered, the earth shook, the raging fires of hell were consumed, showers of petals cascaded

Infant Buddha. "In keeping with the usual Indian convention, Siddhartha appears as a fully grown figure, simply reduced in scale, rather than as a child." (P. Pal) Detail, gilt bronze, Ming Dynasty, China, 16th–17th century.

The Education of Siddhartha. This silk panel and the one on the opposite page, drawn in ink and colors, are from the Dunghuang Cave Temples in China, 9th century.

from the heavens, prisoners were freed, blind men regained their sight, the mute spoke in praise, and celestial trumpets heralded the child's glory.

Seven days after Siddhartha's birth, Queen Maya died. As was the custom, King Suddhodana then married Queen Maya's sister Prajapati, who raised Siddhartha as her own at the king's palace in Kapilavastu.

The king summoned renowned seers from their mountain abodes to predict the child's destiny. Pointing to auspicious constellations in the skies and to the child's perfect features, they agreed that Siddhartha was destined for immeasurable greatness, that his influence would be boundless and his benevolence unsurpassed.

But his astrological chart also held an ambiguity: If the prince succeeded to the throne, he would become India's greatest king, a universal monarch. On the other hand, if he pursued a spiritual path, his magnificence would be unmatched, for if he looked inward, he would conquer himself. The king was pleased to have fathered a great future king, but he was worried for the kingdom, should his son become a sage.

The eminent hermit and prophet Asita was beckoned to Kapilavastu by the miraculous signs that heralded the birth of a future Buddha. When he arrived at the palace, Asita told the king, "For the welfare and happiness of the world, your son will teach the dharma."

prince siddhartha's life

In response to the prophecies, King Suddhodana embarked upon futile attempts to beguile the prince with worldly pleasures.

The palace was surrounded by exquisite gardens, lotus ponds, pleasure groves, and forests. Within his residence the most beautiful maidens, cultivated in the arts of singing, dancing, making music, and making love, were provided for his pleasure. Splendid banquets, the finest steeds, and all earthly delights were designed to win Siddhartha's heart. He also received an extensive education in mathematics, astronomy, and metaphysics under prestigious Brahman tutors. And, as the future leader of the warrior caste, he became skilled in archery, horsemanship, and the arts of battle. He met his challengers on the tournament grounds and defeated even the most accomplished opponents with precision, strength, and grace.

At the age of sixteen, Prince Siddhartha married Yasodhara, a young woman of great beauty and intelligence. Although his life was rich and full, the prophecies of Siddhartha's spiritual awakening remained unrefuted.

Siddhartha devised a means to leave the palace in order to satisfy his curiosity about life outside. When King Suddhodana heard of this, he ordered that the kingdom be whitewashed and that every sign of suffering, old age, and sickness be removed. But Siddhartha's destiny had to be fulfilled.

Riding out from the eastern gate in his chariot, the prince caught sight of an old man, bent, withered, and toothless. Astonished, he asked his driver to explain the disturbing sight. He learned that all beings are subject to the ravages of old age.

Riding out from the southern gate, Siddhartha saw a man disfigured by bodily rot and covered with sores. Dismayed, he again requested an explanation and learned that all beings fall prey to the calamity of disease.

Riding out from the western gate, he discovered a bloated and maggot-ridden corpse deserted by the roadside. Shaken, he learned that all beings inevitably succumb to the finality of death.

Transformed by these experiences, Siddhartha realized the futility of worldly affairs. "Why pursue that which is doomed to such a fate?" he wondered.

Finally, riding out from the northern gate, he saw a gaunt recluse standing with quiet dignity, staff in one hand and an alms bowl in the other. Impressed, he was told, "That is a holy man—one who seeks to break the bonds of becoming and attain a mind that abides nowhere."

While these encounters were certainly pivotal in the transformation of Siddhartha, he also noted how others on the periphery appeared oblivious to the suffering in their midst. He was determined

The Four Encounters of Prince Siddhartha
outside the palace of his father,
King Suddhodana.

43

Palace Scenes. Upper panel: The enthronement of Prince Siddhartha and his wife Yasodhara. Lower panel: Life in the palace of King Suddhodana, father of the Buddha. Gandhara stone carving, Pakistan, 2nd century.

to do differently and to live a life of discipline structured by acute, unflinching awareness.

Leaving the palace once again, Siddhartha requested his driver to take him to the countryside, where he slipped off in search of solitude. His driver later found him seated in deep meditation under a fruit tree. This taste of meditation awakened the young prince to his birthright and destiny.

A new situation opened up for Siddhartha. His wife, Yasodhara, had given him a son, assuring that there would be an heir to the throne. This event accomplished Siddhartha's primary mundane responsibility, enabling him to set his mind free to pursue the life necessary for his liberation. The spiritual goals he set himself led to an ultimate renunciation and provided an example for all those who would follow.

leaving the palace: the great renunciation

Thus, at the age of twenty-nine, forsaking his wife and child, Siddhartha left the palace by chariot and rode into the night. Reaching the edge of the forest, he removed his golden ornaments, cut his long hair, and changed from silks into cotton clothes. His driver attempted to dissuade him, but Siddhartha vowed not to return to the palace until he had attained enlightenment.

Siddhartha's decision, on the one hand, was inevitable. Not only did the laws of karma predetermine his decision, but his own insight took him beyond equivocation. Yet to idealize his determination in ways that diminish the degree of his sacrifice is to miss the point. He left a life of luxury to sleep on the forest floors; he forfeited security to beg for his food; he left the love of family and wife for a solitary life; he gave up a throne to live like a beggar. At the same time, the great renunciation his life continues to exemplify was that of the self.

siddhartha's life as an ascetic

Living the life of a renunciate, Siddhartha went from house to house asking for alms, then retired to the forest to eat what he had gathered. There he began a search for the renowned Arada-Kalama, a hermit sage who lived in the mountains north of Rajagriha.

Siddhartha's noble demeanor escaped no one's notice, and the news of an aristocratic mendicant soon reached the palace of King Bimbisara, the most powerful sovereign of central India. Curious to know who this might be, King Bimbisara sought out the mendicant and was astonished to discover that he was none other than Siddhartha, royal heir of the Shakyas. The king offered him a kingdom, but Siddhartha rejected it and returned to his spiritual quest.

At last locating the master Arada-Kalama, Siddhartha learned methods for developing concentration through breathing techniques and yogic postures. He was soon able to enter the "sphere of nothingness," but he was still not free of his passions. His questions about the source and cessation of suffering had not been resolved.

Miracle Below the Jambu Tree. Upper panel: Siddhartha leaves the palace in search of the cause of suffering. Lower panel: Siddhartha meditates beneath a fruit tree, with his father and step-mother on either side. White limestone carving, India, 3rd century.

The Great Departure and Mara's Assault. Upper panel: Siddhartha is enticed by Mara, Lord of Temptation, and his daughters. Lower panel: Prince Siddhartha's departure from the palace. 56" high stone relief, India, 4th century.

Therefore, since the master had no more to teach him, Siddhartha bowed respectfully and left the hermitage to seek further.

From there he went to study with another hermit sage, Udraka-Ramaputra, a great philosopher whose doctrine guided Siddhartha past the sphere of nothingness to the "sphere of neither perception nor nonperception." Although he quickly mastered this more difficult doctrine, he knew there was more to be learned and again left to continue his search. He determined thereafter not to seek help from teachers but from within himself, and to practice on his own.

One day, as he was meditating on Mt. Gaya, he was joined by five ascetics who so respected his attainments that they were convinced he would become enlightened. They wandered together, living in the forests and subsisting on alms as they pursued their contemplative life.

Siddhartha practiced austerities to the most extreme degree, eventually eating only one grain of rice per day. But he found that this did not lead to liberation; after six years, his body had become so emaciated that he lacked the strength even to meditate. On the verge of collapse, he encountered the maiden Sujata, who recognized how majestic he was despite his appearance and, inspired to help him, offered him a bowl of rice milk. Acknowledging Sujata as a providential presence, Siddhartha accepted the sustenance, at the same time recognizing the futility of his extremist practices.

Taking this as a sign of weakness, Siddhartha's five companions abandoned him. He understood, however, that the nourishment of food and the warmth of compassion were necessary to make him fit for the revelation of the sacred mysteries that had so long eluded him. Now he was able to walk on to the place which would later become known as Bodhgaya, his strength restored and his resolve to attain enlightenment renewed.

Ascetic Shakyamuni. The emaciated Siddhartha during his ascetic period. Bronze with silver inlay, China, 16th century.

enlightenment

Siddhartha was now prepared to meditate until true transformation had been achieved. He seated himself upon some kusha grass, beneath a tree that would be known thenceforward as the bodhi (*bodhi* is the Sanskrit word for "awake") tree, with the determination not to rise until his goal had been attained.

Now he was beset by passions embodied by the demonic forces of Mara, the Lord of Temptation. It could be said that Mara is only a state of mind and that Siddhartha was at that point engaged in a final challenge to let go of his own internal attachments. If Siddhartha were to slip and fall back into believing that his enemies were outside his own mind, then Mara would win and delusion would reign. However, recognizing the true nature of these forces, he defeated them.

At last, meditating with equanimity, wisdom, and total attentiveness, his enlightenment came in four successive stages, which came to be known as the Four Watches, after a term used by ancient peoples to denote division of time during the night.

the four watches

During the First Watch, the details of all of Siddhartha's past lives came to him with such clarity and presence that he actually relived, rather than recollected, his previous incarnations. Filled with empathy for all creatures, he watched the repeated cycles of creation and dissolution of all forms and states of animate and inanimate existence. Here he saw the suffering caused by our clinging to forms as if they were substantial instead of accepting the fact of impermanence.

During the Second Watch, he turned his attention to the cycle of death and rebirth, or *karma.* He saw how the degree of pleasure or pain a sentient being experiences in each moment is determined by that being's prior actions, whether they be committed in the present lifetime or in one from the past; that is, he recognized the underlying causes of the quality of life. And he saw as well that lifetime after lifetime, because of the ever-present fear of death, there is no end to suffering.

The opening of the wisdom eye occurred during the Third Watch. This allowed Siddhartha the inner vision to penetrate the coarseness of the physical world and to experience the true nature of mind.

The Temptation of Mara. "In this beautiful ivory relief from Kashmir, Shakyamuni sits in meditation with downcast eyes. He remains calm and serene, although two of Mara's attendants blow conchs into his ears and a third beats a drum nearby. Simultaneously, several ferocious-looking demons attack him with weapons. Two handsome figures, each smiling and holding a flower, flank the meditating Shakyamuni. The male figure on the left of the relief may represent Mara, and the figure on the right, who appears to be female, may be one of his daughters. The position of Shakyamuni's right hand touching the earth in the gesture known as *bhumisparsamudra* indicates the moment of his Enlightenment and hence his victory over Mara." (P. Pal)

In the Fourth Watch, he focused on the apparent reality of present time. He saw that people are born, they age, they experience good health and sickness, they die and are reborn—that, fettered by desire and ignorance, they live like moles, scurrying blindly through successive lifetimes, too entrapped by their attachments to find their way to the light.

Shakyamuni Buddha at the moment of enlightenment. Brass with colors, Tibet, 11th-13th century.

Thus, having experienced his own prior incarnations and the cycle of death and rebirth, Siddhartha entered the great emptiness. Now, although he searched from the summit of the world downward, he could find no trace of himself. As the morning star rose, he proclaimed, "How wonderful! How wonderful! All things are enlightened exactly as they are."

What he had so arduously pursued had been there all along—obscured by illusions, but still there, within, and waiting to be discovered. The journey of the Buddha had been completed.

Siddhartha continued to sit under the bodhi tree. He had no desire to take action. For seven days he looked into his own mind. Having fully comprehended the essence of suffering, he now formulated the Four Noble Truths. At last, having achieved his goal, he was motivated to leave his "diamond seat" (the diamond being symbolic of the indestructible mind), and he set out to teach. He became known as the Buddha.

The Buddha's decision to teach after attaining his own enlightenment helped to define Buddhism as a path of compassion. Furthermore, his return to the world, to work helping others, was not a descent from Buddhahood to a less elevated realm but was, in fact, the next stage of enlightenment.

the missed opportunity

As the Buddha made his way from Bodhgaya toward the town of Sarnath, the first person he encountered was the mendicant, Upaka. Awed by the countenance of Shakyamuni and perceiving that his senses had been tamed while those of all others were as restless as wild horses, Upaka asked, "What are you?"

Shakyamuni answered, "I am a Buddha."

Upaka then made a deep bow in recognition of Shakyamuni's attainment and proceeded on his way. By not asking to be instructed, he missed an opportunity to become the Buddha's first disciple and to become enlightened.

the buddha's teaching

When the Buddha arrived in Sarnath, he met his former companions, the five ascetics, in a park full of gentle deer thereafter to be known as Deer Park. Remembering his acceptance of the rice milk given him by Sujata, they planned among themselves to reject him when they saw him approach. Yet as he drew nearer, they were awed by his enlightened presence and rose up in unison to take his robe, receive his bowl, arrange a place for him to sit, and prepare water for bathing his feet. However, they addressed him by his former name, Gautama.

The Buddha explained that this name was no longer appropriate, since he was no longer who he had been. He then drew a parallel between those who are hopelessly attached to satisfying their senses and those who are determined to deny them. Both, he told the astonished ascetics, are equally distant from the path of deathlessness, the perpetually enlightened state of mind that is without beginning or ending.

This was the teaching of the Middle Way.

But, asked the ascetics, what alternatives are there to austerities? The Buddha then elucidated the Four Noble Truths.

Shakyamuni Buddha began his first sermon by analyzing the root symptom of the human predicament: suffering. The First Noble Truth is often stated as "Life is suffering," meaning that suffering and dissatisfaction are symptomatic of the human condition. In truth, life is just life, neither good nor bad, neither pleasurable nor painful. It is our attachment to life—how we relate to it, try to control it, manipulate it, regulate it, and cling to it—that causes suffering. The Buddha thus located the problem of suffering within the human mind, not in the external phenomena of life.

In the Second Noble Truth, the Buddha identified the cause of suffering: desire and attachment. Its source is unreal, impermanent, changeable, and insubstantial. Nonetheless the experience of suffering is concrete and cannot be ignored; that is, the Buddha does not dismiss suffering just because it is manifested by the ephemeral self. Since the cause of suffering does not reside outside the mind, suffering cannot be alleviated with material solutions. ("You can't buy happiness" is a familiar Western expression of this essential insight.)

Constructing his teaching the way a doctor assesses the needs of a patient, Shakyamuni offered his diagnosis in the Third Noble Truth. Suffering can come to an end; its cessation is possible and within the

The Buddha Teaching. His hands are in the gesture of "turning the wheel of the law," or teaching the Buddha dharma. Gandhara, 29", stone carving, Pakistan, 2nd century.

reach of everyone. The debilitating disease of self-interest can be transformed and transcended.

In the Fourth Noble Truth, the Buddha outlined the way to bring about this end, known as the Eightfold Path:

1. *Right View:* that which is directly in line with the teaching, or the dharma.
2. *Right Intention:* the motivation of altruism, or being of benefit to others, rather than selfishness.
3. *Right Speech:* verbalizing things which are related to the teachings or otherwise express Right Intention. Buddhists place great importance on speech, seeing it as a direct reflection of mind.
4. *Right Conduct:* action which directly expresses the heart, exemplary of the teachings of the Buddha.
5. *Right Livelihood:* a vocation which is of service to others, advancing oneself and one's fellow beings along the path of virtuous activity.
6. *Right Effort:* the diligence or energy with which one practices the dharma and engages in activity leading to the realization of the enlightened state of mind.
7. *Right Mindfulness:* being attentive, present, and focused in all one's daily activities of body, speech, and mind.
8. *Right Concentration:* "one-pointed" attention, developed through meditative practice, on the teaching of the Buddha.

The Eightfold Path defines the basis of all Buddhist studies.

Upon listening to the Buddha's teaching, the five ascetics attained liberation and became *Arhats,* which literally means "foe destroyers." Thus began the *sangha,* the community of ordained practitioners. They stayed together in Sarnath for the three months of the summer rainy season. Then the Buddha instructed his many disciples to travel throughout the land and teach for the benefit of all living beings.

The Buddha was thirty-five years old at this time. He spent the rest of his long life walking through northern India and teaching the dharma. Disregarding the conventions of caste, he welcomed anyone

Circumambulating the massive stupa at Sarnath, India, the site where the Buddha first taught. A small stupa within this one contains relics of the Buddha.

to come and converse with him and, if they wished, receive the teachings and enter the order. Almost all who came to the Buddha did not merely receive the teachings but were completely enlightened and became exemplary members of the sangha.

The Buddha emphasized practice and training rather than superstition, ritual, and worship. Part of the significance of the Eightfold Path is that it affirms the capacity of human beings to change and to take responsibility for their own well-being. He rejected dependency on anything other than the mind in attaining enlightenment and advocated verification of truth through personal experience.

He denied the existence of an individual, independently existing self. He did acknowledge that an apparent self exists, but only as a projection of the mind shaped by human convention. It is this mind that experiences suffering, he said, and the mind can be transformed. And he taught a way of viewing the world and of behaving in it that would achieve that aim.

The Buddha was always skillful in presenting the teachings. One day, while meditating under a tree, he was approached by a group of young noblemen. They were chasing a prostitute who had run off with their wives' jewelry as they dozed after a picnic. The Buddha asked, "Which is more important, to seek the woman or to seek yourselves?"

Sensing the profundity of this question, the men gathered around to hear more. And the Buddha, seeing that the men were ready to listen, talked about the nature of human existence, and of the senselessness of seeking outside oneself and of blaming others for one's own unhappiness. The men and their wives thus became students of the Buddha.

On another occasion, a jealous priest with great psychic powers put a poisonous snake in the hut where the Buddha was to spend the night. When the Enlightened One emerged unharmed the next morning, the priest asked him to explain his powers.

"Having tamed myself, there is no difficulty taming others," he replied. In the face of such tranquillity, the snake had curled up beside him in the night.

Overcome with remorse, the priest confessed that he himself was more vicious than the venomous snake. He bowed down at the Buddha's feet and requested instruction in taming the mind. The news of this powerful priest's conversion spread quickly and attracted others to the path.

In another version of the story, the Buddha was pitted against the priest in a four-day competition of supernormal powers. Although he easily beat his opponent, resulting in the priest's initiation into the sangha, the Buddha explained that such powers do not lead to enlightenment and are trivial compared to true wisdom. Nonetheless, tales of the Enlightened One's great powers inevitably spread far and wide.

The defeat of the priest also resulted in the conversion of his thousand disciples. The event was monumental historically, because it marked the end of animal sacrifice and supernormal feats as representing the pinnacle of spiritual practice. In their place the Buddha conceived of a path to the ultimate in spiritual experience—enlightenment—through ethical conduct, meditation, and wisdom.

To disciples who had previously been fire worshipers, he taught that the flames of suffering are fanned by greed, anger, and ignorance, and that the fire within the mind must be controlled. To Bimbisara, king of the realm of Magadha, he dared to challenge the concept of caste. With regard to both society and religion, his views were thoroughly revolutionary.

Philosophical debate and metaphysical abstraction were highly popular in the India of that day. Pragmatic mystic and universal healer, the Buddha asked, "Suppose a man were shot by an arrow. What should he do? Should he ask who shot it, of what clan, what complexion, what caste? Or should he busy himself with the task of removing the arrow?"

With such logic he revealed the profound immediacy of spiritual life. One day, when questioned, he simply held a flower aloft. Kashapa the Great, who understood the significance of the silent gesture, was designated the Buddha's first successor.

In time the Buddha returned to Kapilavastu, to the palace where he had spent his youth. His wife Yasodhara, still grieving for the loss of her former husband, eventually dried her tears and bowed down at his feet. The king, at first shocked to see his son seeking alms with his retinue of monks, was filled with awe upon meeting him. No one was quite prepared for the effect of the great presence of the Buddha. His half-brother Ananda and a number of childhood friends became monks, and the ranks of the ordained swelled. Wherever he went, the Buddha affected people profoundly, and his attainments elicited their devotion.

He later visited Kapilavastu again to see his dying ninety-seven-year-old father. After receiving the comforting words of his son, the king died serenely. Queen Prajapati, recognizing the truth of impermanence and the suffering that follows from desire and attachment, requested admission to the order.

At first the Buddha was reluctant to admit her in view of the prevailing attitudes of the day. But eventually he agreed, and asserted the potential of women for achieving enlightenment, thereby defying social norms by allowing them to abandon household life. Prajapati, along with 500 noblewomen, formed the first order of Buddhist nuns. Eventually Yasodhara herself joined the community. Shakyamuni continued to expound the spiritual equality of the sexes, again contradicting one of the most cherished beliefs of his day.

Not all of the Buddha's disciples became immediately enlightened. One among them, his own cousin Devadatta, slandered him and attempted to kill him out of jealousy. But even when Devadatta rolled a huge boulder down a mountain and released a mad elephant to crush him, the Buddha remained unscathed.

The Lotus Sutra, considered to be the final sermon, proclaims the potential for anyone to attain enlightenment, from any point in his or her life, thus neutralizing any fixed definition of good and evil. Making specific reference to Devadatta, the Buddha implied that even he, after all his "evil deeds," had as good a chance for enlightenment as any upstanding member of the sangha.

Buddhists summarize the accomplishments of the Buddha by presenting them as the Twelve Deeds:

1. Life and teaching in Tushita heaven, a pure land
2. Entering the womb of his mother
3. Taking birth
4. Displaying his skill in the worldly arts—mathematics, fencing, archery, horsemanship
5. Fully experiencing the pleasures of women
6. Renouncing worldly life and ordination as a monk
7. Engaging in arduous ascetic practices
8. Meditating under the bodhi tree
9. Defeating the host of demons
10. Attaining full enlightenment
11. Teaching, or turning the wheel of doctrine
12. Passing into the state of peace.

the death of the buddha

By the age of eighty-one, after more than forty years of traveling the length and breadth of northern India disseminating his teachings, the Buddha's physical body began to fail. Upon completing the annual rainy-season retreat, he began a journey to Kusinagara to prepare for his *parinirvana,* or passing away.

The group stopped in the village of Pava for lunch. The meal caused the Buddha immediate discomfort and he became very sick. Sympathetic to the cook Chundra's despair at the inadvertent result of his meal, the Buddha called Ananda to his side and said, "Tell Chundra that the two meals in my whole life that I am most grateful for are Sujata's and his. One helped me to attain enlightenment, and the other is helping me to enter nirvana. Tell this to Chundra and ease his mind."

At Kusinagara the party entered a grove, and there the Buddha instructed Ananda to prepare a place for him to lie down with the head of his bed facing north. Some interpret this to mean that his teachings would spread to the north—to Tibet, Mongolia, and beyond.

The Buddha lay on his right side, resting his head on his right palm, and put one leg over the other in the posture of parinirvana, also known as the lion's pose. He instructed his grief-stricken disciples to take the dharma as their teacher, relying neither on any person nor on anything else, and to work diligently for enlightenment.

Legend has it that the earth trembled, celestial music filled the air, and the Buddha's golden form was showered with a rain of heavenly flower petals. The Enlightened One had fulfilled his mission. He died an ordinary death, demonstrating the impermanence of life and the suffering of cyclic existence to his followers.

Nehan (paranirvana), the passing of the Buddha. Detail from hanging scroll, color on silk, Japan, Mouromachi Period, 14th century.

57

The Early History *of* Kalachakra

INDIA AND SHAMBALA

OUT OF HIS GREAT compassion and desire to benefit
all beings, the Buddha manifested himself as the
Kalachakra deity to confer the initiation at Dhanyakataka
in South India. Some say this happened one year after he
attained enlightenment, while others say it was one year
before he shed his body. However, the Dalai Lama says
that logically it must have been at the end of his life, because the Kalachakra Tantra reflects so many of his life-
long accomplishments.

A three-dimensional
Kalachakra Mandala of
gold, 12 feet in diameter,
in the Potala Palace,
Lhasa, Tibet.
18th century.

The Buddha gave the Kalachakra Initiation at the request of Suchandra, the king of Shambala. Shambala is said to be to the north of India and Tibet, a land where an enlightened society

King of Shambala. King Suchandra, who requested the Kalachakra Initiation from the Buddha. Suchandra is identified by his golden crown and wheel pendant, symbolic of the Kalachakra Tantra. Thangka painting, Tibet, 18th-19th century.

lives in a perpetual state of bliss consciousness. Some say it is as close as northern Tibet while others believe that it is in Siberia or as far north as the North Pole. Others believe it is still farther, as far as North America, or even the stars. Early texts do, however, contain elaborate, graphic descriptions known as guidebooks to Shambala.

During the lifetime of the Buddha, when ordinary communications and travel were extremely limited, a pure land beyond the treacherous northern mountains of India was presented as an almost unattainable goal and only those with the highest realization attempted it. But no matter where the believer lived, Shambala has existed as a goal symbolic of the enlightened state of mind.

the setting

The Buddha taught the Kalachakra Tantra inside the *stupa* known as Sri Dhanyakataka. This sacred place, existing from the time of the third historical Buddha (the incarnation of the Buddha prior to Shakyamuni, the Buddha of our time), is located in a southern Indian village currently known as Amaravati.

During the time of the third Buddha there had been a famine, so the people had prayed to the appropriate deities and were saved by a shower of rice that came down from the sky and formed itself into the stupa at Dhanyakataka. This came to be known as the Rice Pile Shrine, the sacred and miraculous place where Shakyamuni Buddha came to confer the Kalachakra Initiation.

Shakyamuni, although he led the life of a monk, manifested himself as the Kalachakra deity in union with his consort Vishvamata, and the entire mandala of Kalachakra. The palace of Kalachakra, replete with deities, offerings, balconies, and portals, was projected in its fully three-dimensional form resting on the cosmic Mt. Meru.

Placed upon the mountain were several different "seats" for the mandala, such as the seat of the sun and the seat of the moon. And at the top center of the mandala itself stood the Buddha in the form of the deity Kalachakra with his consort, surrounded by the many hundreds of other deities who reside within the mandala, deities who are all manifestations of Kalachakra, or the Buddha's realized mind.

But this was not all. There was a second mandala projected upon the ceiling of the stupa, in a configuration that we might imagine to appear like a canopy of stars and comets. This was called the

Miniature Chorten. A stupa (*chorten* in Tibetan) is a monument composed of a square, a circle, a triangle, a crescent, and a flame, one on top of the other. Consecrated to have the power of the Buddha, a stupa may be a small personal reliquary or as large as several acres. 7⁵/₈" high, cast brass, Tibet, 12th century.

Glorious Constellation Mandala, or Great Integration Mandala, which consisted of ten different mandalas. The initiation took place within the time it takes to snap one's fingers.

Above: *Meru Mandala*. Abstract representation of the universe, including the elements earth, water, fire, and wind, which are depicted by squares, circles, triangles, and crescents. On top of the cosmic Mt. Meru is the palace of Kalachakra. Kalachakra and his consort are symbolized by a crystal on a lotus flower. 12" in diameter, gilt copper, eastern Tibet, 18th century.

Right: The cosmic Mt. Meru as it is traditionally visualized, in the shape of a cone composed of earth and water, and resting on the other elements of the universe. On top of the mountain is the five-story palace of Kalachakra. All of this is depicted in the Kalachakra Sand Mandala.

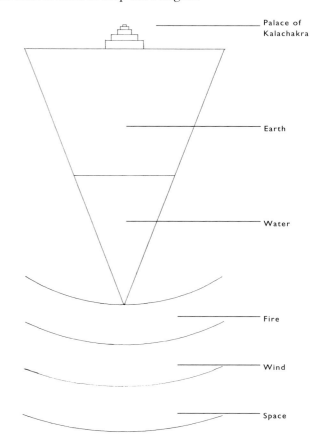

Manjusri, the Buddha of Wisdom, is always present at the transmission of the Kalachakra Tantra. He hovered over the mandala in the stupa, in luminous colors, to assist Shakyamuni.

How could it be that something as great as a mountain could fit inside a mere stupa? We can begin to understand this if we think of what it is like to look through the eye of a needle. If we hold the needle up close to our eye and look through it, we can actually see things that are very large, even something as large as a mountain. It is a matter of perspective, and what is happening in the mind of the person who is having the experience.

king suchandra and the ninety-six generals

King Suchandra, a member of the Shakya clan of which Shakyamuni Buddha was the prince, was king of Shambala. He was more than an ordinary human being—he was an emanation of Vajrapani, the deity who represents the power of all the Buddhas. This is how he knew of the Kalachakra teachings and had the wisdom to request them.

In fact, there were many advanced disciples in Shambala who were particularly suited to this initiation. The people of Shambala already had a strong faith in the Buddha dharma and in spiritual practice in general. They were a peaceful people, unlike the peoples around them and unlike those of today.

Suchandra was accompanied by his retinue of ninety-six generals, governors, and chieftains of provinces, or minor kingdoms within Shambala, most of whom were ready for these highest of teachings. The people were so advanced that they were adept in the practice of the *powa*—that is, at the time of death they could transfer their consciousness at will and with great dexterity from this physical body to the next so that they could be reborn in a pure realm.

They had other powers as well, which is why they were able to make the journey from Shambala to south India to witness the miraculous appearance of the Kalachakra Mandala and of the deities that the Buddha manifested.

the guidebooks

In Tibetan literature there are many guidebooks to Shambala. The most famous was written in 1775 by the Third Panchen Lama, Lobsang Palden Yeshe, and was called *Shambhalai Lamyig* [the Description of the Way to Shambala]. An older one is found in the *Tengyur,* the commentary section of the Tibetan Canon, which the Panchen Lama's version closely follows. To succeed in the journeys described, one needs to be particularly adept at the path and practice of meditation, and must possess great physical stamina. Many have actually attempted this arduous journey northward, without success.

The Buddha Vajrapani, who represents the power of all the Buddhas. This Sanskrit name literally means, "holder of the vajra." Ink on paper, Nepal, 19th century.

The guidebooks were written to serve those who, in degenerate future times, might wish to seek enlightenment for the benefit of others. Those qualified for the journey would have received an initiation into the Kalachakra Mandala, learned the science of the ritual, and become students of the tantra.

According to Edwin Bernbaum in *The Way to Shambhala*, Shambala "symbolizes the end of a journey into the hidden depths of the mind." The unexplored lands between India (or Tibet) and Shambala represent the subconscious mind, that which lies just beneath the familiar realm of surface consciousness. The pure land of Shambala, on the other hand, is the superconscious mind. The fact that there is such a great distance between Shambala and the known world tells us that the superconscious is "hidden deep in the mind, far from the illusions of the surface consciousness." According to the guidebooks, this subconscious area, which stands between the surface consciousness (or self-conscious mind) and the superconscious mind, contains a mixture of different elements, "bright rays of illuminating awareness" mixed with "dark clouds of delusion."

description of shambala

Shambala is said to lie within a ring of snow-covered mountains, which serves to bar all those not fit to enter. These mountains are forever hidden in mist, so that there is no way the kingdom's presence might be detected. Inside this outer ring lies another ring of even higher mountains. And the area between the two rings is divided into eight regions by rivers and smaller mountain ranges. These regions are shaped like petals symmetrically arrayed around the center of a flower; the entire shape is often compared to a lotus blossom. Each of these regions is in turn divided into twelve principalities, making up the individual domains of the ninety-six chieftains or governors.

Despite its great northerly location, the land of Shambala is lush with parks and meadows, flowers and abundant vegetation. Peace reigns throughout the land, and there is no need for harsh laws. Even the poorest of inhabitants has more material wealth than he or she can use and lives free of sickness. These qualities coincide with those in the description of Shangri-la, the idyllic realm created by James Hilton in his romantic novel *Lost Horizon.* However, the people of Shambala devote their entire lives to the study and practice of the Kalachakra Tantra.

The inner-ring mountains surrounding the central portion of the "lotus blossom" shine brilliantly because they are made entirely of

ice crystals. At the very center of this complex system lies Kalapa, capital of Shambala.

In *The Way to Shambhala*, Bernbaum has translated descriptions of Shambala from Tibetan texts. He writes, "The jeweled palace of the King at the center of Shambhala shines with a glow that lights up the night like day, reducing the moon to a dim spot in the sky. The palace's pagoda roofs gleam with tiles made of the purest gold, and ornaments of pearl and diamond hang from the eaves. Coral molding carved with dancing goddesses decorates the outside walls. Emeralds and sapphires frame the doorways while golden awnings shade windows made of lapis lazuli and diamonds. Pillars and beams of coral, pearl, and zebra stone support the interior of the palace, which is sumptuously furnished with carpets and cushions of fine brocade. Different kinds of crystal embedded in the floors and ceilings control the temperature of the rooms by giving off cold and heat."

At the center of the palace is the king's golden throne, supported by eight carved lions encrusted with jewels. "As long as the King remains on this seat of wisdom and power, a magic jewel given him by the serpent deities who guard hidden treasures enables him to satisfy all his wishes."

The inhabitants of Shambala also possess extraordinary powers. They have developed an advanced science and technology, which has been put to the service of spiritual ends. "Tibetan medical texts believed to have come from the kingdom describe human anatomy and physiology, sophisticated theories and methods of diagnosis, and ways to cure and prevent serious diseases such as smallpox. Other Kalachakra texts from Shambala have provided Tibetans with their systems of astronomy and astrology, as well as one of the calendars they use today." Their "stone horses with the power of wind" sound very much like our modern aircraft. The skylights in the king's palace are fitted with lenses that reveal life in other solar systems, and the king has a mirror that enables him to view whatever is happening in his country. But all of these sciences are only important to the Shambalans to the extent that they aid them in mastering "the highest science of all—the science of mind, or meditation." Through

Opposite: The mythical kingdom of Shambala, with the king at the center surrounded by the domains of the ninety-six governors. Shambala is known for its development of advanced sciences and technologies used in the practice of enlightened activity. Thangka painting, Tibet, 19th century.

the direct awareness and control of their minds and bodies which they have developed, they are able to heal their own bodies. They are also capable of telepathy and clairvoyance and can walk at very high speeds. Despite these marvelous abilities, the people of Shambala are not yet fully enlightened. "They still retain some human failings and illusions, but many fewer than people of the outside world. They all, however, strive to attain enlightenment and bring up their children to do likewise. Theirs is the closest to the ideal society that can be reached in this world."

The Kalachakra Tantra was given to the people of Shambala to help them make that final step to the highest enlightenment.

kalachakra in shambala

Following the empowerment at the Rice Pile Shrine, King Suchandra returned to Shambala and, over the next two years, transcribed all that had been imparted to him by the Buddha in what is known as the *mula,* or root text. It comprised twelve thousand verses, written in an unknown tongue called "twilight language." The Buddha had instructed Suchandra to teach the Kalachakra Tantra and also to build a three-dimensional Kalachakra Mandala in his native land. Some scholars say that certain realized practitioners who went to Shambala in search of these teachings returned to India without the texts in written form, but with an understanding of them. They then wrote the teachings in Sanskrit.

It is, of course, impossible to provide scholarly documentation or irrefutable proof of the history of Kalachakra in Shambala before these written teachings. But Tibetans feel that this kind of concern derives from too linear a way of seeing the world, or of having only one level, or interpretation, of reality. By getting beyond this limited perspective, we are able to see and understand how the world functions on subtler levels. Those who believe in Shambala and practice the Kalachakra Tantra, for example, feel that they have become one with the deity and have been given instant and nonlinear access to a vast resource of knowledge, wisdom, and compassion.

the seven shakya kings
and the twenty-five kalkis

King Suchandra was the first of seven Shakya kings of Shambala, each of whom was an emanation of a different *bodhisattva,* and each of whom reigned for 100 years. The seventh was called Yashas, "the Renowned." King Yashas was an emanation of Manjusri, the Buddha of Wisdom, and predicted the coming of the "barbarian dharma" after 800 years. Through his skill as a teacher, he converted the Brahman sages to Buddhism, combining all the castes of Shambala into one "vajra family" (those who have received the tantric empowerment from the same lama). Henceforth, the rulers were known as *kalki,* meaning governors who hold the lineage.

In this way, Yashas ensured that Shambala would remain outside the range of the barbarian influences and that his descendant, the prophesied twenty-fifth kalki, Raudra Chakri, and his army would perform their role at the end of the *Kali Yuga,* or the current Age of Darkness or Strife. Shakyamuni Buddha prophesied that the twenty-fifth kalki would usher in the Age of Enlightenment, transforming all the ignorance of the barbarians to Buddha mind. This age was prophesied to last for 1,000 years before the cycle returned to a degenerative process.

We are currently in the reign of the twenty-first kalki king, which would place the new Golden Age some 400 years from now; but such calculations are futile, in any case, since we cannot even be sure of what constitutes 100 years of Shambala time. Many scholars think the Golden Age will occur during the twenty-first century.

The first of the kalki was named Manjukirti, again an emanation of Manjusri. He was also the first to put the Kalachakra Tantra into condensed form, called the *Sri Kalachakra,* which became the basic tantra for the Kalachakra system.

At the time the Buddha gave the first initiation, he prophesied that 600 years later Manjusri would take the form of the first of the kalki and would compose a Kalachakra text. Then he said that Manjusri would appear again much later in the form of the twenty-fifth

kalki, whose particular task would be to overcome the forces of the barbarians.

Manjusri represents the mind that realizes emptiness. Because the disputes that arise between nations and between individuals come about through wrong views, only Manjusri, emanation of the Wisdom Mind of the Buddha, can truly overcome the "barbarians."

It was King Yashas's son, Pundarika, who wrote the commentary on *Sri Kalachakra* entitled the *Vimalaprabha*. These two texts comprise our basic written sources for the Kalachakra system as a whole.

Of the twenty-five kalkis, the seventeenth, Shripala, is especially important to us because it was through him that the Kalachakra lineage reentered India in A.D. 1027. The kalki system continues even to this day in Shambala.

kalachakra returns to india

The Kalachakra Tantra was nurtured and kept alive in Shambala for at least 1,500 years before it reemerged in India. However, there is evidence that there was at least an aspect of it still known and practiced in India. For example, the Indian yogi Mahasiddha Ghanda had composed a *sadhana* (meditation practice) of Kalachakra.

There are many different accounts of how the transmission of Kalachakra reentered India from Shambala. One commonly accepted version states that the tantra reappeared in India during the 10th century and was first brought to Tibet in 1027 by a Kashmiri scholar named Somanatha. However, others insist that 1027 was the year in which the tantra returned to India through King Shripala and that it was brought to Tibet later.

king shripala

This version of the story comes to us from Buton, a Tibetan scholar, writer, spiritual master, and historian of the early 14th century. He says that the tantra was brought to India by the seventeenth kalki of Shambala, Shripala himself, in 1027.

Opposite: *Kingdom of Shambala,* detail. This painting depicts the defeat of the "barbarians" (those who hold wrong views) through the application of the wisdom mind of the Buddha. Following this conflict, the present Age of Darkness will be transformed into an enlightened society. Kalachakra appears in a colored circle at the top right. 44"x44" thangka, Mongolia, 19th century.

It had become psychically known to Shripala that a young man who had a burning desire to master the profound tantras would attempt to cross the dangerous terrain from India to Shambala. Concerned for the youth's welfare on a journey across a waterless desert that would have taken four months, the king, through an emanation body, went to meet him. He initiated the young man and over the next four months taught him all the highest yoga tantras. The youth remembered all he heard and attained a mastery of the knowledge he received. When he returned to India, he became famous as an emanation of Manjusri and was known as Kalachakrapada.

tsilupa

Another version of the coming of the Kalachakra Tantra from Shambala to India tells of Tsilu, a great 11th-century *pandit* (highly realized scholar) who was born in Orissa, in the eastern part of India. He studied all the available texts on Buddhism at several of the great Buddhist universities of his day, but he did not find what he was seeking. He realized that in order to become self-liberated within a single lifetime, he needed further clarification of the doctrines associated with the Vajrayana, or tantric Buddhism. Hearing that the kings of Shambala had kept the sacred and secret doctrine intact in their distant reaches to the north, Tsilu resolved to visit them.

According to one version, Tsilu's intuitive voice told him to seek the protection of a group of traders who dealt in pearls and other jewels from the sea. However, in his restlessness he could not wait for them, so he set out on his own to traverse the mountains. Luckily, as he was climbing a mountain one day, he met a man who offered to spare him the long and arduous journey by giving him the teachings he sought then and there.

Recognizing the man to be an emanation of Manjusri, Tsilu prostrated himself before him, and Manjusri imparted to Tsilu all the initiations, the tantric commentaries, and the oral instructions of the Kalachakra Tantra through mind transmission. At last, he placed a flower on the top of Tsilu's head, blessed him, and said, "Realize the entire bodhisattva corpus," whereupon Tsilu realized the entirety of

Opposite: A *Mahasiddha*, or great yogi master, most likely sculpted by one of his students. Somanatha of Kashmir, who was such a Mahasiddha, translated the Kalachakra Tantra from Sanskrit and brought it to Tibet in the 11th century. Brass, 12 1/2", Tibet, 11th-12th century.

Naropa, an Indian Mahasiddha of the 11th century who was instrumental in the transmission of Kalachakra. Naropa was the abbot of Nalanda University. Detail from *Virupa, Naropa, Saraha, and Dombi Heruka,* thangka, eastern Tibet, 18th century.

the transmission. Henceforth, he was known as Tsilupa, as a title of respect.

Some think that Tsilupa was in fact Naropa, a well-known master of Kalachakra and other tantras of the 11th century.

kalachakrapada
the elder and younger

Tsilupa's foremost student, Pindi Acarya, taught the bodhisattva corpus to Kalachakrapada the Elder. He in turn taught it to Kalachakrapada the Younger, who taught it openly at Nalanda University, the great seat of Buddhist learning in northern India.

There Kalachakrapada the Younger, intent on establishing and propagating the Kalachakra Tantra, issued a challenge to the five hundred pandits residing at Nalanda, who immediately engaged with him in a debate. With the knowledge of the highest tantra on his side, Kalachakrapada the Younger proceeded to defeat them all. They became his disciples, and the Kalachakra Tantra was firmly established in India.

Nalanda University, the great Buddhist learning center of northeastern India, had as many as 20,000 students during the 11th century, when Kalachakra became part of its curriculum. It was destroyed by foreign invaders in the 12th century.

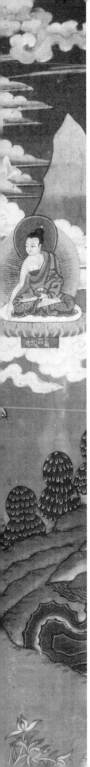

Kalachakra Comes *to* Tibet

mountains and mystery

TIBET IS A NATURAL setting for a highly developed spiritual culture. It is ringed by lofty peaks, some the highest on the planet, which have traditionally formed borders with China, Mongolia, Ladakh, Pakistan, India, Nepal, and Myanmar (Burma). Tibet is the size of Texas and Alaska put together, and has an astounding average altitude of more than 12,000 feet. This challenging geography limited outside influences, while the harsh climatic conditions kept the population low and the country sparsely settled. Although the stark magnificence

Padma Sambhava, known to the Tibetans as Guru Rinpoche, or "Precious Teacher," was instrumental in establishing Buddhism in Tibet in the 8th century. Thangka painting, Tibet, 18th century.

of the so-called roof of the world offers little creature comfort, it seems to inspire spiritual pursuit. In fact, the mythologies of both India and China place the mountains of Tibet as home to their gods and immortals.

Archeological excavations date Tibetan civilization back 3,600 years to the Late Stone Age. Tibetan mythology holds that all Tibetans are descended from the union of the deity Avalokiteshvara in the form of an ape and the goddess Tara (who is the Mother of All Buddhas) in the form of an ogress. Their progeny gave birth to the Tibetan people in the Yarlung Valley of central Tibet. Long before the common era, Tibetans developed a particular brand of shamanism that was ideal preparation for the arrival of Buddhism in the 7th century.

The goddess Tara. 28' high, solid cast copper with jewel inlay, Nepal, 17th century.

the early kings and the bon tradition

As legend tells it, the first nine kings of Tibet descended a rope ladder from the sky onto a mountain top. Other stories suggest that the first king, Nyatri, was fleeing defeat in the Mahabharata war in India when he was seen coming down from the mountains by the peasants who inhabited the Yarlung Valley. The peasants took him for a mountain god and gave him the title of *tsenpo,* meaning "mighty one." Nyatri Tsenpo reigned around 127 B.C., approximately 400 years after the birth of the Buddha, although other accounts place him as far back as 400 B.C. He was the first of what became known as the Yarlung dynasty, whose kings united the many tribes scattered over this wide area.

The shamanic practices of the ancient Tibetans, which may have originated in Iran, were performed to appease local gods and demons. They placed great emphasis on divinations and sacrifices conducted by priests, primarily at burial sites, as well as rites concerning oracles, magical possessions, and healings. The word *Bon,* meaning "priest," was eventually applied to the religion as a whole, which began to take shape about the time of the ninth Yarlung king. The heavenly ladder was said to have been cut during this king's reign, so that he died on earth instead of returning to heaven. The kings, or tsenpos, were worshiped as divine, so the coronation or burial of a tsenpo provided a great occasion for religious ceremonies. The practice of Bon continues in contemporary Tibet, with the strong influence of Buddhist doctrine and monasticism.

the coming of buddhism to tibet

Although Buddhism may have been introduced to Tibet as early as A.D. 173, during the period of the twenty-eighth Yarlung king, Lha Thothori Nyantsen, it was not well known until the reign of the thirty-third king, the great Songtsen Gampo. He began his reign in the year 627 at the age of thirteen and succeeded in unifying Tibet by defeating the armies of Tang China, extending his empire to the trade cities of the Silk Route to the north and into what is now Ladakh, Nepal, and northern India, and to Myanmar to the south. He was introduced to Buddhism through his exposure to the cultures of India and China as his kingdom expanded. He decided it would be useful to incorporate elements of Buddhism into his own culture to establish a greater governmental stability based on its teachings of nonviolence. He introduced several aspects of Indian culture to Tibet, including writing and a legal system.

Songtsen Gampo's conquests brought him two foreign wives, one from the Chinese imperial family and the other from the royal family of Nepal. Both princesses brought Buddhist scholars and artists, as well as dowries which included gifts of great Buddhist art. The king built temples based on Indian architecture, with wooden doorways and pillars adorned with celestial carvings reminiscent of the 5th-century Buddhist cave temples of Ajanta in western India. The temples built by Songtsen Gampo housed images of the Buddha, including one statue, the famous Jowo (or "Precious Lord") Buddha, brought by the Chinese queen, which is said to contain relics of Shakyamuni Buddha. The Jowo Buddha is now housed in the Jokhang temple, also built by Songtsen Gampo, in Lhasa, the capital of Tibet, and it remains the holiest image in Tibet to this day.

King Songtsen Gampo, who unified Tibet in the 7th century. This detail is from the oldest known Tibetan thangka, from the 11th century, which was discovered by Mokotoff Asian Arts.

Songtsen Gampo is thought by Tibetans to have been an incarnation of Avalokiteshvara. He was the first to assume the role of "Dharma King," although he continued to support the indigenous religion which assured his theocracy.

During the reign of the thirty-seventh king, Tri Song Deutsen (755-797), Tibet became a very powerful country and Buddhism spread throughout the kingdom. Tri Song Deutsen brought the abbot Shantarakshita, known in Tibet as the "Bodhisattva Abbot," from India to teach and to establish the first monastery at Samye. But there

Above: This sculpture of the Jowo Buddha was part of the dowry of Songtsen Gampo's Chinese wife, Princess Wencheng. It was bejeweled by Tsong Khapa in the early 15th century and contains relics of Shakyamuni Buddha. Housed in the Jokhang Temple in Lhasa, *Jowo Rinpoche* is Tibet's most revered icon.

Right: The roof of the Jokhang Temple in Lhasa, Tibet. The eaves terminate with decorative guardians to ward off evil spirits.

was great resistance from the Bon deities, so Shantarakshita suggested that Tri Song Deutsen call upon a practitioner from the land of Swat (an area of Pakistan) known for his great power, to subdue the warrior deities. This extraordinary man was Padma Sambhava.

Padma Sambhava is revered by Tibetans even today as Guru Rinpoche, or "Precious Master." His success became known far and wide, along with his great strength and vision. Thought to be an incarnation of the Buddha himself, Padma Sambhava was skilled in magic and mysticism; he performed many miracles, such as flying and emanating his presence into several places at once. He buried precious teachings known as "treasures" throughout the country for later generations to discover and put into practice.

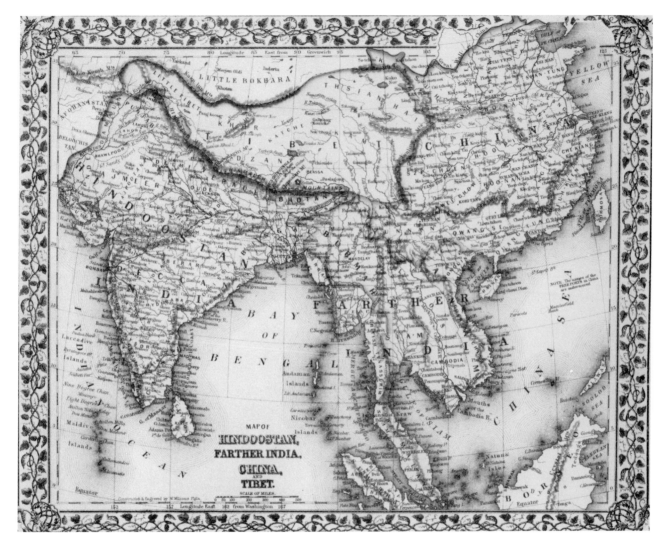

Official map from the United States Library of Congress showing India, China, Southeast Asia, and Tibet, published in 1872.

Samye Monastery was eventually completed based upon a mandala made by Padma Sambhava. A great philosophical debate was held at Samye in about 792 to determine whether Tibetans would follow Indian or Chinese Buddhist practice. The Indian scholars argued the Mahayana position, which held that it was necessary to accumulate merit and knowledge over many lifetimes to attain

Buddhahood. The Chinese position was that any practitioner could achieve Buddhahood by establishing a state of complete repose, regardless of his morality or intellectual endeavor. The Indian view prevailed, and it was determined that Tibetan Buddhist doctrine would thereafter come from India.

Buddhism entered a golden age in Tibet during which it grew in popularity as monasteries appeared throughout the land, along with serious Buddhist practitioners and renowned scholars. The fortieth king of the Yarlung dynasty was Tri Relwajen. Songtsen Gampo, Tri Song Deutsen, and Tri Relwajen are known as the Three Great Religious Kings of Tibet. Tri Relwajen was a great supporter of Buddhism, responsible for the construction of many temples. He issued a decree that every monk should be supported by seven households, which made him highly unpopular with the nobles and the Bon priests. A reaction set in, and Tri Relwajen was assassinated. His brother, Lang Darma, came to power in 836, bringing the golden era and the Yarlung dynasty to an end.

During the violent persecution in the reign of Lang Darma, Buddhist temples and texts were destroyed. Monks who didn't flee to safety in eastern Tibet were forced on pain of death to choose between marriage, the army, and adoption of the Bon religion, which they proclaimed by ringing a bell wherever they went. Bon flourished for a century, acquiring a written canon, and Buddhism went underground. After the death of Lang Darma, the country was split into small kingdoms and many outlying territories were lost.

Samye Monastery, Tibet's first monastery, which was built in the 8th century according to a plan by Padma Sambhava. Thangka painting, Tibet, 18th-19th century.

a new wave of buddhism

In the 10th century, interest in Buddhism was revived by a religious king in western Tibet named Yeshe O, who came from the Guge dynasty, which was descended from the Yarlung kings. Monks began returning from exile in the East, monasteries were rebuilt, and ancient texts reappeared, some of which had been buried by Padma Sambhava. Many important teachers came from India, including the great Atisha in 1042, whose arrival heralded the Buddhist revival in Tibet. A disciple of Atisha's named Drom Tonpa founded the

Kadam school, which is the historical predecessor of today's Gelug order of Tibetan Buddhism.

During this new wave of enthusiasm, many Tibetans traveled to India to study directly with Indian masters, transforming themselves into the embodiment of the teachings. Several of them, in-

The Three Great Religious Kings of Tibet: Songtsen Gampo, Tri Song Deutsen, and Tri Relwajen. These kings implemented the development and practice of Buddhism in Tibet during the 8th and 9th centuries.

cluding Rinchen Sangpo, Drokmi, and Marpa, brought back new Tibetan translations of many sutras and important tantras. Rinchen Sangpo's translations, which included the Kalachakra text, added a new vitality to the science of mind the Tibetan Buddhists were so eager to perfect.

The translator Drokmi is considered to have been among the first to bring the Kalachakra Tantra to Tibet. Drokmi's descendants founded the first Sakya Monastery in 1073. In the 13th century, the high lama of Sakya Monastery, named Kunga Gyaltsen, was invited to visit the camp of the Mongol prince who was threatening to invade the Tibetan states. He impressed the warlord with his spiritual wisdom and he became honored as a teacher of Buddhism. In 1253, the lama's nephew Pagpa became the court priest of Kublai Khan. When he became emperor, Kublai Khan granted Pagpa the administration of all Tibet, establishing a "patron and priest" relationship that would maintain the Sakyas in power for a hundred years.

Another important translator was the famous teacher Marpa, who studied with the Indian master Naropa. Among Marpa's thousands of disciples was the yogi poet Milarepa, who became a legend for having attained Buddhahood, or complete enlightenment, within one lifetime. The teachings of Naropa, Marpa, and Milarepa were responsible for the establishment of the Kagyu lineage of Tibetan Buddhism. In the 12th century, Dusum Kyenpa was the head of one of the twelve sub-orders of the Kagyu lineage, known as the Karma Kagyu. He became known as the Karmapa, the first line of religious leaders to be recognized as reincarnations of their predecessors. This tradition of recognizing incarnations, called *tulkus,* became widespread in Tibetan Buddhism and is the system used to determine the succession of the Dalai Lamas.

kalachakra enters tibet

As many as sixteen translations of the *Sri Kalachakra* from Sanskrit to Tibetan appeared between the 11th and 14th centuries. Because there is no written proof that the Kalachakra Tantra was originally transmitted by the Buddha himself, some scholars speculate that it was invented in invasion-plagued India in the 11th century. But its message of compassion and loving kindness clearly overshadows any debate concerning its origin.

The translations of Rwa and Dro, dating from the latter part of the 11th century, emerged as the primary sources for the Kalachakra Tantra as it exists today. The Tibetan Rwa Lotsawa (*lotsawa* is Tibetan for "translator") was a student of Samantashiri of Nepal, while Dro Lotsawa was a student of Somanatha, the first Indian to travel to Tibet to spread the Kalachakra system. The Rwa translation became the most widespread.

The two lineages of Rwa and Dro continued through successions of teachers and eventually flowed together. Buton (1290-1364) played a key role in the preservation of all the different

The 11th-century yogi master Milarepa, who sang his expressions of enlightenment in verses known as the *100,000 Songs*. He was the foremost student of Marpa, a great scholar who traveled to India and translated many important Buddhist texts into Tibetan, including the Kalachakra. 4 3/8" high, cast brass with inlay, Tibet, 16th century.

types of tantras and produced a close study of the Kalachakra in particular.

Born in the Tsang province of central Tibet, the infant Buton had a very shriveled and ugly appearance. But, according to legend, when Buton was very small he would say to his mother that she

The oldest known image of Kalachakra, this bronze came to the Newark Museum from eastern Tibet, though its style indicates that it probably traveled from Kashmir with the spread of the teachings. The faces of Kalachakra on the front side were decomposed by a blessing substance commonly applied by practitioners in Kashmir; the back face is intact. Front and back. 8$^{1/2}$" high, solid cast bronze with silver inlay, 10th-11th century.

shouldn't be afraid to show her unattractive child to others, since he was prepared to deal with them. In fact, *Bu* means "son" and *ton* means "to show."

Buton became a superior scholar and, unlike other scholars, an expert in all the teachings, from sutra to tantra to poetry, astrology,

Buton, a master Kalachakra scholar of the 14th century, consolidated many translations of Kalachakra into one text, which has been used by the major traditions of Tibetan Buddhism since that time. Thangka, central Tibet, 17th century.

Opposite: Tsong Khapa (1357-1419) was an incarnation of Manjusri, the Buddha of Wisdom. A realized master and important reformer of Tibetan Buddhism, he was a pivotal figure in the transmission of Kalachakra. Thangka painting, Tibet, 18th century.

Tsong Khapa had a student named Khe-Drub. Khe-Drub's life work on Kalachakra resulted in the text used by the Dalai Lama today, called *The Mandala Rite of Glorious Kalachakra: Illumination of Thought.*

and grammar—a kind of Renaissance man of Tibet. He became a master of Kalachakra with a powerful personal practice and even saw Kalachakra himself. He wrote about thirty volumes, most of which deal with the Kalachakra Tantra, and he is regarded as the most reputable of all the scholarly sources on the subject.

It was under Buton's hand that the Rwa and Dro traditions became consolidated into one. His translation was incorporated into many different Kalachakra traditions, including the Sakya, Kagyu, and, later, the Gelug.

from tsong khapa to the dalai lamas

The great Tsong Khapa Lo Sang Drakpay (1357-1419) was one of the most eminent of Tibetan meditation masters, scholars, and writers. Although he was ordained as a novice monk by the fourth Karmapa, he studied with Nyingma, Kadam, and Sakya, as well as Kagyu, masters for twenty years. He is believed to have achieved complete enlightenment in 1398.

Concerned with the decline in monastic discipline, Tsong Khapa convened a great council of all of the major orders of Tibetan Buddhism, which resulted in a movement of monastic renewal. The Gelug (meaning "virtue") order arose out of Tsong Khapa's reform of the Kadam tradition originated by Atisha in the 11th century, as well as from his own teachings.

Tsong Khapa's hermitage was built up to become Ganden Monastery, which, with a community of more than 10,000 monks, was one of the largest monasteries in the world until its destruction during the Chinese takeover of Tibet in the 1950s. Drepung and Sera Monasteries were built after Ganden, forming the three major monasteries of the Gelugs. Those who continued to follow the form of Buddhism introduced by Padma Sambhava had become identified as the Nyingma (meaning "ancient") order. Thus the four great orders of Tibetan Buddhism were now fully established: the Nyingma, Kagyu, Sakya, and Gelug.

It is important to note that these lineages, orders, or schools of Tibetan Buddhism share the same teachings of the Buddha and the same goal of attaining liberation from suffering for the benefit of others. The orders distinguish themselves from one another by the

Ganden Monastery, built on the hermitage site of Tsong Khapa, the largest monastic universities in the world. This photograph was taken in 1932 by Sir Hugh Richardson, who served as the representative of British India to Lhasa.

different teachers who are the lineage holders and by the particular teachings they emphasize at various stages of the path.

Tsong Khapa studied Kalachakra under Chokyi Palwa, a disciple of Buton. He conducted an extensive meditation retreat on the completion stage of Kalachakra. At the conclusion of the retreat, he received a vision of Kalachakra who, laying his hands on Tsong Khapa's head, said he would do the same work as King Suchandra of Shambala in propagating the teachings of Kalachakra.

Near the end of his life, Tsong Khapa transmitted the entire Kalachakra teachings to Khe-Drub, who, as a result of extensive meditation practice, wrote a comprehensive text and commentaries on Kalachakra.

A successive lineage of eminent lamas transmitted the Kalachakra Tantra directly from Khe-Drub to the Seventh Dalai Lama, who in-

troduced it to Namgyal Monastery in the early 1700s. By the middle of the 14th century, the *Sri Kalachakra* had been translated into Tibetan at least fifteen times, and the *Vimalaprabha* commentary more than ten times, which represents the most attention given any Buddhist text by the great Tibetan translators. Even though the Kalachakra Tantra was widely practiced throughout Tibet in hundreds of monasteries, it was Khe-Drub's text which was retained by Namgyal Monastery over the centuries and which is followed by the Dalai Lama of today.

Ganden Monastery after its destruction in 1959. The buildings that are still relatively intact were reconstructed after 1981.

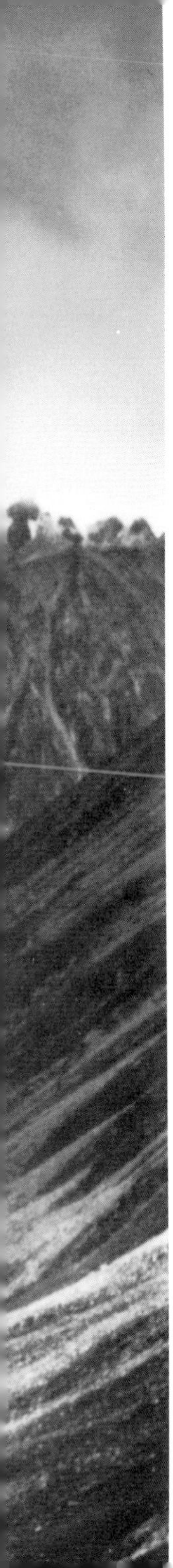

The
Dalai Lamas
and
Namgyal
Monastery

TENZIN GYATSO, the warm and charismatic man we
know today as the Fourteenth Dalai Lama, has been a
major influence in the preservation and practice of Tibetan Buddhism throughout the world. The attention
he has placed on the Kalachakra Tantra is appreciated
by the thousands of students who have entered into
its practice, and the tens of thousands of museum visitors who have been drawn to demonstrations of the
Kalachakra Sand Mandala and other ritual arts. The
Dalai Lama, with his great energy and devotion to the

The mystical lake, Lhamo
Latso (Lake of Visions),
located in central Tibet at
an altitude of 12,000 feet.
Since the early
15th century, auspicious
signs have been seen in its
waters, such as clues to
the birthplace of a newly
reincarnated Dalai Lama.

teachings of the Buddha, is the holder of a lineage that is part of a rich tradition unique to Tibet.

The Dalai Lama is seen as an early manifestation of Avalokiteshvara, the Buddha of Compassion. Since 1642, the Dalai

Shadakshari Avalokiteshvara. The Dalai Lamas are said to be manifestations of this four-armed form of the Buddha of Compassion. Detail from a huge 11th-century wall fresco in Alchi Monastery, Ladakh.

Lamas have been both the political and spiritual leaders of their country, a situation that was interrupted by the Chinese takeover of Tibet in the 1950s.

Buddhists believe that all sentient beings are reborn again and again. However, until we have achieved liberation (sometimes referred to as *nirvana*) from this cyclic existence, we are automatically reborn as a result of our actions in previous lives. This is an aspect of the law of karma. "Tulkus" such as the Dalai Lama are beings who have achieved a level of realization where they can choose rebirth at any time or place, or in any form, in order to help or liberate all sentient beings.

When a Dalai Lama dies, a committee is formed to find his reincarnation. Search parties go out into the countryside, following the prophecies of the State Oracle, predictions of high lamas, visions from the sacred lake Lhamo Latso, and mystical clues left behind by the previous Dalai Lama. When the likely candidates, usually very young children, have been found, they are subjected to extensive tests, one of which requires them to identify possessions of the previous Dalai Lama from among an assortment of similar items.

the training of the dalai lamas

Historically, once the incarnation had been identified, the boy would be taken to Lhasa, the capital of Tibet, and installed as the Dalai Lama on his throne in the Potala Palace, often at a very early age. He would be raised and educated by highly qualified teachers from many of the great monasteries of Tibet, representing all the major traditions of Tibetan Buddhism. Many of these teachers might have been trained themselves by the previous Dalai Lama. This education program lasts more than twenty years and includes learning to read and write Tibetan and becoming adept in all the various traditions of Buddhist philosophy and ritual arts. During the Dalai Lama's minority, affairs of state would be handled by a regent appointed by the previous incarnation of the Dalai Lama or by the government.

In order to achieve his Master of Metaphysics degree, the young Dalai Lama would be expected to prove himself through extensive debate with Tibet's most respected scholars, before an assembly of thousands of highly educated monks from Sera, Drepung, and Ganden monastic universities.

the first dalai lamas

Gyalwa Sonam Gyatso was given the title Dalai Lama by the Mongolian king Altan Khan in 1578. Detail from an 18th-century Tibetan thangka.

The title Dalai Lama was first given by the Mongolian king Altan Khan to Gyalwa Sonam Gyatso during his visit to that country in 1578. Altan Khan granted this honor in recognition of Sonam Gyatso's great knowledge and compassion and out of gratitude for the positive influence he had had in turning the minds of the Mongolians toward Buddhism. *Dalai* is the Mongolian word for "ocean," while *lama* means "teacher of wisdom." Thus the title means "Teacher (like the) Ocean," or "Ocean of Wisdom." Interestingly, until only recently the Tibetans themselves never used this term but referred to him instead as Kundun ("in the Presence of"), Yishin Norbu ("Wish-fulfilling Gem"), or Gyalwa Rinpoche ("Precious Victorious One"). Today, due to the influence of Westerners, he is also referred to as His Holiness.

Even though Sonam Gyatso was the first person to receive the title, he was actually the Third Dalai Lama, since he was an identified reincarnation of the great Gyalwa Gendun Gyatso, who was himself a reincarnation of the renowned Gyalwa Gendun Drub. These two were posthumously named the First and Second Dalai Lamas.

The First Dalai Lama, Gyalwa Gendun Drub, was born in 1391 to nomadic peasants in central Tibet. After the death of his father when Gendun Drub was seven, his mother placed him in a monastery to be educated. He became a disciple of Tsong Khapa in 1415 and soon demonstrated his greatness as both a scholar and a realized master, as his fame spread throughout Tibet. In 1447, he established Tashi Lhunpo Monastery at Shigatse, which eventually became one of the most important monastic universities of southern Tibet. The First Dalai Lama died in 1474 at the age of eighty-three, while sitting in meditation.

The next year Gyalwa Gendun Gyatso was born, also in central Tibet, to a renowned yogic practitioner of the Nyingma tradition. He was recognized at the age of four as Gendun Drub's reincarnation, and, posthumously, was called the Second Dalai Lama. He studied at Tashi Lhunpo and Drepung Monasteries and became the abbot of Drepung. There he built a residence for himself called Ganden Phodrang (Joyous Palace).

Gyalwa Gendun Gyatso's principal focus was on practicing the tantric tradition, but he wrote extensively on practices from several distinct major lineages of Tibetan Buddhism. He discovered the power of Lhamo Latso (Lake of Visions), which is reputed to foretell events, and he established an important monastery on its shore.

It was at this time that what was to become Namgyal Monastery became associated with the Dalai Lamas. One account tells us that there was a monastery called Phende Gon, which was struck by an epidemic that left only eight monks alive. These monks then abandoned their home and set out on a pilgrimage to perform rites at various holy sites for their deceased companions. En route, they encountered Gyalwa Gendun Gyatso, who was also traveling for religious purposes. They found their meeting propitious and decided to stay together.

In 1542, at the age of sixty-seven, Gendun Gyatso passed away, like his predecessor, while sitting in meditation.

the establishment of namgyal monastery

In 1543, Gyalwa Sonam Gyatso, the Third Dalai Lama, was born near Lhasa and was soon recognized as the reincarnation of Gendun Gyatso. Educated at Drepung Monastery, he was renowned throughout Asia for his great wisdom as a teacher and scholar. The Namgyal monks, though they were not yet so named, received formal recognition as being associated with the Dalai Lama only when Sonam Gyatso established Phende Lekshe Ling in 1564-65. This monastery had its roots in Phende Gon and later became known as Namgyal Monastery. From that time on, the monks became responsible for performing prayers for the welfare of the land and people of Tibet and for assisting the Dalai Lama in performing religious rites.

The name Namgyal came into use in 1571 when the king Altan Khan of Mongolia became very ill and requested that his teacher, the Third Dalai Lama, perform long-life prayers for his recovery. The Dalai Lama instructed his monks to perform the sacred long-life prayer of the goddess Namgyalma, and from that moment on, Phende Lekshe Ling Monastery was also known as Namgyal Monastery.

Famous for combining the teachings of the Nyingma and Gelug lineages of Tibetan Buddhism, Sonam Gyatso established monasteries throughout Tibet, Mongolia, and western China, bringing their traditions to these lands as he traveled and taught. When he finally left Mongolia, he promised to return sometime in the future. He passed away in 1588 at the age of forty-five.

The next year, Gyalwa Yonten Gyatso was born in Mongolia to a direct descendant of Altan Khan. He was the only Dalai Lama to have been born outside of Tibet, demonstrating that the selection of the Dalai Lama goes beyond race, country, or class. It was not until he was twelve years old that he was brought to Tibet as the Fourth Dalai Lama. He spent his short life in study, practice, and teaching, declining, as did his predecessor, the many invitations sent by the emperor of China to visit the Manchu court. He passed away in 1617 at the age of twenty-eight.

the great fifth

The Fourth Dalai Lama's relatively modest contribution was amply made up for by the Fifth Dalai Lama, Gyalwa Ngawang Lobsang Gyatso, who is known as "The Great Fifth." Born in 1617, soon after the passing of the Fourth, this great leader reunited the three regions of Tibet in 1642, areas that had been functioning as separate kingdoms since the mid-9th century.

For the first time, a Dalai Lama became the spiritual and political leader of all of Tibet. In that same year, he declared Lhasa to be the capital of Tibet, and he named the government *Ganden Phodrang* in honor of the Second Dalai Lama, whose residence bore this name. This government, which divided authority equally between the

The goddess Namgyalma, protector deity of Namgyal Monastery. Cast bronze, 20th century, India.

clergy and the laity, continues today in exile under the leadership of the Fourteenth Dalai Lama. As the new head of Tibet, the Fifth Dalai Lama appointed governors for the various districts and ministers to form a new government. He also instituted a national system of medical care, as well as a program of national education. He visited Peking in 1652, at the request of the Ching emperor, in order to restructure the Buddhist monasteries there.

The Great Fifth Dalai Lama, Gyalwa Ngawang Lobsang Gyatso (1617-1682). 7⁷/8" high, gilt bronze, central Tibet, 17th century.

The Great Fifth was known for his spiritual power and personal dynamism. He produced voluminous scholarly writings, especially on history and classical Indian poetry, and he clarified and improved upon the rituals of Namgyal Monastery. He introduced to Namgyal many meditation practices influenced by the Nyingma order, added forms of sacred dance and chanting based on ancient Tibetan traditions, and established Namgyal as one of the few Tibetan monasteries to include the ritual practice and study of all four lineages of Tibetan Buddhism. The Great Fifth Dalai Lama traveled and taught widely, and he became famous for his statesmanship.

The Fifth Dalai Lama had a beloved teacher named Lobsang Chokyi Gyaltsen, who was a fourth-generation disciple of Tsong Khapa. When Lobsang Chokyi Gyaltsen died, the Dalai Lama recognized a boy as his incarnation and gave him the title of Panchen Lama (*panchen* means "great teacher"). The Panchen Lama became the abbot of Tashi Lunpo Monastery, founded by the First Dalai Lama, and all subsequent Panchen Lamas have held this position. The Panchen Lama is considered second in stature only to the Dalai Lama, and whichever one is the elder has traditionally served as teacher to the younger. The Panchen Lamas have figured strongly in the transmission of the Kalachakra Tantra, and the Third Panchen Lama is especially noted for having written one of the most important guidebooks to Shambala.

an astonishing palace is built

It was the Great Fifth Dalai Lama who began construction of a magnificent palace in Lhasa in 1645 upon the ruins of Tritse Marpo (Red Mountain), which had been the castle of King Songtsen Gampo, built in A.D. 636. It became known as the Potala, named for the pure land where the bodhisattva Avalokiteshvara resides.

Knowing that he would die soon, the Great Fifth Dalai Lama called Desi Sangye Gyatso, prime minister of the government, to his side to tell him to keep his death a secret and to give him instructions on how to govern the country once he had passed on. In 1682, at the age of sixty-eight, the Fifth Dalai Lama passed away, leaving the entire responsibility of temporal and spiritual administration in the hands of his prime minister.

An important part of the Dalai Lama's instructions dealt with the necessity of completing the Potala Palace. In order to accomplish this, Desi Sangye Gyatso had to keep the Dalai Lama's death secret lest the people stop working on the palace, as well as to avoid occupation of Tibet by the Mongolian and Manchu emperors.

The First Panchen Lama (1567-1662) was recognized by the Fifth Dalai Lama as the reincarnation of his teacher. Thangka painting, Tibet, 18th-19th century.

Overwhelmed by his new responsibilities, the young Desi despaired of fulfilling them without the Dalai Lama's direction, and he cried out in great grief at his ruler's death. Suddenly the Dalai Lama came back to life for a moment and said, "Don't worry. The

small things you can decide for yourself. Whenever there are important matters that you don't know how to handle, just stand before this image of (the protector deity) Penden Lhamo and ask her for direction."

Desi Sangye Gyatso told the people that the Great Fifth was in a prolonged meditation retreat. When high Mongolian officials, who

The Potala Palace in Lhasa, Tibet, photographed during the New Year's celebration in 1890. A great appliquéd banner is displayed down the front face of the palace. This first known photograph of Tibet was taken by a Buryat Mongolian with a camera brought from Russia.

had traveled a great distance, demanded to see him, one of the monks of Namgyal Monastery, who looked somewhat like the Dalai Lama, was placed on the throne to impersonate him.

In this manner, the Desi kept the Great Fifth Dalai Lama's death a secret from the people of Tibet for thirteen years, until the completion of the Potala Palace in 1695. Today a prayer that the young prime minister secretly carved into a stone palace wall still proclaims his wish for the Dalai Lama to be reincarnated.

the poet dalai lama

The Sixth Dalai Lama, Gyalwa Tsangyang Gyatso, who was born in 1682 in southern Tibet, was unique in his own way. He was the only Dalai Lama who chose not to follow monastic discipline, preferring sports and social life instead. Enthroned in 1697 at the age of fifteen, he refused to become a monk and moved out of the Potala when he was twenty years old. He was much loved by the people of Tibet for his lightheartedness, romantic poetry, and disregard for authority. But the Mongolian ruler Lhasang Khan invaded Lhasa in 1705, seized the young Dalai Lama and, according to one version of the story, murdered him in 1706. Tsangyang wrote to one of his lady friends as he was being forcibly taken away:

White crane!
Lend me your wings
I will not fly far
From Lithang, I shall return.

Lithang proved to be where the next incarnation was found.

Another version of the story says that the Sixth Dalai Lama, after being deposed by Lhasang Khan, sought refuge among the people, fading into anonymity, choosing life as a beggar, and making many pilgrimages to the holy places of Tibet, India, and China.

Gyalwa Tsangyang's unorthodox behavior with women came to be thought of as a tantric teaching of the greatest wisdom. It is said that even his love life followed advanced yogic practices. He wrote:

The goddess Penden Lhamo, protector deity of the Dalai Lamas and the Tibetan government. 6", cast silver inlaid with semiprecious stones, 17th century, Tibet.

Never have I slept without a sweetheart
Nor have I spent a single drop of sperm.

Lhasang Khan enthroned a puppet ruler in place of the Sixth Dalai Lama and kept Tibet under his rule by occupying it with his Quogshot Mongol troops. But in 1717 the Dzungar Mongols entered Lhasa and killed Lhasang Khan. The monks of Namgyal Monastery were either executed or sent away. Those who survived founded a new monastery.

the seventh dalai lama

Gyalwa Kalsang Gyatso, who was to become the Seventh Dalai Lama, was born in 1708 in Lithang, as the Sixth Dalai Lama predicted, and was identified soon thereafter. Troubles with Mongolia prevented his official recognition until 1720, when Lhasa became liberated from Mongolian rule by popular rebellions aided by the Manchurians, who hoped to extend their interests in central Asia.

Despite this background of political intrigue, the Seventh Dalai Lama lived the life of a pure and simple monk and was dearly loved by his people. He composed a new liturgy for the *tsok,* or feast offering, and is noted for introducing the Kalachakra Tantra to Namgyal Monastery. Realizing that the complete lineage of the Kalachakra Tantra propagated by Tsong Khapa was in danger of extinction, the Seventh Dalai Lama requested the aged and noted scholar Ngawang Chokden to collect and master it. Having done this through extensive retreat, Ngawang Chokden passed the endangered lineage on to the Seventh Dalai Lama, who in turn passed it on to thousands of monks.

When he was at last admitted back to the Potala in 1735, he reassembled the monks and reestablished Namgyal Monastery. Concerned with preventing any further degeneration of the tradition, the Seventh Dalai Lama appointed a group of monks to maintain the rituals of the Kalachakra Tantra—and they continue to be practiced to this day.

The Seventh Dalai Lama also instituted the practice among the Namgyal monks of performing a one-day Kalachakra prayer ritual

The Seventh Dalai Lama, who brought the Kalachakra Tantra into the lineage of the Dalai Lamas. This image, measuring 2³/8" high, is made of dark, reddish-brown clay, Tibet, 19th century.

on the eighth day of every Tibetan month. In addition, in the third month of the Tibetan year the monks perform the entire Kalachakra ritual in commemoration of the Buddha giving the Kalachakra. This includes the construction of the sand mandala and performance of ritual dances, as well as other rituals, including a concluding fire offering purification ceremony. The Kalachakra Earth and Offering dances were also taught to Namgyal monks by monks from Shalu Monastery in south central Tibet at the request of the Seventh Dalai Lama.

In 1751, Kalsang Gyatso regained full spiritual and temporal authority over Tibet, but he died just six years later at the age of forty-nine. In spite of the political turbulence of his time, the pious Seventh Dalai Lama's religious and scholarly achievements were extraordinary.

a quiet period

No major changes were made in the routines or traditions of Namgyal Monastery during the lives of the Eighth through Twelfth Dalai Lamas, whom the Namgyal monks continued to serve.

The Eighth Dalai Lama, Gyalwa Jamphel Gyatso, was born in central Tibet in 1758 and was brought to Lhasa in 1762. He exhibited spiritual qualities, along with a strong distaste for the political maneuverings going on around him. It was during his reign that the British Empire was making itself known in Asia, and Tibet's isolationist policy developed as a defense. This policy continued until the 1950s. In 1783, he built the Norbulingka (Jewel Park), the magnificent summer palace of the Dalai Lamas. He died in 1804 at the age of forty-six.

The Ninth through Twelfth Dalai Lamas—Gyalwa Lungtok Gyatso, Gyalwa Tsultrim Gyatso, Gyalwa Khedrup Gyatso, and Gyalwa Thinley Gyatso—each had a very short life, ranging from ten to twenty-one years. Whether they died from disease due to increased exposure to foreign contacts, natural causes, or as a result of political intrigue, the lack of effective leadership made the country weak and unstable. Many Tibetans believe that the degeneration of virtuous activities may have been responsible for the short and ineffective lives of these four Dalai Lamas.

a modern dalai lama

Gyalwa Thubten Gyatso, born in 1876 and enthroned as the Thirteenth Dalai Lama in 1879, was to become the strong spiritual and political leader Tibet needed in the changing times ahead. The coming of the 20th century brought to Tibet political pressure from its three great neighbors: Russia, China, and British India.

Around the turn of the century, the Thirteenth Dalai Lama became acquainted with a Buryat Mongolian monk named Agwan Dorjiev, who was not only a renowned scholar but also strongly connected to Czar Nicholas II of Russia. The Russians, like the British and the Chinese, were interested in Tibet for its strategic

The beloved Thirteenth Dalai Lama, Gyalwa Thubten Gyatso (1876-1933), who brought Tibet into the modern world. This photograph was taken in 1911 in Darjeeling, India.

location. Dorjiev tried to convince the Dalai Lama that the Romanov czar was the king of Shambala, and a book published in Russia in 1913 supported this theory. The Dalai Lama believed that Russia could be an important ally, and gave Dorjiev a large financial contribution with which he had the Kalachakra Temple built in St. Petersburg between 1909 and 1915, and he sent artistic treasures with which to decorate it as well. It functioned as a Tibetan Buddhist monastery until it was closed by Stalin's secret police in 1935. It was used as a radio-jamming station during World War II and as a zoological garden by the Communist Party until 1989, when it began to function as a monastery and Buddhist learning center once again. Today it is headed by a young Buryat monk named Samaydev, who journeyed to New York in 1991 to receive the Kalachakra Initiation from the Fourteenth Dalai Lama.

Concerned over Tibet's friendship with Russia, Great Britain sent an expeditionary force to Lhasa in 1904, which resulted in peaceful trade agreements. This prompted the Manchus ruling China to send an army in 1910, and the Dalai Lama fled to safety in India. The Manchus ruled Tibet for only about a year. China became engaged in a revolution, and by 1912 the Tibetans had managed to defeat and deport any remaining foreign soldiers. With the aid of the British, the Thirteenth Dalai Lama negotiated the Simla Agreement of 1914, which placed Tibet under the suzerainty of China, meaning that China would be responsible for its foreign affairs but would not have the right to colonize Tibet or maintain troops there. China never ratified the agreement.

The Thirteenth Dalai Lama was the first to have extensive contact with the West. He traveled for long periods of time and was deeply loved everywhere he went. He prophesied in 1932 that Tibet would soon be invaded from the East, and he urged his people to prepare themselves. He sought to modernize the country but met with considerable resistance from the established power structure.

The Kalachakra Temple in St. Petersburg, Russia, built in 1909 under the patronage of the Thirteenth Dalai Lama with the permission of Czar Nicholas II. Above the entrance are two deer facing a dharma wheel, signifying Deer Park, where the Buddha first taught.

the recognition of the fourteenth dalai lama

Strange cloud formations were noticed in the sky northeast of Lhasa some time after the Thirteenth Dalai Lama died in 1933. According to custom, his body had been seated in state on the throne in his summer residence, the Norbulingka, facing south. After a few days, the face had turned eastward. In addition, a large star-shaped fungus suddenly appeared on a wooden pillar on the northeastern side of the

The Norbulingka (Jewel Park) was the summer palace of the Dalai Lamas in Lhasa. It was built by the Eighth Dalai Lama in 1783. This photograph was taken during the Cutting Expedition in 1937.

shrine. The regent consulted the magical lake of visions, Lhamo Latso, where, after several days of prayer and meditation, he saw an image of a monastery with roofs of green and gold and a house with turquoise tiles. The regent also saw the Tibetan syllables *AH KA MA*, which were interpreted as clues to go to the Amdo region of eastern Tibet.

Portrait of the Fourteenth Dalai Lama at his installation on the Lion Throne in 1940. Oil on canvas, 34³/4"x25", by Indian painter Kanwal Krishna.

Parties set out in various directions throughout Tibet to search for a very special child. The party that traveled to Takster in eastern Tibet arrived without fanfare at a farmhouse with turquoise tiles, and the two-year-old baby of the family went immediately to the high abbot, who was disguised as a servant, and sat on his lap.

The child pulled at the abbot's rosary, which had belonged to the Thirteenth Dalai Lama, and said he wanted it. The abbot said he'd give it to him if the child could tell him who he was. The boy called him Sera Aga, which means "a lama from Sera Monastery." He also correctly identified the two others in the party. When they left, he wanted to go with them, and afterward he would often tell his mother he was packing to go to Lhasa. When the delegation returned, they showed the child a series of articles, some of which belonged to the Thirteenth Dalai Lama, including two rosaries, a small hand drum, and a walking stick. Each time, the boy chose the article belonging to the Dalai Lama. He did hesitate over the wrong walking stick, but it turned out that this stick had indeed belonged briefly to the Dalai Lama as well. Further testing proved remarkably successful, and the little boy was officially recognized as the Fourteenth Dalai Lama.

The only obstacle that remained was the warlord who ruled western China at the time. When he found out the new Dalai Lama had been discovered in Amdo, he demanded an enormous ransom to let the child be taken to Lhasa. After the ransom was paid, the warlord demanded even more money and valuable objects, and again the Tibetans were forced to comply.

The present Dalai Lama, who was born Lhamo Dhondup on July 6, 1935, with his eyes wide open, was finally brought to Lhasa to be enthroned in 1940. When he was given his novice vows at the Central Cathedral, he was renamed Tenzin Gyatso, which means "Ocean Which Protects the Dharma." He gave his first teaching in 1946 at the age of eleven.

china invades tibet

The signs were ominous: warnings from the State Oracle, a comet with a bright tail (signifying war), and a tremendous earthquake that rocked southeastern Tibet, changing the course of the Brahmaputra River, which swallowed up hundreds of villages in the province of Kham. The sky glowed an unholy red.

Then in 1950 the newly formed People's Republic of China announced its plans to "liberate" Tibet. The Tibetan government, which was led by a regent due to the Dalai Lama's youth, knew that

Gyalwa Tenzin Gyatso, the Fourteenth Dalai Lama, took over the Tibetan government in 1950 at the age of fifteen.

The Dalai Lama greeted by Mao Tse-tung during a visit to Beijing in 1954.

Ling Rinpoche, the senior tutor to the Fourteenth Dalai Lama, gave him the Kalachakra Initiation in 1955. Here he holds a vajra and bell at the Kalachakra Initiation in Bodhgaya, 1973.

its limited military resources would be no match for the Red Army. Delegations were sent to Britain, India, Nepal, and the United States to appeal for support, to no avail. Chinese troops moved into the eastern provinces of Amdo and Kham, overwhelming the Tibetan army in spite of the able assistance of the renowned Khampa warriors. The United Nations refused to take up the question of Tibetan independence. The long years of isolation were taking their toll.

The Tibetan government consulted the State Oracle, whose message was clear: the Dalai Lama must take charge. Although he was only fifteen years old, he was the only leader who could unite the country during this time of crisis. On November 17, 1950, Tenzin Gyatso was granted full authority in a traditional ceremony at the Potala.

The Dalai Lama sent a delegation of officials to Beijing, where they arrived early in 1951. Zhou Enlai, the foreign minister of China, forced them to sign the Seventeen-Point Agreement agreeing that Tibet was part of China, under threat of both personal violence and large-scale retaliation against Tibet.

Chinese troops marched into Lhasa, taking over houses, fields, and animals, and demanding large quantities of food, soon overtaxing the fragile economy. Rumors of violent oppression were pouring in from the eastern provinces.

The Dalai Lama felt that the only hope for his people against such a powerful enemy was to persuade the Chinese peaceably to fulfill the promises they had made in the Seventeen-Point Agreement, such as not to interfere with the Tibetan political system or religious practices and to aid in agricultural development. He committed himself to the Buddhist path of nonviolence, advocating cooperation when it was possible and passive resistance when it was not. And he accepted an invitation to visit Mao Tse-tung in China in 1954.

The Dalai Lama returned to Lhasa with Mao's plan for a committee to prepare for regional autonomy, but soon realized the committee would be a mere puppet of the Chinese. The situation in the eastern border provinces was becoming desperate, and by mid-1957 the Khampas were in full-scale revolt. Chinese Communist troops were obliterating entire villages, publicly executing hundreds of people by the most brutal means imaginable, and destroying monasteries and their inhabitants. Refugees poured into Lhasa.

In spite of the political turmoil, the young Dalai Lama was already dedicated to the Kalachakra teachings. He gave the Kalachakra Initiation in Lhasa for the first time in 1956, and again in 1957, to more than 10,000 people. As was traditional, forty-four Namgyal monks assisted him with the rituals of the empowerment. In 1959, the Dalai Lama took his examinations upon completion of the highest level of religious training, undergoing intense day-long philosophical debates before thousands of onlookers in order to receive his degree.

Associated Press report from 1959: "FEARFUL JOURNEY: Pursued by Red Chinese troops, struggling against the harsh elements of the Himalayas, the God-King of Tibet—the Dalai Lama—is shown here on the fourth day of his flight to freedom. The 23-year-old ruler, wearing spectacles, is aboard the white horse. At this point, the escape party is crossing the Zsagola Pass in Southern Tibet on March 21st—four days after the Dalai Lama fled Lhasa, the sacred city, dressed in the drab robes of an ordinary monk."

A Chinese general invited the Dalai Lama to attend a theatrical presentation on March 10, insisting that he come without any government officials or armed escort. The rumor spread throughout Lhasa that the Chinese were planning to kidnap the beloved Tibetan leader. By the morning of March 10, 30,000 Tibetans had gathered in front of the gates of the Norbulingka to keep him from going. The general exchanged several letters with the Dalai Lama, finally asking where in the palace he would be, so that he would not be harmed. This confirmed popular fears that the palace would be shelled, and on March 17 the first shots were fired.

The Dalai Lama decided he must leave Tibet. Dressed in unfamiliar soldier's clothes, and without the glasses by which he would be immediately recognized, he left the Norbulingka preceded by his family and followed by a small party of ministers and attendants. Thus began a long and dangerous journey across Tibet and the Himalayas to asylum in India.

The Chinese occupation of Tibet resulted in the deaths of a reported 1.2 million Tibetans from military activities, starvation, torture, and the hard labor inflicted during long-term prison sentences. Six thousand monasteries were systematically destroyed, with their sacred art sent to Beijing, destroyed, or desecrated. Some of the Chinese policies which continue to oppress Tibetans into the 1990s are the dumping of nuclear waste in the eastern provinces, significant deforestation, forced population control, and a massive transfer of the Chinese population into Tibet.

The Dalai Lama has remained in exile since 1959 in an attempt to keep the Tibetan civilization alive and to maintain a sense of Tibetan identity for those Tibetans born outside of their native land. He has said repeatedly, "It's best for me to stay outside Tibet where I can speak on behalf of the six million people inside Tibet." The government of an independent Tibet continues to operate from Dharamsala, India, supervising an expatriate population of more than 150,000 in India and Nepal, with other Tibetan settlements in Switzerland, Canada, and the United States. All of the principal cultural institutions have been rebuilt in India, including the medical college and numerous monasteries, where texts are being painstakingly reproduced and art forms relearned. A prime example has been in the work of Namgyal Monastery.

the reestablishment of namgyal monastery in exile

Fifty-two Namgyal monks managed to escape to India with nothing, not even their sacred texts or begging bowls. The Dalai Lama encouraged them to preserve the monastery's traditions, which they did by writing down texts from memory. They found themselves in a land where the climate was quite foreign, where they did not speak the language or know the customs, and where most of their time was taken up by menial work on road construction crews, which was the only livelihood available to them.

The Dalai Lama was eventually able to bring all the exiled Namgyal monks together again in 1961 at his newly established residence in Dharamsala, India. In Tibet, there were as many as sixty-five Namgyal monks trained in the rituals of Kalachakra, and the annual routine of the monks included rituals that were performed in conjunction with the lunar phases, the third month of the lunar year being devoted to Kalachakra. The traditional sacred practices were gradually reestablished, and by 1963 the Namgyal monks were once again able to perform the Kalachakra rituals.

Also in 1963, an abbot was appointed by the Dalai Lama to be in charge of Namgyal Monastery for the first time. His name was Samten Chophel. In 1969, the Tibetan government-in-exile began to build the Central Cathedral of Thekchen Choling in Dharamsala, and twenty-eight boys were admitted to Namgyal Monastery as student monks.

Innovations were made in the course of study as well. The Dalai Lama introduced the teaching of philosophy, which was added to the regular studies in mandala construction, sacred dance, and other rituals. The monks' attire was simplified to a standardized robe of burgundy and yellow cotton or wool, doing away with the silk brocades often worn in Tibet.

Thekchen Choling, the Central Cathedral in Dharamsala, India, decorated for the Dalai Lama's return from Oslo, Norway, where he received the Nobel Peace Prize in 1989.

kalachakra beyond tibet

In 1971, in addition to many other Buddhist teachings and cere-monies, the Dalai Lama conferred the Kalachakra Initiation—for the first time in exile—on the Dharamsala community at the newly completed Central Cathedral. Two thousand people attended.

The night the empowerment was completed, the Dalai Lama dreamed he was in the center of the mandala, with the assisting monks reciting certain verses from the principal Kalachakra text, when the mandala dissolved into and remained in a clear-light state. Speaking of the dream, he said, "When I woke up, I knew that in the future I would perform this ritual many times. I think in my previous lifetimes I had a connection with the Kalachakra teaching. It's a karmic force."

In 1972, he gave the initiation in south India, at the Bylakuppe Tibetan settlement. He performed it again at the end of 1973 in Bodhgaya, India. This time, 50,000 people attended, of whom 1,000 were Westerners. The Kalachakra Initiation was again given by the Dalai Lama in 1976 at Leh in Ladakh, India.

In 1980, when the Chinese opened up the borders of Tibet for travel, a number of the former monks of Namgyal Monastery who were still in Tibet were invited to visit Dharamsala at the monastery's expense. They participated in special prayer ceremonies and some remained.

In July 1981 the Dalai Lama traveled to America to confer the first Kalachakra Initiation in the West, in a cornfield outside of Madison, Wisconsin, on 1,200 people, including many Tibetans now living in North America. Sixteen Namgyal monks also jour-neyed from Dharamsala to Madison to assist in the many different aspects of performing the Kalachakra Initiation ritual. In the Dalai Lama's effort to simplify the rituals since coming into exile, he has reduced the standard number of ritual assistants for the Kalachakra Initiation from forty-four to sixteen.

In 1983 the Dalai Lama gave the Kalachakra Initiation twice in India. In 1985 he gave it in Switzerland, and then in Bodhgaya for

150,000 people in December. This included more than 100,000 Tibetans living in exile and 50,000 who had journeyed from Tibet. In July 1988 he gave the initiation in Zan-skar, which is in Ladakh, India. At the same time, four Namgyal monks constructed, for the first time in history, the Kalachakra Sand Mandala as a cultural offering at the American Museum of Natural History in New York City.

The Dalai Lama returned to the United States in July 1989, and gave the Kalachakra Initiation to 3,000 people at the Santa Monica Civic Auditorium overlooking the Pacific Ocean. At the same time, four more Namgyal monks created the Kalachakra Sand Mandala at the Natural History Museum of Los Angeles County, again as a cultural offering.

Since 1978, the Kalachakra Initiation has been given throughout the world by a number of eminent lamas in exile who represent the various traditions of Tibetan Buddhism. Among them are Sakya Trizin Rinpoche, the present head of the Sakya lineage, the beloved Kalu Rinpoche, Chutgye Trichen Rinpoche, and Jamgon Kongtrul Rinpoche.

They and many other great Tibetan lamas of modern times, including the Dalai Lama himself, encourage the nonsectarian dissemination of the teachings of the Buddha. The Kalachakra Initiation, because it is often given to large groups, serves as an excellent vehicle for sharing the path of compassion with people of all backgrounds.

the international year of tibet

In 1991, as part of the International Year of Tibet, the Dalai Lama, assisted by the monks of Namgyal Monastery, conferred the Kalachakra Initiation on 4,000 students in the newly renovated Paramount Theater at Madison Square Garden in New York City, while another group of Namgyal monks demonstrated the Kalachakra Sand Mandala as part of the *Wisdom and Compassion* exhibition at the IBM Gallery.

The Kalachakra Sand Mandala was also demonstrated in conjunction with the *Wisdom and Compassion* show at the Museum of Asian

Kalu Rinpoche (1905-1989), renowned meditation master of the Shangba Kagyu lineage, conferred the Kalachakra Initiation to many thousands of students in North America, Europe, and India. His profound expression of the teachings of the Buddha and his ever-present compassion attracted students from all four schools of Tibetan Buddhism. He established numerous meditation and retreat centers throughout the world and taught at Namgyal Monastery at the request of the Dalai Lama.

Art in San Francisco and other cultural institutions across America, including the Mingei International Museum of World Folk Art at La Jolla, California; the Field Museum of Natural History in Chicago, Illinois; the Buffalo Museum of Science in New York State; and the Royal Academy in London, England. The Namgyal monks demonstrated other mandalas at the Newark Museum in New Jersey; Windstar Foundation in Aspen, Colorado; the Herbert F. Johnson Museum in Ithaca, New York; the St. Louis Art Museum in Missouri; and the Virginia Museum of Fine Art in Richmond, Virginia. They have also constructed mandalas in art galleries at Kutztown University and Dickenson College, and at Samaya Foundation and the Open Center in New York City.

Many people believe that the destruction of Tibet and the spread of Buddhism to the West were prophesied by the great Padma Sambhava in the 8th century, who said, "When the iron bird flies and horses run on wheels, the dharma will be carried to the land of the Red Man." And, in fact, there is an ever-growing interest in Buddhism in the West, evidenced by the growing number of Tibetan studies programs in major universities and the many Tibetan Buddhist meditation centers and monasteries that have emerged in recent years, as well as impressive museum collections of Tibetan art.

In 1992, Namgyal Monastery established a branch in Ithaca, New York, where the monastery's religious practices and sacred arts will be continued in conjunction with a new Institute of Tibetan Buddhist Studies. The institute, open to men and women from around the world, will provide an opportunity for the systematic education in English of Tibetan Buddhist studies in a traditional monastic setting.

religious activities of namgyal monastery

The Namgyal monks' main responsibility is to assist the Dalai Lama in his spiritual activities and to perform ritual prayers for the Tibetan government. Daily, the monks perform prayers of Penden Lhamo, the main protector deity of the government as well as of the

His Holiness the Dalai Lama at the sunrise meditation celebrated in Central Park, New York City, October 1991.

left: Appliqued scroll painting of the diety Kalachakra with his consort Vishvamata. above: Ritual dance protecting and consecrating the site. Performed during the Kalachakra Initiation outside Madison, Wisconsin, 1981.

above: Western Tibetan Buddhist monks lead a Long Life prayer for the Dalai Lama following the Kalachakra Initiation at Bodhgaya, India, 1973. left: Kalachakra Initiation at Bodhgaya, India, 1973.

above: 1,200 North Americans gather in a corn field outside Madison, Wisconsin to receive the Kalachakra Initiation in 1981.

right: Mountain people of Zanskar in northern India await the arrival of the Dalai Lama for the Kalachakra Initiation in 1989.

Sue Byrne

Vajra Vega, the wrathful emanation of Kalachakra.

monastery. In addition, they perform elaborate ritual ceremonies for one to two weeks out of every month for different deities and protectors, to request prosperity for the Tibetan government and nation.

A high point of the year at Namgyal is the New Year, which usually occurs during the Western month of February or March. After two weeks of preparatory ritual offerings, the Tibetan New Year celebration begins at 3:00 A.M. with special prayers and meditations in a private ceremony on the roof of the Central Cathedral in Dharamsala, with the Dalai Lama presiding. Later that morning, the Namgyal monks offer long-life prayers to the Dalai Lama, and approximately 500 government staffers join the ceremony to receive blessings from him. On the second day of the ceremony, as many as 10,000 people come to receive individual blessings from the Dalai Lama.

the course of study at namgyal monastery

Boys are admitted for entry into the monastery between the ages of ten and seventeen. The youngest and least educated among them are first schooled in reading and writing. To be admitted as trainees, they must be well behaved and able to learn to recite ten pages of religious text from memory in one month. The main study program includes memorization of ritual texts and philosophical dialectics, in addition to Tibetan and English grammar.

The Dalai Lama blessing children during the Tibetan New Year celebration near his residence in Dharamsala, India, 1985.

The memorization takes three to seven years, depending on individual ability. The students are tested upon completion of each of the ritual texts. Debating philosophical dialectics is learned in an ongoing process, with an exam at the end of each year. Once they have completed the memorization program, the young students begin to practice the ritual arts, becoming members of the monastery and advancing to the level of junior monk.

The junior monks are entitled to join in the daily prayers of the monastery as part of their training. At the completion of the thirteen-year program, the monks receive a Master of Sutra and Tantra degree.

the dalai lama in today's world

Philosophical debate at Namgyal Monastery is an important part of the curriculum and often has moments of great humor.

Inspired by the leadership of the Fourteenth Dalai Lama, Tibetans in exile have worked tirelessly with the growing support of the international community to stop the devastation of Tibet and to free their homeland. The Dalai Lama made his first visit to the West in 1973, with an eight-country tour of western Europe, including a special

stop in Switzerland where he officiated at several Buddhist ceremonies for the 2,000 Tibetans living in Rikon, outside Zurich.

In 1979, the Dalai Lama visited the United States and the American Buddhist community for the first time. Subsequently, he has traveled widely throughout the world, to Russia, Mongolia, Japan, Central and South America, Mexico, Canada, and throughout Europe and the United States several times. He presented his Five-Point Peace Plan in 1987 before the Human Rights Caucus of the United States House of Representatives and elaborated on that plan before the European Parliament in Strasbourg shortly thereafter.

In the face of China's continuing campaign to discredit his efforts, he has received countless awards—including the Albert Schweitzer Award, the World Management Council Award, the Raoul Wallenberg Award, and the 1989 Nobel Peace Prize—in recognition of his unremittingly nonviolent struggle for Tibetan independence, as well as his humanitarian effort on behalf of both inner peace and global survival. In spite of his protest that he is "only a simple Buddhist monk," Gyalwa Tenzin Gyatso has come to be recognized as that rare kind of leader who can inspire and change the world.

Tenzin Gyatso, the Fourteenth Dalai Lama,
receives the 1989 Nobel Peace Prize
in Oslo, Norway.

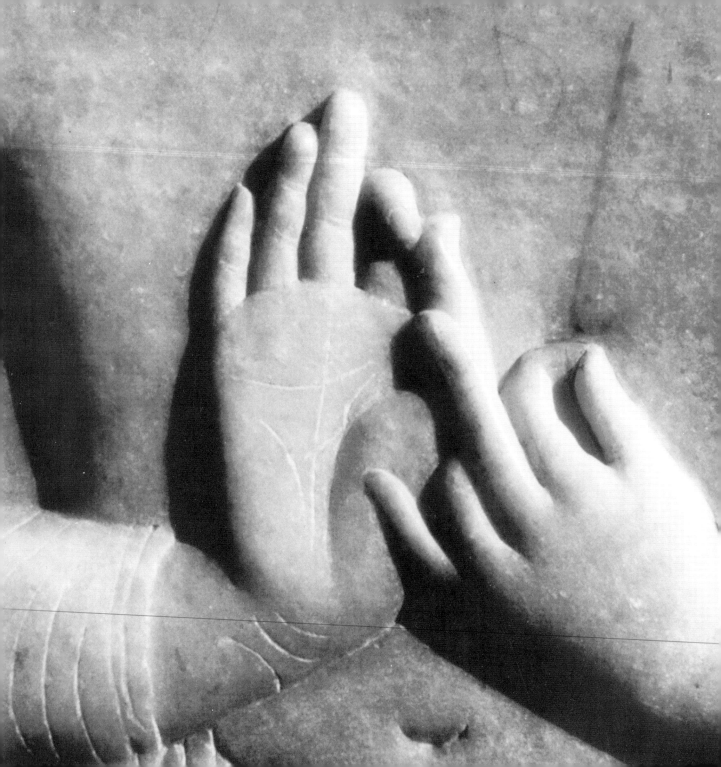

Tibetan

I vow to work for the benefit of all sentient

Buddhist

beings until the last one is enlightened.

Philosophy

The Bodhisattva Vow

the seed of compassion

IT IS SAID at the beginning of every Tibetan Buddhist text, at the start of every Vajrayana Buddhist initiation, by teachers both ancient and contemporary, that in order to practice and make progress on the path, the most important thing is to develop *bodhicitta,* the mind of compassion. This is the state of mind that wishes to attain enlightenment for the purpose of liberating all sentient beings from the cycle of suffering—birth, old age, sickness, death, and rebirth.

The hands of the Buddha teaching the dharma. Detail of life-sized statue of polished stone at Bodhgaya, India.

The word *bodhi* means "awake" and the word *citta* means "mind." Thus the word *bodhicitta* has been translated as "enlightened attitude," "awakened mind," "awakened heart," and, most commonly, "mind of compassion." It is interesting to note that the English word *compassion* means "feeling with." This sense of complete one-

Bodhisattva Avalokiteshvara. The eleven-headed, one thousand-armed form of the Bodhisattva of Compassion. 29 1/2"x20", silk and silver metal cord appliqué on silk, China, circa 1780.

ness with others is precisely what is at the core of the mind that is awake.

Another word that is relevant here is *bodhisattva,* or "awakened knowledge." A bodhisattva is a being who has attained enlightenment and has taken a vow to work for the benefit of all sentient beings. Moses, Jesus, and Mohammed are acknowledged by some Buddhists to be bodhisattvas—beings who have achieved liberation from the Wheel of Life but have chosen to be reborn in order to help others. Mahatma Gandhi, Martin Luther King, and Mother Teresa are recent examples. In addition, sincere Buddhist practitioners are considered to be aspiring bodhisattvas, on the path to realizing the state of perfection, or selflessness. The vow of the bodhisattva is renewed again and again.

To prepare students for the Kalachakra Initiation, the ritual master begins by giving teachings on developing compassion. This is regarded not only as a necessary foundation for receiving the empowerment, but also the essential requirement for attaining the enlightened state of mind. The Dalai Lama offers these preparatory teachings over a period of four days, before the students enter the initiation site.

The spirit of compassion is symbolized by Avalokiteshvara (known as Chenrezig in Tibet, Kuan Yin in China, and Kwannon in Japan), who is one of the best known and most beloved bodhisattvas of the Buddhist pantheon. Like all bodhisattvas, he can be visualized in colorful detail but is at the same time understood to be a symbol for an essential quality of Buddha mind.

According to legend, Avalokiteshvara vowed to free all beings from the cycle of suffering before his own liberation. The enormity of this task caused him to emanate in a form with one thousand arms reaching out in limitless compassion.

a question of motivation

Tibetan Buddhists examine and purify their intentions before every practice so that they may be fit vessels to receive the empowerment or instruction. On a daily basis, practitioners attune their motivation to the Buddha ideal of conducting all activities in a spirit of altruism.

The purpose of Tibetan Buddhist practice is not only to gain enlightenment in order to rise above earthly struggles. Practitioners must also have the sincere desire to alleviate the suffering of all sentient beings, placing others' well-being above their own just as a mother would that of her child.

Westerners, raised from birth on principles of individuality and independence, on values of personal achievement and material success, may find this very fundamental precept of compassionate motivation to be an alien concept. We may agree intellectually with the principle, which is in accord with the Judeo-Christian ethical tradition ("Love thy neighbor as thyself") and other spiritual practices. But the reality is that most of us, when we examine ourselves earnestly, find that we act primarily from self-interest.

How, then, can this most basic teaching of the Mahayana Buddhist path be considered to be even remotely attainable, as something relevant for each of us and not just for saints and spiritual devotees? The answer was given by the Buddha when he said that all beings possess the seed of enlightenment, and if that seed is properly cultivated it is only a matter of time before it will ripen. The path of the Buddha is intended to hasten that development.

The Buddha speaks of extremes as a cause of suffering, and therefore Buddhism is called the Middle Path, or Middle Way. Equanimity and moderation set the course by which to steer.

Ironically, some see the Buddhist ideal itself as an extreme. How can I work for the benefit of others, one might think, when it's all I can do to take care of myself? But working for the benefit of others is actually a path to personal satisfaction. If we have a genuine attitude of loving kindness and generosity toward others, people respond to us and we are rewarded in many ways. This is what the Dalai Lama calls being "wise selfish."

By taking the attention away from ourselves and thinking of others, we expand our vision of what the world holds for us, which benefits us and those around us.

Shakyamuni Buddha seated in his teaching posture. 24³/₄" high stone carving, Pala period, India, late 10th century.

the core of the teaching

The subject of altruistic motivation, or bodhicitta, brings us to the fundamental precepts of Mahayana Buddhism. The Buddha's first teachings after he achieved enlightenment were the Four Noble Truths, and the first of these is a recognition of the existence of suffering. This is often a stumbling block to people who are curious about Buddhism. They ask, quite understandably, "What relevance does this have to my life? If I'm happy, why do I need this? If I'm unhappy, how can this benefit me?"

However, the Buddha's observation cuts right to the root of the human condition: No matter what our station in life, no matter how favored we may be with material comfort or loving family and friends, no matter how powerful we are, we all grow sick, we all grow old, and every one of us will die. In addition, there are all the smaller, psychological sufferings of every moment—the frustration of not being able to do what we think we want, the anger of experiencing what we don't like, the envy of desiring what we don't have, the fear of losing what we do have.

The fact is, there isn't one of us who doesn't long for happiness, yet we are all constantly laying the foundations of unrest through the habitual conditioning of our own minds. When we realize that we are in a seemingly endless cycle of rising and falling emotions which offers no sense of fulfillment or lasting happiness, many of us look for a remedy.

Seeking another way to live, we listen to the words of the Buddha, and he begins with the subject of suffering. At first, this seems a rather bleak confirmation of our own painful experience. However, once we recognize that what we thought was a normal condition is in fact suffering, we begin to understand that there may be a remedy.

The Buddha explained the cause of suffering (the Second Noble Truth) to be attachment (desire) and aversion. Both causes of suffering are rooted in our belief in a self that exists apart from others, in the concepts of "me" and "mine." But underlying that is our fundamental ignorance of the nature of reality. Taking everything at face value and naively presuming that "I" am the most important object in "my" world, we blind ourselves to the truth that everything is interdependent. "My" very existence hangs upon a multitude of fortunate occurrences and could be extinguished at any time as easily as a candle flame in the wind.

Furthermore, what exactly is this "I" that is mistakenly conceived to be so permanent? Is the "I" that is reading this sentence, possibly relaxed and focused, the same "I" that is so upset when wrongly accused or threatened with physical danger? And does the enemy who wrongs me exist independently, or do I perceive him to be my enemy by virtue of some words or action which I interpreted as malicious? Were my enemy to speak kindly and offer assistance, would I not then call him my friend? This "I" that I am aware of "in here" is merely an illusion, an entity fabricated by the mind out of what is in fact a never-ending flow of consciousness. It is our self-centered attitude, at the root of which is our ignorance, that causes these reactions of liking/attachment versus not-liking/aversion.

understanding emptiness

When Buddhists speak about "emptiness," they do not mean the void. Emptiness is what is experienced in the absence of concepts, emotions, and other obstacles that prevent us from realizing the essential, pure nature of mind.

The Buddhist concept of emptiness is extremely difficult to translate into Western thought, because our philosophies have no comparable concept. Emptiness is not produced by the mind. One of the reasons we have so much trouble understanding it is its nondualistic nature; that is, it is beyond the references that we use to comprehend the world.

Buddhism encourages the cultivation of "mindful" activity, as a means to develop awareness of our actions. We are thus able to penetrate the ordinary mind, or what Buddhists call the coarse levels of consciousness, and begin to experience the subtle mind. Only the intuitive or subtle mind is capable of perceiving emptiness.

the elimination of suffering

This takes us right back to the Buddha's fundamental insight into the existence of suffering and its causes. Attachment and aversion, which are manifestations of the ordinary or coarse mind, stem from a distortion of our basic need for survival and result in our possessive drive to satisfy our desires. In daily life, we are continually grasping and clinging as a result of this conditioned reflex to possess. As this grows unchecked, what results is a cumulative effect in which we experience desire as a fixed, concrete condition of life. This perpetuates the basic delusion that getting what we want will make us happy; in reality, it only results in our wanting something else. While we are in a state of desire, reaching for something outside ourselves, we cannot find happiness. As long as we grasp at life to make it conform to our ideas of what it should be, and as long as we reject whatever we deem at the moment to be painful or unpleasant, we will always be in a state of suffering. We create our own suffering by the way we choose, consciously or unconsciously, to perceive the moment.

Our responses of attachment and aversion are deeply conditioned within us and, according to the Buddhist view, even continue through many lifetimes. So unless and until we "wake up" from our delusion, we are compelled to repeat our suffering throughout an endless cycle of rebirths.

All Buddhist disciplines are aimed at eliminating suffering. Often, we do not see how our emotions contribute to our suffering. Emotions are a construct of our own minds, which we alone are responsible for creating and for changing. The primary afflictive emotions, known to Buddhists as "the Three Poisons," are ignorance, greed, and hatred. They superimpose judgments of good and bad—judgments that exist entirely in our minds—and prevent us from experiencing the true nature of reality.

What we perceive as a problem often contains hidden opportunities. The Vajrayana path actually teaches us to use our disturbing emotions as a catalyst for transformation. As we actively replace our unwholesome mental activity with consciously chosen enlightened

responses, our old habits are naturally and spontaneously transformed. This is a way out, using suffering as a path to freedom.

the importance of practice

To understand the importance of setting out upon the path of liberation, practitioners reflect deeply upon the four fundamental meditations that turn the mind toward the dharma, or the teachings of the Buddha. These are (1) the preciousness of a human birth and the difficulty of attaining it; (2) the impermanence of all things; (3) the inexorability of karma; and (4) the pervasiveness of suffering in cyclic existence (life, death, and rebirth), or *samsara*.

Imagine the world as one large ocean with a single, doughnut-shaped, wooden yoke floating on its surface, blown about by the winds. Now suppose that a blind sea turtle were to poke its head out of the water once every hundred years in an attempt to find the center of that ring. While that may seem impossible, it is suggested that to obtain a human birth in the vast sea of beings is even more difficult.

According to Buddhist thought, there are six realms of existence. For instance, there is the hell realm, where conditions are so miserable that all one can think about is one's pain; there is the realm of hungry ghosts, in which beings hunger and thirst incessantly; and there is the animal realm, where stupidity and ignorance reign. Only human beings are in a position to realize the truth of the dharma and to choose to act on it.

Therefore a human lifetime is regarded as being precious indeed. Death may occur at any time; stability and certainty are mere illusions, so there is not a moment to waste. We must practice now.

The second meditation that turns one's thoughts toward the dharma is the reality of impermanence. This is the fact that nothing is static: everything is constantly in flux, being born and dying, even this very moment. Everything that seems so solid—monuments, empires, the earth itself—is in a continuous process of growth, decay, and dissolution. One season changes into the next; the body that once seemed so strong and healthy becomes weakened and eventually dies.

Opposite: *The Wheel of Transmigration.*
The center circle illustrates "the three poisons,"
represented by a cock (greed), a pig
(ignorance), and a snake (hatred). They are
surrounded by the six realms of existence,
which are in turn surrounded by
the perpetual cycle of life, or *samsara.*
43" thangka, Tibet,
19th century.

The third thought to contemplate is inexorable karma—the accumulation of our actions as governed by the law of cause and effect—which determines our destiny. Our actions, both positive and negative, accumulate in our karmic "account" throughout every second of our lives. No matter how insignificant they may appear, these actions affect not only the present but the future. We are, in fact, the summation of our collective karmic experience, all of which comes to bear on each moment. If we pay attention to our daily actions of body, speech, and mind, we can improve our karma in our present life and for our future lives.

The final subject for meditation that leads us to see the wisdom of practicing the Buddhist path is that the suffering of samsaric, or cyclic, existence is everywhere and it is unending. Whatever is left undone at the time of our death, whatever cravings or resistances were not recognized as illusion, whatever debts were owed in the course of harming others—all these things will coalesce once more into yet another lifetime of ignorance, greed, hatred, and continued suffering.

training the mind

The Buddha was recognized as the "Great Physician" not only because he was explicit in his diagnosis of the disease of suffering, but also because of his presentation of the Third and Fourth Noble Truths. The good news is that there is a cessation to the endless round, and the antidote, like the disease itself, resides in the mind.

Just as samsara arises from ignorance, so enlightenment is born of awareness. And this awareness is developed through hearing the teachings, contemplating them, and putting them into practice. Thus we can purify our minds of the "defilements" that keep us from perceiving the true nature of all beings and things. This purification allows the "pristine consciousness" of the enlightened state to enter into our lives, like sunlight shining through a newly washed window.

In the Kalachakra Tantra it says that "sentient beings are Buddhas, defiled by adventitious stains. When those are removed, all beings are Buddhas." Elsewhere in Buddhism, it is taught that the true nature of the mind is as clear as the sky. The stains or defile-

ments are not intrinsic to our nature but are collected through many lifetimes of ignorance and unwholesome actions. They are like clouds which, when removed, leave only the clear sky of our essential Buddha nature.

Monks seated beneath the bodhi tree during the 1973 Kalachakra Initiation at Bodhgaya, India. They are listening to a teaching by Ling Rinpoche, the senior tutor to the Dalai Lama.

To remind us that it is really quite possible (though not easy) to accomplish this, the Dalai Lama has said, "We human beings have a developed brain and limitless potential. Since even wild animals can gradually be trained with patience, the human mind also can gradually be trained, step by step. If you test these practices with patience, you can come to know this through your own experience. If someone who easily gets angry tries to control his or her anger, in time it can be brought under control. The same is true for a very selfish person: first that person must realize the faults of a selfish motivation and the benefit in being less selfish. Having realized this, one trains in it, trying to control the bad side and develop the good. As time goes by, such practice can be very effective. This is the only alternative."

one human family

We must recognize the necessity for a compassionate attitude. The Dalai Lama explains that each of us is essentially the same human being, made up of the same basic flesh, blood, and bone. Our differences of language, culture, social status, and gender are merely superficial and secondary. The most important thing we have in common is our desire for happiness, and we each have an equal right to its pursuit.

We are joined together in the same family, on the same planet, so it is necessary that each of us be a good and loving member of that family. Everyone, even the most cruel, appreciates being shown kindness and consideration. This is intrinsic to human nature, just as infants need to be held and touched with love in order to grow strong and survive. Thus we can all understand the basic human need for compassion.

Furthermore, we can recognize that we always feel friendlier toward someone who shows good will toward us, and we naturally feel uneasy when shown hostility, anger, or jealousy. Since we are in essence so much like one another, our own experience can tell us what everyone else must feel. The simple fact is that without a

The Dalai Lama teaches on compassion to prepare the students to receive the Kalachakra Initiation in Rikon, Switzerland, 1985.

newly affirmed sense of responsibility and cooperation toward one another, our very survival as a species is threatened.

realizing inner peace

The conclusion is unavoidable: The most important thing in life, the most primary and basic, is the practice of kindness and honesty. If we have the right attitude, we can face even tragedy with the inner peace that comes from love and trust in others, whereas a negative attitude will bring us only unhappiness.

What the Dalai Lama calls "universal responsibility" comes from the courage and self-confidence that replace fear and insecurity. Forgiveness, tolerance, and patience are evidence of true inner strength, and we can practice these states of mind consciously. Selfishness, on the other hand, is the result of ignorance and can only lead to suffering. The Dalai Lama teaches us that the cultivation of our positive human qualities takes constant effort, time, and mental determination. The need is urgent—more urgent than ever. And we must never give up hope.

The Tibetan Buddhist path is in no way passive. It is very much an active process of continually purifying the activities of one's body, speech, and mind, by reciting prayers and performing various ritual practices and meditations daily.

The accumulated thoughts, habits, and behavior patterns of this and former lifetimes are transformed by the continuous application of mental discipline. The effect of practicing altruism is a calming of the mind, which allows one to experience its finer, subtle levels. Eventually, this altruistic intention becomes a way of life from which love, compassion, and wisdom arise spontaneously. Through such continued and prolonged practice, the atomic structure of one's very being is altered.

The practice of Buddhism leads us beyond suffering by turning our attention away from our self-absorption and toward the ultimate truth.

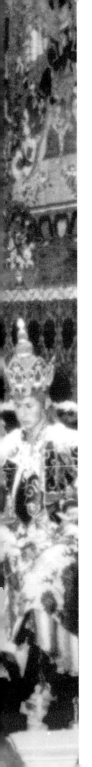

The Kalachakra Initiation

THIS CHAPTER is a general introduction to the Kalachakra Initiation based primarily on *The Mandala Rite of Glorious Kalachakra,* written by Khe-Drub in the 16th century. For the practitioner, it will be useful in understanding the proceedings. For the reader who is unfamiliar with tantric Buddhism, it is intended to introduce a magnificent ritual that has remained a secret for centuries, until now. An attempt has been made to simplify and clarify this extremely complex, esoteric initiation, which has resulted in the omission of many details. Also, certain descriptions have been abbreviated. The reader wishing a complete

Monks perform a celebratory dance upon the completion of the Kalachakra Sand Mandala on the eighth day of the Kalachakra Initiation at Madison Square Garden, New York City, 1991.

description of the Kalachakra Initiation is referred to the Dalai Lama's book, *The Kalachakra Tantra,* translated by Jeffrey Hopkins, Ph.D. (London: Wisdom Publications, 1985).

Every step of the initiation process and every gesture of the ritual master is intended to help develop an attitude of compassion and generate the wisdom mind of pristine awareness. Even the simple act of reading about the initiation may provide a useful glimpse into the compassionate teachings of the Buddha.

tantric initiation

The Tibetan word for initiation is *wong-kur,* which literally means "giving permission," or granting the authority to practice the tantra. The person conferring the initiation is known as the ritual master or *vajra* master, because the vajra is the ritual implement that cuts through illusion and represents the indestructible mind. Since the tantra itself lives through direct transmission by the vajra master, the initiation fulfills the vajra master's pledge to pass on the tantra without diminishing it in any way, always for the benefit of all sentient beings.

During the initiation, the student makes a similar pledge to respect and uphold the teachings. In this way the student enters into the lineage.

Students may choose to take on different levels of commitment. One who maintains the commitment to a conscientious daily practice will achieve greater results, and the lineage will be strengthened. Or, the initiation may be received simply as a blessing.

introduction to the kalachakra initiation

The Kalachakra Initiation allows the student to hear the tantra, practice its path, and teach its wisdom. The goal is to attain the mental and physical "accomplishments" of the deity, the ultimate goal being bliss consciousness. The initiate who incorporates the tantra into his or her daily practice will achieve new levels of such

The second day of preparation for the initiation begins with the performance of a ritual dance "protecting and consecrating the site," in which interfering forces are summoned to protect and bless the mandala site. Madison, Wisconsin, 1981.

mental accomplishments as generosity, patience, and concentration. The physical accomplishments, which are achieved by opening the subtle energy channels of the body, include relaxation, strength, and healing.

The student, by generating himself (or herself) as the deity, is introduced to new mental patterns which help him to abandon old, destructive conditioning, thus bringing him closer to the experience of bliss consciousness of Kalachakra.

The entire Kalachakra Initiation comprises fifteen separate initiations. The first seven, referred to as the Seven Childhood Initiations, are described in this chapter. The Four High Initiations and the Four Higher Initiations are considered to be more secret in nature and are given to advanced practitioners.

The Kalachakra Initiation is generally given over twelve days. There are ten days of preparation rituals, of which the first eight are conducted by the vajra master, assisted by monks, without the students present. The students participate in the ninth day's activities, known as "The Preparation of the Disciple." The actual initiation ritual takes two days. There is one final day of ceremonies, which includes the dismantling of the sand mandala.

Throughout the initiation, the students repeat many basic steps, including the purification of body, speech, and mind; the verbal expression of one's intent to become enlightened for the benefit of others; and the rejection of one's misdeeds. This repetition plants the teachings clearly and firmly in the minds of the students.

the assumption rite

The assumption rite is the student's request for empowerment on the first day. The vajra master performs the rites to protect the initiate from external and internal obstacles until the time that the actual initiation begins. The assumption rites can be received by one or more representatives for the students.

The vajra master, who has generated himself as the embodiment of Kalachakra, makes torma (barley flour and butter) offerings to

Five *torma* offerings made of *tsampa* (roasted barley flour and butter) represent deities. In the foreground are offering bowls containing rice and incense.

spirits of the local area. He explains to the students the importance of their developing the three principles of the path: renunciation, bodhicitta, and the perfect view of emptiness.

The vajra master is visualized not as an ordinary person but as the deity Kalachakra. The empowerment site is seen not as an ordinary place but as the mandala itself, which is the manifestation of Kalachakra's nondualistic mind. The students prostrate themselves and recite verses to the vajra master, requesting the assumption rite.

The students or their representatives repeat the following vow after the vajra master:

> *To remove the nonvirtuous deeds which I have accumulated in many lives, I will maintain all individual vows. Do allow me to become Kalachakra.*

Above: The tooth stick divination.
Right: On the first day, Richard Gere, as the representative of the students, makes the traditional supplication of the vajra master to bestow the Kalachakra Initiation. New York City, 1991.

Next, it is determined which of the six afflictive emotions requires particular attention for keeping the vows. This identification is made by dropping the "tooth stick" (a smooth twig from the fig tree, used in India as a tooth brush) on a specially painted mandala board indicating the six families of Buddhas. Each family is repre-

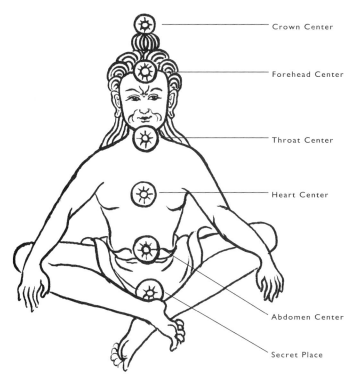

Crown Center

Forehead Center

Throat Center

Heart Center

Abdomen Center

Secret Place

Left: The six *chakras,* or energy centers.

Below: *Purbas,* or ritual daggers, protecting the mandala site.

sented by a different color which stands for an aspect of the wisdom of the Buddha.

The students request that the vajra master bestow the initiation upon them. In response, the vajra master spiritually anoints the students' six energy centers, or *chakras* (the crown of the head, the forehead, the throat, the heart, the abdomen, and the "secret place" or reproductive organs), with the Sanskrit "seed" syllables of the Buddhas, each syllable representing the essence of one of the six Buddha families. A seed syllable is the essence of a deity, the syllable from which the deity arises.

rituals of the site

Once the vajra master has agreed to confer the Kalachakra Initiation, the mandala site is prepared through a series of special rituals known as the Rituals of the Site:

On the second day, the vajra master prepares the "five substances" for the initiation. Here the Dalai Lama consecrates the ten vases filled with saffron water. New York City, 1991.

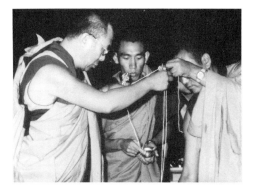

The chalk strings used to draw the mandala lines are blessed. The sand for the mandala is also consecrated at this time.

The vajras and bells used in the initiation are consecrated.

- "Testing the site" gains assurance through divination that the site is worthy of being home to the mandala.
- "Requesting the site" involves special prayers, which ask permission of divine beings and earthly authorities for use of the site.
- "Purifying the site" clears the area of extraneous objects and purifies it ritually. This includes invoking a circle of protective, ritual daggers known as *purbas,* which have been generated into protective deities, around the mandala site. The vajra master then meditates on emptiness. (This concludes the activities of the first day.)
- "Taking possession of the site" elicits permission from various divine beings to draw the mandala.
- "Protecting and consecrating the site" summons and then disperses interfering forces through an elaborate ritual dance by the monks.

rituals of preparation

"Preparation of the earth goddess" consists of offering flowers and perfume to Tenma, the earth goddess, at the center of the mandala site.

"Preparation of the five substances" consecrates substances which will be used throughout the initiation. These are:

1. vases filled with purified water in which the deities will be generated
2. strings dipped in liquid chalk, to be used for drawing the mandala
3. the sand with which the mandala will be constructed
4. the vajra, a ritual implement symbolizing the indestructible mind
5. the bell, symbolizing the realization of emptiness.

The vajra master and his assistants visualize the Kalachakra Mandala, and then visualize lifting this meditated mandala into the space above the mandala base for its protection.

Above: Snapping the chalk string. The vajra master and his assistants begin the construction of the sand mandala by defining the major axes and four base lines with the blessed chalk strings. New York City, 1991.

Left: At the completion of the second day's rituals, the four gatekeepers draw the sand mandala. Los Angeles, 1989.

During the "preparation of the chalk strings," the vajra master and his assistants draw the major axes and four base lines of the mandala, and then place grains of sand at each point where a deity will be represented.

In the "preparation of the deities," the vajra master summons the 722 deities from their abodes through prayer, asking them to take up residence in the mandala.

"Marking the mandala" calls for the preparation of a special thread made of five strands. The vajra master and his attendants visualize the five Dhyani Buddhas (the principal Buddhas of the five Buddha families) dissolving into the five strands and empowering them. The thread is then placed along the main outlines of the mandala, marking both the square mandalas and the outer, concentric circles into which the entire mandala is divided.

This completes the second day.

The four gatekeepers spend the next day drawing the mandala, including details such as the animals and the cushions for the deities, according to the Kalachakra text. The vajra master begins the third day by generating himself as Vajra Vega, the wrathful emanation of Kalachakra. He visualizes the syllable *HI* in each of his

Top: Vajra Vega, fierce emanation of Kalachakra. Detail from a contemporary Tibetan thangka.

Right: At the beginning of the third day, the vajra master generates himself as Vajra Vega and performs a dance opening the gates of the mandala. Here, a similar protective dance at the western gate is shown.

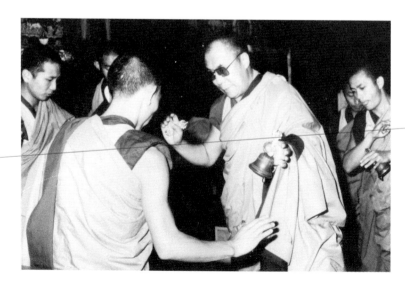

eyes. The syllable on his right eye becomes the sun and the one on his left eye the moon. He stands with his hands at his waist, reciting *HI HI* while smiling fiercely and turning his head from right to left. While doing this, the vajra master opens the three eastern gates by rubbing out the excess lines with saffron water. He repeats this process in the south, north, and west.

Now that the blessed string has fulfilled its function, the vajra master recites prayers to request that the five Dhyani Buddhas that had been invoked into the wisdom string now return to their sacred homes.

The next important step is known as the "preparation of the deities." The vajra master, reciting the mantra of all the deities, marks the seats of the 722 deities with scented water. First, the seat of Kalachakra in the center of the mandala is marked with a square, and the same is done for the four "ones gone thus" deities of the four directions, who sit at the cardinal points of the Mandala of Enlightened Wisdom. The seats of the other deities are symbolically marked with a circular motion. Then a grain of wheat is placed at each point to represent the flower that acts as a cushion for each deity.

The vajra master, facing the eastern gate, generates the entire mandala and all of the deities. He invokes the meditated mandala that was lifted into space on the second day and dissolves it into the generated mandala. Rituals are performed and blessings and offerings given.

The vajra master blesses his own tongue and generates it into the form of a one-pointed vajra, purifying his mouth and speech, and kneels on his right knee to supplicate Kalachakra as follows:

> *Oh, victorious Kalachakra, lord of knowledge, I prostrate myself to the protector and possessor of compassion. I am making a mandala here out of love and compassion for my disciples and as an offering of respect to you. Oh Kalachakra, please be kind and remain close to me. I, the vajra master, am creating this mandala to purify the obstructions of all beings. Therefore, always be considerate of my disciples and me, and please reside in the mandala.*

Grains of wheat indicate the placement of the 722 deities of the Kalachakra Sand Mandala.

Top: The vajra master holds the five colored strings representing the five Dhyani consorts.

Above: He and five assistants invoke the exalted wisdom of the consorts to enter the principal lines of the sand mandala.

Sitting on the eastern quadrant of the mandala, the Dalai Lama applies the first sand. New York City, 1991.

The vajra master repeats this supplication three times and visualizes that Kalachakra grants him his authority by replying:

I grant your wish to you, holder of awareness.

The vajra master then visualizes lifting the deities and the mandala of preparation into the space above the mandala base. Thus the Rituals of Preparation are completed.

main ritual

The main ritual consists of "snapping the wisdom string," applying sand, placing vases around the mandala, decorating the thekpu, generating the entire mandala, making offerings, and, finally, conferring the initiation.

The ritual assistants begin by collecting the grains of wheat from the mandala. They clean the mandala and anoint it with sacred water. The vajra master, as Kalachakra, stands in the west, facing east, while five assistants stand opposite him, generated as Vajra Vega. They make offerings to the lifted mandala, which they visualize suspended in front of them.

The wisdom string is made of five strings of different colors representing the five Dhyani consorts. The vajra master wraps the strings around the forefinger of his left hand, starting with green at the bottom, followed by black, red, white, and yellow. Holding the string above his navel and reciting a mantra, he passes the other end of each of the colored strings to the assistants.

Reciting prayers, the vajra master and his assistants dissolve the common wisdom string into emptiness, from which they generate the five strings into the five consorts. The vajra master visualizes that each of the Dhyani consorts of the lifted mandala has at her heart center a Sanskrit syllable. These syllables are transformed into the five wisdom strings in the form of five colored rays. The vajra master generates rays from his heart center into the hearts of the consorts, activating the syllables in the form of wisdom string rays. These rays dissolve into the five colored strings held by the vajra master and his assistants.

The vajra master and his assistants twist the strands of the strings together, and then make offerings to the consecrated colored strings. The vajra master and an assistant, standing at the west and the east, hold the colored string above the mandala and recite a special mantra for snapping the string. They visualize the sound of the snap alerting the deities, who then descend from the lifted mandala and dissolve into the colored string in the hand of the vajra master. The vajra master and his assistant now hold the string directly over the east/west axis, and snap it while reciting the same mantra. At the sound of the snap, they visualize the nature of exalted wisdom in the form of the five colored strings descending and dissolving into the east/west axis line and blessing it. This procedure is repeated over the north/south axis, and then over the two diagonal lines, the four root lines, and all

The Dalai Lama holds a five-strand, multicolored string called a "mantra string" and a vajra at his heart while he recites mantras during the construction of the sand mandala. The other end of this string is tied around the top of each of the ten vases, so that the mantras may travel along the string to bless the vases.

Above: Sand is applied to the mandala from the third through the eighth days of preparation for the Kalachakra Initiation. The monks begin by sitting on the mandala base and working from the center outward.

Right: Toward the end of the sand painting process, the monks stand on the floor, leaning forward to work near the perimeter of the mandala. Other monks join in to ensure that the mandala is completed by the end of the eighth day.

the other lines of the mandala until they have all been blessed. The vajra master then makes offerings to the colored wisdom string and requests that all the wisdom deities of the string return to their sacred home.

Now the applying of the sand begins. The vajra master sits on the east side of the mandala facing the center, reciting prayers. He applies white, red, and black sand to the foundation wall of the Mandala of Enlightened Mind. He paints a small portion of this wall to signify beginning the construction of the mandala with its foundation. The four gatekeepers and their assistants continue the process of sand painting in their respective directions.

By the conclusion of the third day, the assistants have completed applying the sand to at least the Mandala of Enlightened Great Bliss and the Mandala of Enlightened Wisdom.

The fourth through eighth days consist of ritual prayers and the construction of the sand mandala. Each day, before working on the mandala, the artisan monks generate themselves as the deities of the mandala, led by the vajra master.

After the mandala has been completed on the eighth day, it is blessed by the vajra master. Purbas symbolizing the ten directions are set around the completed mandala to protect against any obstacles which might disturb the initiation. Also set around the mandala are ten ritual vases, each representing a female deity (called a *shakti*) who personifies the wisdom of enlightenment. The shaktis will help in the empowerment of the student. A curtain is lowered on all four sides of the mandala house (thekpu) to shield it from the uninitiated, and a celebratory dance of the offering goddesses is performed by the monks. The students are invited to see the dance.

Two of the ten decorated vases, connected by the mantra string, which are placed around the completed sand mandala. The vases represent female deities called *shaktis*, who will aid in conferring the initiation.

preparing the students

Before the students are initiated, they undergo a process of preparation which involves approximately forty steps. This process is generally referred to as "enhancement," or sometimes "enhancement of the students."

Throughout the enhancement and initiation processes, the student uses visualization to become acquainted with the mandala, and

to experience the qualities of the deities. "Becoming" the deities in this way helps students to transform their ordinary consciousness to that of Buddha mind. This practice is known as "deity yoga."

On the afternoon of the ninth day of the ceremony, the students officially enter the mandala site for the first time. In preparation, they have cleansed their bodies, dressed in clean clothes, and readied themselves for spiritual rebirth. Before entering the site, they rinse their mouths as a gesture of purification and respect, and they make prostrations to the vajra master, whom they visualize as the deity Kalachakra. The site itself is likewise visualized as a manifestation of the nondualistic mind of Kalachakra. The students recite a verse while making a *mudra,* or hand gesture, requesting that the vajra master give them the preparation for the initiation. This is called a "mandala offering."

The vajra master acts as a guide for the students. He introduces them to the deities of the mandala, the commitments, and the procedures for realizing the purpose of the initiation—achieving the enlightened mind of Kalachakra.

The vajra master explains the history, lineage, and essence of the Kalachakra Tantra, and he instructs the students on how to correct their motivation. He explains that for the initiation to be effective, vows bodhisattva—with the motivation to achieve enlightenment for the benefit of all sentient beings—must be taken.

internal initiation

The students now ask for spiritual rebirth as a child of the vajra master. They envision the vajra master as Kalachakra in union with his consort Vishvamata. Light rays radiating from the vajra master's heart draw the students into his body. They imagine themselves entering the vajra master's mouth, dissolving into light at his heart, moving downward through his body and into Vishvamata's womb, where they are dissolved into emptiness and, finally, reborn as Kalachakra with one face and two arms. Having emerged from the womb of the consort, the students visualize returning to their seats.

Above: This *mudra,* or hand gesture, is an offering mandala representing the universe. The students make this gesture as a supplication many times during the Kalachakra Initiation.

Opposite: The Dalai Lama as vajra master at the Kalachakra Initiation in Bodhgaya, India, in 1973. He sits on a throne facing the students in front of a thangka painting of Kalachakra. On the ninth day of the ritual, the students enter the mandala site for the first time. They will visualize the vajra master as the deity Kalachakra.

The students then supplicate the vajra master to bestow the pledges and the mind of enlightenment. The vajra master encourages them to keep a firm faith in the tantric teachings by implementing them in their daily practice.

the vows and blessings

Led by the vajra master, the students take two levels of vows: the common, or bodhisattva vow, and the uncommon, or mantra vow. Through meditation, the students generate the state of mind to practice the altruistic path of the bodhisattvas, and they repeat the verses of the bodhisattva vow three times:

I go for refuge to the Three Jewels,
individually repent all misdeeds,
rejoice in the virtue of others,
will hold the mind of enlightenment.

The uncommon mantra vow is taken by repeating additional verses three times.

The students are then symbolically blessed. First, they generate themselves as Kalachakra, and then they visualize six Sanskrit syllables, which represent the six Dhyani consorts (of the five Buddha families, plus the consort of Vajrasattva, Vajradhatvishvari) appearing on their six chakras. The six syllables, or goddesses, are imagined to dissolve into the chakras to purify them. The vajra master recites the syllables *HUNG AH OM* many times, and then touches the student with scented water, visualizing these syllables at the student's heart, throat, and forehead. This will bestow the seeds of Kalachakra and bless the three places, which represent mind, speech, and body. The vajra master then makes offerings to the students with flowers, incense, butter lamps, and food.

dropping the tooth stick

The student prays that his or her body, speech, and mind be cleansed in preparation for divination. The tooth stick ritual is repeated, with the same or different results, to determine the student's spiritual tasks to be accomplished.

the scented water and the kusha grass

A small amount of scented water is placed in the hands of the students. Three sips are taken to remove subtle defilements. They each then receive two reeds of kusha grass, because the Buddha was sitting on kusha grass when he became enlightened. The longer of the

two reeds is to be placed under each student's mattress to clear the mind of obscuring thoughts, while the shorter reed, to be placed under the pillow, aids in the generation of clear dreams.

As one of the forty steps in the "enhancement of the students," each student receives two reeds of kusha grass, used to clear the mind of obscuring thoughts and aid in the generation of clear dreams.

the protection cord

Students receive a red thread with three knots, which is tied around the upper arm. The Dalai Lama has explained its meaning: "As love increases in your mind, harmful forces do not affect you. Thus, the most effective method for protecting against harm is the cultivation of love. The best protection from one who is trying to harm you is to think, 'This being wants happiness just as I do. May this being attain happiness.' The thread is a reminder of this."

instructions for sleeping

The vajra master teaches about the cycle of suffering, which includes birth, death, and rebirth. He explains that it is a rare opportunity to be born in human form, even rarer to hear the teachings of the Buddha, and rarer yet to have access to the practice of the Kalachakra Tantra. Even if the student can't practice the secret mantra teachings, to see the sand mandala and circumambulate it will be of great benefit and is cause for rejoicing.

He then gives the students special instructions for the night. He suggests that they recite the Kalachakra mantra as many times as possible before going to sleep. They are to sleep on their right sides with their heads toward the mandala, or, if that is not possible, to imagine that the mandala is in the direction of their heads. This will keep them mindful of Kalachakra, of bodhicitta, and of the inherent emptiness of all things. It is explained that students are most likely to be in contact with the mysteries of subtle consciousness at dawn, and they are reminded to place the kusha grass under both mattress and pillow.

dream analysis

The next day, which is the tenth day of the ritual, the vajra master offers guidelines for analyzing the dreams of the previous night. He describes positive and negative symbols, and offers a means of overcoming troublesome dreams through meditation on emptiness, and by the generation of compassion and love toward the threatening forces in the dream. The vajra master then scatters water while recit-

On the tenth day, the vajra master gives representatives of the students the costume of the deity to instill in them a sense of the dignity of Kalachakra.

ing a special mantra associated with the "wisdom which understands emptiness," to disperse any negative effects.

The actual initiation begins here.

The vajra master gives general teachings on Kalachakra. The students, visualizing themselves as Kalachakra with one face and two arms seated at the eastern gate of the mandala, supplicate the vajra master to bestow all the vows, which are repeated.

To instill a sense of the dignity of the deity, the vajra master, reciting a mantra, gives several representatives of the students the costume of the deity: brocades, multicolored garments, and a three-tiered, red-haired knot crown.

Since they are not spiritually ready to see the mandala before receiving the initiation, the vajra master gives the students a red blindfold. The blindfold is placed symbolically over the forehead.

The vajra master recites a mantra and gives students a garland of flowers to offer to the deity upon entering the mandala.

The vajra master again gives the students three sips of scented water, as on the preparation day, to cleanse the three gates of body, speech, and mind.

Sprinkling scented water, the vajra master recites a mantra to remove obstacles and dissolve the students into emptiness. Within that emptiness, he generates them as Kalachakra with two hands embracing blue Vishvamata. The students focus on the emptiness of all phenomena. Then, imagining that all the Buddhas and bodhisattvas are present as witnesses, the students take the bodhisattva vow.

Students are then given further teachings on the five Buddha families and the Kalachakra Initiation itself. They take special tantric vows to train in the path of highest yoga tantra, in order to bring help and happiness to all sentient beings.

Students wear red blindfolds given to them by the vajra master to protect them from seeing the mandala before being initiated.

the twenty-five modes of conduct

The vajra master introduces the twenty-five modes of conduct, which include refraining from harming others, lying, killing, stealing, gambling, engaging in sexual misconduct, and becoming attached to sensory pleasures. A four-line verse expresses the students' motivation to maintain the rules of conduct forever.

The vajra master leads the students through an essential visualization process called the "generation of the all-encompassing

Flowers are given to the students by the vajra master to offer to Kalachakra upon entering the mandala.
After the flower divination is performed, the flowers are placed on their foreheads to generate the wisdom of bliss and emptiness.

yogic mind." This mind of bodhicitta and emptiness generates the powerful seed which is the basis of the tantric path to achieve enlightenment. Students make pledges of secrecy—not to speak of the initiation to the uninitiated—as well as other pledges of correct behavior.

entering the mandala

While the vajra master recites mantras, the students (or their representatives) enter within the curtained mandala area. With blindfolds still on, they circumambulate the mandala while visualizing themselves as the Buddhas of the four entrance gates to the mandala. They repeat the secrecy oath. A complex visualization is led by the vajra master, in which deities are dissolved into blessings within the students, to the accompaniment of cymbals, bells, drums, and incense.

the flower garland initiation

Each student holds a single flower between his hands in prayer and visualizes offering a garland of flowers to Kalachakra. A representative drops a flower onto a painted mandala board that represents the sand mandala, which is too fragile to use for this purpose. As in the tooth stick divination, where the flower lands determines what the students' Buddha lineage is and what feats they will accomplish. The flower is also an offering to the deities of the mandala, who purify and transform the offering, "blessing it into magnificence." The flower, placed on the student's head afterward, help to generate the wisdom of bliss and emptiness. This initiation creates an auspicious bond between the deity and the students.

The vajra master exhorts Kalachakra to open the students' eyes. They remove their blindfolds, which symbolizes that the darkness of ignorance has been removed. They can now see the entire mandala.

The vajra master guides the students through a visualization of the three-dimensional mandala, introducing them to the deities.

explanation of the mandala

The outermost circle of the mandala is seen as brightly colored lights, representing a mountain of fire that completely surrounds and protects the mandala. This is the element of bliss consciousness, or pristine awareness. The green circle inside it represents space, with a chain of vajras seen as the indestructible mind protecting the mandala. Next comes the gray wind element, then the pink-red fire element, then the white water element, and finally the yellow earth element with its pattern of green swastikas, a benevolent symbol derived from ancient Indian culture representing stability and well-being.

These six concentric circles represent the six elements. Within the concentric circles is a square which represents the five-story palace of Kalachakra. Surrounding the palace are four crescent-shaped gardens, which contain various offerings to the deities.

The colors of the four quadrants of the square palace correspond to the four faces of Kalachakra. When the black quadrant is at the bottom, the viewer is facing the black face of Kalachakra, who resides at the center facing east.

The outermost square is the Mandala of Enlightened Body. Inside that are the Mandala of Enlightened Speech, the Mandala of Enlightened Mind, the Mandala of Enlightened Wisdom, and the innermost square, the Mandala of Enlightened Great Bliss.

description of the principal deity kalachakra

At the center of the mandala, on a green, eight-petaled lotus, is the principal deity—glorious Kalachakra with his consort Vishvamata. Kalachakra has a dark blue black body with three necks, the middle one black, the right red, and the left white. He has four faces, of which the main one is blue black and wrathful, with a fang. His right face is red, with an expression of desire; his left face is white, and very peaceful; and his rear face, depicted at the extreme left, is yellow, with a look of meditative concentration. Each face has three eyes. His hair is in the form of a topknot. The three-tiered crown is topped with the seal of the bodhisattva Vajrasattva and other ornaments, including a crescent moon and a double vajra. His body ornaments include vajra, jewels, vajra earrings, vajra necklace, vajra bracelet, vajra belt, vajra ankle bracelets, flowing vajra scarves, vajra garlands, and a skirt of tiger skin.

Kalachakra has six shoulders, three on each side, which are black, red, and white. Connected to each shoulder are two upper arms, and connected to each upper arm are two lower arms, each with one

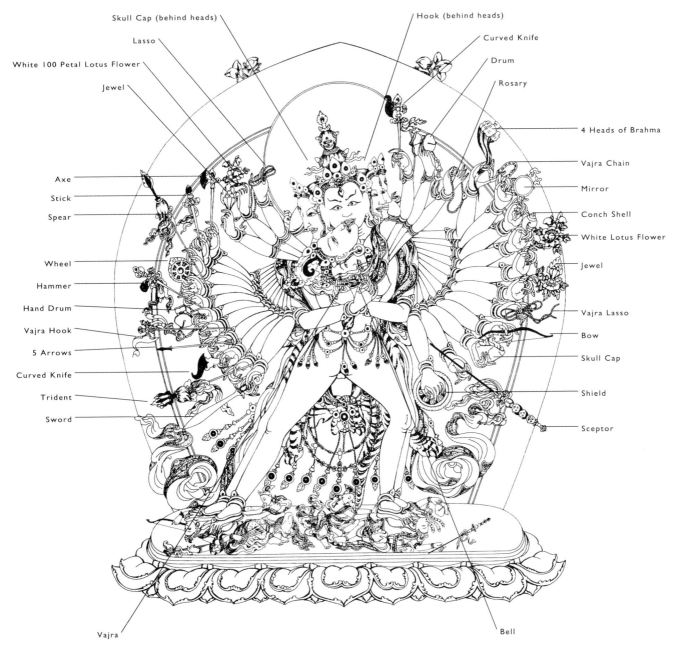

Skull Cap (behind heads)
Lasso
White 100 Petal Lotus Flower
Jewel
Axe
Stick
Spear
Wheel
Hammer
Hand Drum
Vajra Hook
5 Arrows
Curved Knife
Trident
Sword
Vajra

Hook (behind heads)
Curved Knife
Drum
Rosary
4 Heads of Brahma
Vajra Chain
Mirror
Conch Shell
White Lotus Flower
Jewel
Vajra Lasso
Bow
Skull Cap
Shield
Sceptor
Bell

The implements of the deity Kalachakra and his consort Vishvamata.

hand, for a total of twenty-four hands, twelve on each side. The different-colored arms, symbolizing the purified and enlightened body, speech, and mind of Kalachakra, are used to remove the obstructions of body, speech, mind, or wisdom. The lower eight arms are black, the middle eight are red, and the upper eight arms are white. The back side of all the thumbs are yellow, the first fingers white, the second red, the third black, and the fourth green.

The inside of the hand and the lower joints of all his thumbs and fingers are black; the middle joints are red, and the upper joints white. Each finger is adorned with a shining, radiant ring.

the implements of kalachakra

Kalachakra holds twenty-four implements. In his four right black hands are a vajra, a sword, a trident, and a curved knife. His four right red hands hold five arrows, a vajra hook, a hand drum, and a hammer. His four right white hands hold a wheel, a spear, a wooden stick, and an axe. The implements of his four left black hands are a vajra bell, a shield, a scepter, and a skull cup filled with blood. His four left red hands hold a bow, a vajra lasso, a jewel, and a white lotus flower. The implements of the four left white hands are a conch shell, a mirror, a vajra chain, and a four-faced head of Brahma. Kalachakra stands on the cushion of a lotus and moon, sun, Rahu, and Kalagni disc, with his outstretched red leg pressing on the heart of the demonic god of the desire realm and his left white leg pressing on an afflicted deity. The consorts of the demonic god of the desire realm and the afflicted deity prostrate themselves at the feet of Kalachakra.

kalachakra's consort vishvamata

Vishvamata has a yellow body with four faces. Her main face is yellow, and the other three faces are white, blue, and red. Each face has three eyes. She has eight yellow arms. The implements of her four right hands are a curved knife, a hook, a small drum, and a rosary. The implements of her four left hands are a skull cup, a lasso, a white, eight-petaled lotus flower, and a jewel. Her head is crowned by Vajrasattva, indicating her lineage. Her body is adorned with the five seal ornaments of the Buddha families. With her left leg out-stretched, Vishvamata is in embrace with Kalachakra.

Kalachakra, represented by a vajra, and Vishvamata, represented by a dot of colored sand, reside at the center of the lotus of the sand mandala. Inseparable from them are green Akshobhya with his consort Prajnaparamita, and blue Vajrasattva with his consort Vajradhatvishvari, who are present on layers of sand below Kalachakra and Vishvamata. Thus, there are a total of six deities residing at the center of the lotus flower. On the eight petals of the lotus flower surrounding Kalachakra and Vishvamata reside eight shaktis. These eight, plus Prajnaparamita and Vajradhatvishvari, who are both in the center, make ten shaktis. The ten shaktis represent the ten perfections: generosity, morality, patience, effort, concentration, wisdom, method, spiritual aspiration, spiritual power, and transcendent wisdom.

The place where Kalachakra, Vishvamata, and the eight shaktis reside is the Mandala of Enlightened Great Bliss, which is at the center of the Mandala of Enlightened Mind.

The students visualize having seen and met the 70 deities of the Mandala of Enlightened Mind, the 116 deities of the Mandala of Enlightened Speech, and the 536 deities of the Mandala of Enlightened Body.

The vajra master states that all of the 722 deities have the nature of light rays and the great bliss of the enlightened state of mind. The 722 deities of the Kalachakra Mandala also correspond to specific parts of the human anatomy and their functions. Kalachakra himself corresponds to the heart, which is the seat of the subtle mind.

Having entered the mandala and having been introduced to the deities, the students sing a song of joy. This concludes the ceremony of the tenth day.

the seven childhood initiations

The students prepare for the Seven Childhood Initiations with offerings and supplications. The vajra master dispels all inauspiciousness.

Each of the seven initiations empowers the student to attain a particular spiritual goal, and each is analogous to an event in the development of a child.

The first two initiations, known as the Water and Crown Initiations, are associated with purification of defilements of the body. They take place at the northern gate of the Mandala of Enlightened Body.

The Silk Ribbon and Vajra-and-Bell Initiations are associated with purification of defilements of speech. They take place at the southern gate to the Mandala of Enlightened Speech.

The Conduct and Name Initiations are associated with the purification of defilements of mind. They take place at the eastern gate of the Mandala of Enlightened Mind.

The last initiation, that of Permission, is associated with the attainment of bliss, and takes place at the western side of the innermost Mandala of Enlightened Great Bliss. Thus, the Seven Childhood Initiations introduce the students to the deities of each mandala, allowing them to penetrate to the heart of Kalachakra.

There are twelve stages of realization to the practice of the Kalachakra Tantra. These are known as "bodhisattva levels," with the twelfth level corresponding to full enlightenment. The Seven Childhood Initiations authorize the students to achieve the first seven levels.

1. *Water Initiation.* Corresponds to a child's first bath.

The vajra master confers this initiation (and the others) as the embodiment of Kalachakra. He leads the students to the northern entrance of the Mandala of Enlightened Body, facing the white face of Kalachakra's "vajra body."

The purpose of the Water Initiation is to purify the student's five elements: wind, fire, earth, water, and space. The students make a mandala offering to the vajra master and then supplicate him to bestow the Water Initiation upon them. Using a conch shell, he sprinkles them with water taken from the ten purified vases. This water is the substance through which the initiation will be conducted. The Water Initiation begins with what is known as the Internal Initiation. The student visualizes being drawn by a ray of light from

The initiation substance "water," represented by a conch shell, used in the first of the Seven Childhood Initiations.

Kalachakra's (the vajra master's) heart center into his mouth, and through the vajra path into Vishvamata's womb. Then, after being transformed into emptiness, the student is generated as Kalachakra's "vajra body." A ray of light from Kalachakra's heart center then draws Kalachakra himself into the student, who becomes one with the deity. The heart rays then invite all the male and female Buddhas and bodhisattvas, known as the "initiators." The vajra master makes offerings and supplications to them. The male and female initiators melt in the heat of great desire into a substance called the "mind of enlightenment," which enters into Kalachakra's body through his crown and penetrates the womb of the consort. There the initiators initiate the student. The student then emerges from the womb of the consort to take his or her initiation seat. This Internal Initiation is repeated at the gate of each mandala entered during the Childhood Initiations.

Next follows a complex visualization wherein the five elements of the students and the water from the vases are generated into deities. The elements and the water, as deities, are then initiated by the five Dhyani consorts of the mandala. The vajra master makes a gesture of touching the conch shell filled with water to the "five places" of the students: head, shoulders, arms, thighs, and hips. This symbolizes the purification of their five elements. He places water in the students' hands to drink. Drinking this water, the students experience a union of bliss consciousness and the realization of emptiness.

There is a concluding water ritual during this and several of the following Childhood Initiations wherein the student is anointed with sacred water in order to experience bliss consciousness.

The Water Initiation instills in the students the power to achieve the first bodhisattva level, known as "The Very Joyous Level," as its fruit.

2. *Crown Initiation.* Corresponds to a child's first haircut.

The students first make a mandala offering and a supplication to receive the Crown Initiation in order to purify each of their five "aggregates"—form, feeling, discrimination (perception), compositional factor (unconscious tendencies), and consciousness. The vajra master describes the auspicious nature of the event and the students envision a rain of flowers falling upon them.

Next follows a complex visualization in which the five Dhyani Buddhas who reside in the mandala confer the initiation while holding the initiation substance, which is the crown.

The vajra master touches the crown to the students' five places. At that moment, the students experience the nonconceptual mind of great bliss. At the conclusion of the Crown Initiation is a water ritual.

The Crown Initiation instills the power to achieve the second bodhisattva level, known as "The Uncontaminated Level," as its fruit.

The Water and Crown Initiations purify the body elements and aggregates of the students. Just as these are initially formed in the womb, these two initiations, which take place at the northern gate to the Mandala of Enlightened Body, establish the seeds in the students for attaining the exalted "vajra body."

3. *Silk Ribbon Initiation.* Corresponds to a child having its ears pierced.

The vajra master, as the embodiment of Kalachakra, leads the students to the southern gate of the Mandala of Enlightened Speech, facing the red face of Kalachakra's "vajra speech."

The students make a mandala offering and then a supplication to receive the Silk Ribbon Initiation. This initiation purifies the ten winds circulating through the body, which are associated with speech. Again, the vajra master describes the auspicious nature of the event and the students envision a rain of flowers falling over them. An Internal Initiation follows, generating the students into the "vajra speech" of Kalachakra. Then, through a complex visualization, the ten shaktis of the mandala, holding the silk ribbons, initiate the students, purifying their ten winds.

The vajra master touches the silk ribbons to the students' five places and then ties the ribbons to their headdresses (crowns). This blesses their ten winds into magnificence and generates the exalted wisdom of great bliss. A concluding water ritual follows.

The initiation substance "crown."

The initiation substance "silk ribbon."

The Silk Ribbon Initiation instills the power to achieve the third bodhisattva level, known as "The Luminous Level," as its fruit.

4. *Vajra-and-Bell Initiation.* Corresponds to a child's talk and laughter.

The vajra represents compassion, or the male principle, while the bell represents wisdom (the realization of emptiness), or the female principle. Thus the vajra and bell together form an important symbol of the essence of Tibetan Buddhism, the union of wisdom and compassion (referred to in the Kalachakra Tantra as the union of "empty form and immutable bliss").

The students make a mandala offering to the vajra master, and supplicate to receive the Vajra-and-Bell Initiation in order to bind the two main (right and left) channels of energy into the central channel.

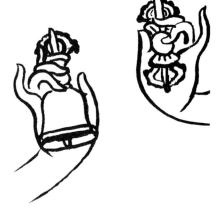

The initiation substance "vajra and bell."

After the vajra master clears away all obstructions, he instructs the students to visualize their right and left channels. The right becomes Kalachakra and the vajra; the left becomes Vishvamata and the bell. Both channels are purified as the two deities confer the initiation. Since all words move through these channels in order to form speech, this is a step of preparation for the students' attainment of enlightened speech.

The vajra master touches the vajra and bell to the students' five places, and the students experience the mind of great bliss that realizes emptiness. A concluding water ritual follows.

The Vajra-and-Bell Initiation purifies the students' internal sun and moon, and authorizes them to achieve the accomplishments of the two principal deities. The power is instilled in the students to achieve the fourth bodhisattva level, known as "The Radiant Level," as its fruit.

Just as the winds and channels are formed in the womb, the Silk Ribbon and Vajra-and-Bell Initiations, given at the southern gate facing Kalachakra's yellow face of exalted speech, purify the defilements of speech. They plant in the initiates the seed of "vajra speech."

5. *Conduct Initiation.* Corresponds to the child's first enjoyment of the five senses.

The vajra master leads the students to the eastern gate of the Mandala of Enlightened Mind, facing the black "vajra mind" face of Kalachakra.

The students begin with a mandala offering and supplicate to receive the Conduct Initiation in order to purify the sense powers and their objects.

The students visualize themselves becoming the deities of the Mandala of Enlightened Mind and are granted an Internal Initiation.

Then the senses and sense objects, along with the vajra thumb ring, which is the initiation substance, are generated into bodhisattvas and their consorts, who reside in the mandala. They grant the initiation, blessing the five internal senses into magnificence. The thumb ring acts as a symbolic restraint upon the finger, and thus the senses.

As the vajra master touches the five places of the students with the thumb ring, they experience the exalted wisdom of the union of emptiness and great bliss. A concluding water ritual is also performed.

The vajra master reminds the students not to allow their senses to obscure their knowledge of the essential emptiness of objects. This is described as being the essence of the Conduct Initiation, wherein the students may achieve the fifth bodhisattva level, known as "The Difficult to Overcome Level," as its fruit.

6. *Name Initiation.* Corresponds to the naming of a child.

The students make a mandala offering to the vajra master, and supplicate to receive the Name Initiation in order to purify their "action faculties" (mouth, arms, legs, and reproductive, urinary and defecatory organs). The wrathful deities in the mandala comply, holding the initiation substance "bracelet."

The students then visualize a rain of flowers falling upon them. The vajra master touches the five places of the students with the bracelet, and they experience nonconceptual wisdom. A concluding water ritual follows.

The Conduct Initiation substance
"thumb ring."

The students then receive names from the vajra master, who stands wearing a special robe. The names, which are derived from the Buddha families, are determined through flower divination. When the vajra master speaks the student's lineage name, the student visualizes becoming enlightened, generating great bliss and exalted wisdom.

It is prophesied that each student will become a Buddha with this particular name in the future.

This initiation enables the students to overcome the four demons through the four immeasurables, which are as follows:

1. Love—the mind that wishes all sentient beings happiness
2. Compassion—the mind that wishes all sentient beings liberated from suffering
3. Joy—the mind that wishes that all sentient beings achieve happiness and not be separated from it
4. Equanimity—the mind that wishes all sentient beings free of discrimination regarding friends, enemies, and others.

The Name Initiation instills the power to achieve the sixth bodhisattva level, known as "The Approaching Level," as its fruit.

The Name Initiation substance "bracelet."

Just as the sense organs and action faculties are formed in the womb, the Conduct and Name Initiations are in the area of the eastern gate of the Mandala of Enlightened Mind, facing Kalachakra's black face, planting in the students the seed to achieve the exalted "vajra mind."

7. *Permission Initiation* (and "Appendages"). Corresponds to a child's first reading lesson.

The vajra master leads the students to the western gate of the Mandala of Enlightened Great Bliss, where they face the yellow face of Kalachakra's "vajra pristine consciousness."

The students present an offering mandala to the vajra master and supplicate to receive the Permission Initiation. The purpose of this initiation is to purify the defilements of wisdom through the initiation substance called "the five hand symbols" (vajra, jewel, sword, lotus, and wheel).

The vajra master dispels obstructions and expresses the auspicious nature of the occasion. The students visualize a rain of flowers falling upon them.

The students are generated into "vajra pristine consciousness" deities in an Internal Initiation. They are then generated, along with the five hand symbols, into Vajrasattva and Prajnaparamita, who reside at the center of the mandala. These deities bestow the initiation, holding the hand symbols. This helps the initiates to teach the tantra according to the interest and disposition of their students.

The vajra master touches each of the five places of the students with the five hand symbols, generating the exalted wisdom of great bliss. A water ritual is performed. The vajra master then gives each initiate an individual authorization to teach, which is done by recitation:

> *Turn the vajra wheel (teach the dharma)*
> *in order to help all sentient beings*
> *in all worlds, in all ways*
> *Turn the vajra wheel*
> *in accordance with the needs of the people*
> *in order to help all sentient beings . . .*

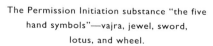

The Permission Initiation substance "the five hand symbols"—vajra, jewel, sword, lotus, and wheel.

As the recitation continues, a conch shell is placed in the students' right hands, the sound of which is symbolic of spreading the sound of the Buddha dharma. Then a Kalachakra text is presented. Next, a bell is placed in their left hands, symbolic of wisdom and emptiness, with the instruction that the teacher of the doctrine should always keep the knowledge of emptiness in mind and view all things as illusion.

A wheel that symbolizes bringing about the welfare of others through the spread of the dharma is given to the students, and instruction is given as to the meaning of emptiness. The students then promise to work for the welfare of all sentient beings by always joining compassion with wisdom. There are several Permission appendages:

1. In the "mantra transmission," the students visualize that a rosary chain of mantras arises from the heart of the vajra master, who is the embodiment of Kalachakra. This chain of mantras comes out of his mouth to enter the mouths of the students and descend to their heart centers. The students repeat the mantra three times after the vajra master, visualizing the mantra circling a syllable at their hearts.

2. The students receive "eye medicine" from the golden spoon of the vajra master, so that they may visualize that the haze of ignorance has been removed and the wisdom eye opened.

3. The vajra master gives the students a mirror, which symbolizes the fact that all phenomena can be perceived as illusions. Like reflections in a mirror, they are empty of inherent existence.

4. A bow and arrow are given to the students. The arrow deeply implants the realization of emptiness into the students' minds, thus encouraging the swift attainment of enlightenment.

5. In the "initiation of the vajra master," the pledges of vajra, bell, and "seal" are given. First, the students and a vajra and bell are generated as deities. The vajra master recites verses and a mantra and places the vajra in the students' right hands. He says that the secret nature of the vajra is the exalted wisdom of great bliss. Holding the vajra will recall the true nature of the ultimate vajra, or what is called "method."

The vajra master places the bell in the students' left hands while reciting verses. The sound of the noninherent existence of all phenomena is proclaimed by the sound of the bell. Holding the physical bell will recall the true nature of the ultimate bell of exalted wisdom realizing the emptiness of all phenomena.

The students are presented with a wheel which authorizes them to spread the dharma for the benefit of all sentient beings. The wheel of the law shown here is a symbol of the office of the Panchen Lama. The sun, with its halo of flames, sets the law of Buddhism in motion. 20" high, silver repoussé, Tibet, 18th century.

Vajrasattva, the Buddha of Purification, is visualized by the students to purify defilements at the conclusion of the Seven Childhood Initiations. 9" high, cast brass, Kashmir, 10th–11th century.

Holding the vajra and bell, the students recite verses after the vajra master and ring the bell at the end of each verse. They make the gesture of embrace with the vajra and bell, contemplating the meaning of great bliss and the meaning of emptiness.

The students visualize themselves as the divine body of Vajrasattva, in whom this exalted wisdom arises.

The vajra master performs a water ritual, again to achieve purification, and offerings are made. He recites instructions on behavior to the students, who are now of the vajra lineage.

The Permission Initiation empowers the students to achieve the seventh bodhisattva level as its fruit. This is known as "The Gone Afar Level."

The Permission Initiation, which takes place at the western gate of the Mandala of Enlightened Great Bliss, facing Kalachakra's face of pristine consciousness, corresponds to the wind of pristine consciousness which circulates just after a child is conceived. It purifies any defilements and plants the seed of Vajrasattva purity, granting the students the capacity to achieve the seventh bodhisattva level.

The students have now completed the Seven Childhood Initiations, the purpose of which are to purify the defilements of nonvirtuous activity. This empowers the initiates to practice the tantra and attain the highest accomplishments. Having received this empowerment, and having become "aspirants of the seventh level," they are now authorized to practice what is known as deity generation meditation, and to actualize all aspects of the mandala. One who can do so in this lifetime is said to become the "lord of the seventh level." One who is not able to actualize the tantra in one lifetime but practices all the virtues will become "lord of the seventh level" in seven lifetimes. The initiates recite a mantra and, while exalting in the glory of being the deity, visualize that they have attained the seventh bodhisattva level.

conclusion of the initiation

The exact time, date, and place of the initiation are determined by the vajra master according to the Kalachakra astrological system.

The vajra master instructs the students to abandon the fourteen root infractions:

1. Disturbing the mind of the teacher
2. Deviating from the teacher's word
3. Harboring anger toward vajra brothers and sisters (those initiated by the same teacher)
4. Abandoning love (for a sentient being)
5. Allowing the mind of enlightenment to degenerate
6. Deriding the tenets of other paths
7. Exposing the secrets to unripened persons
8. Abusing one's mental and physical body
9. Losing faith in the purity of phenomena (emptiness)
10. Associating with deriders of the dharma
11. Not remembering the view of emptiness
12. Influencing against the dharma
13. Forsaking commitments or vows
14. Deriding any woman.

The students express their intention to follow this behavior, and recite the verses:

> *Today my birth is fruitful.*
> *My being alive is also fruitful.*
> *Today I have been born in the lineage of the Buddhas,*
> *Now I have become a child of the Buddhas.*

An offering mandala is made.

The vajra master and all the initiates together recite prayers of wish fulfillment, dedication of merit, and benediction.

The dedication of merit, performed at the end of every Mahayana Buddhist practice, is a request that any merit acquired by the practitioner be given away to benefit all sentient beings.

The vajra master and his assistants remain to perform the conclusion rituals, comprised of making offerings, and praising the deities with additional prayers of wish fulfillment, dedication of merit, and benediction.

This concludes the eleventh day.

A rice offering mandala is made by the ritual assistant at the conclusion of the eleventh day of the Kalachakra Initiation while the students make offering mandalas with a mudra. These are offered to Kalachakra, as embodied by the vajra master, in gratitude for the initiation.

the dismantling of the sand mandala

On the twelfth day, the vajra master and his assistants perform self-generation, as on the other days, and make extensive tsok (food) and torma offerings to the deities. All the initiates are welcomed to circumambulate and view the sand mandala.

People from the surrounding mountain areas form a line to receive a blessing from the Fourteenth Dalai Lama and to view the Kalachakra Sand Mandala at the completion of the Kalachakra Initiation in Leh, Ladakh, India in 1976.

After the viewing of the mandala has been completed, the vajra master and his assistants stand at the eastern entrance of the mandala, bless both the external and internal offerings, and offer them to the deities with prayers and verse. The vajra master circumambulates the sand mandala three times, holding the vajra and bell. Back at the eastern entrance, facing the mandala, he expresses regret to the deities for any mistakes which might have been made during the initiation and recites verses for forgiveness.

The vajra master visualizes and requests the transworldly deities to enter into his heart. Snapping a finger of his right hand, he requests the worldly deities to return to their respective abodes.

The vajra master then meditates on emptiness. Reciting a mantra, starting from the east and going clockwise, he picks up the colored dots and Sanskrit syllables of sand representing the 722 deities and places them on a plate.

Again starting at the east, he cuts the major lines of the mandala with the point of a vajra. All the sand is swept up and put in a vase, which is then covered with the deity's costume and crown. The vajra master places a small portion of the sand on the crown of his head as a blessing, and then offers small amounts of the consecrated sand to the students.

The vase containing the sand is carried with great respect, to the accompaniment of auspicious verses, mantras, musical instruments, banners, and much ceremony, in a procession to a local river, lake, or ocean.

A special altar is placed near the water's edge. The vajra master and his assistants place special torma offerings for the *nagas* (water spirits) on the altar in front of the vase of blessed sand. They recite prayers while an assistant draws a small sand mandala of an eight-petaled lotus flower on a round offering mandala base. The symbols of Kalachakra and Vishvamata are placed at the center and the ten naga kings in the eight petals.

The vajra master generates the dots of sand as Kalachakra and Vishvamata and the ten naga kings, and performs a complex visualization invoking wisdom deities.

The vajra master visualizes that Kalachakra and Vishvamata dissolve into him. He visualizes that all the nagas, with great respect, receive the blessed sand. The sand from the small mandala is then added to the sand from the Kalachakra Mandala, and the vase is carried to the water. The vajra master pours the sand into the body of water, imagining that the blessed sand is being accepted by the water spirits and aquatic life, filling them with joy. The vajra master then rinses the vase, fills it with water, and returns to the mandala site.

After the students have viewed the sand mandala, the vajra master requests that the 722 deities return to their respective abodes. He then meditates on emptiness before beginning the dismantling process. The Dalai Lama at Bodhgaya, India, 1973.

Right: The Dalai Lama picks up the colored dots and Sanskrit syllables representing the 722 dieties of the sand mandala, placing them on a plate. New York City, 1991.

Below: Starting at the center of the sand mandala, the Dalai Lama cuts the energy of the major axes and base lines (eight directions in all) with the point of a vajra. Los Angeles, 1989.

Above and left: Monks sweep up the sand
of the mandala.

Opposite: The Dalai Lama performs a ritual at a special altar at the edge of the Hudson River before pouring the sand into the water. Near the World Trade Center in New York City, 1991.

Above: Rinsing the vase after offering consecrated sand to the local river. Rikon, Switzerland, 1985.

Left: Offering blessed sand to the marine life of Santa Monica Bay. Los Angeles, 1989.

The Dalai Lama pours water brought back from
the river on the mandala drawing.
New York City, 1991.

pacification of the mandala site

The vajra master sprinkles the water around the mandala site and, with his assistants, removes all the chalk lines of the mandala by scrubbing the mandala base with this water.

Then, holding the vajra and bell, the vajra master recites a mantra, removes the ten purbas from their holders, and cleans the point of each with milk, symbolically liberating the spirits subdued by the purbas.

dedication of merit

The vajra master, sitting at the center of the mandala base and facing east, recites concluding prayers of dedication and benediction.

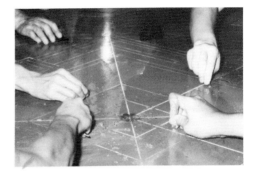

Above: The monks remove all the lines of the mandala drawing by scrubbing the base with river water.

Left: The Dalai Lama sits on the mandala base reciting concluding prayers.
New York City, 1991.

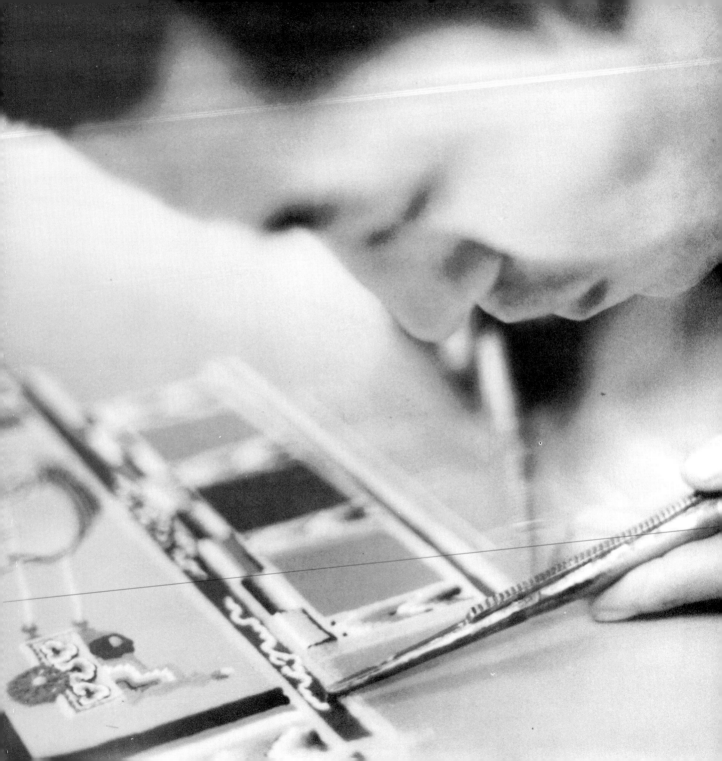

The Kalachakra Sand Mandala

the step-by-step process

THE KALACHAKRA SAND MANDALA is a two-dimensional representation of the five-story palace of the deity Kalachakra. It is one of the colored-particle mandalas, a generic term used to describe any mandala made of crushed materials such as jewels, flower petals, rice, or, most commonly, sandlike stone ranging in density from soapstone to marble.

In ancient times in Tibet, sand ground from brightly colored stone was often used. Today, white stones are ground

Namgyal monk Pema Lobsang Chogyen applies sand to the bottom beam of the western gate of the Mandala of Enlightened Body of the Kalachakra Sand Mandala. The American Museum of Natural History, New York City, 1988.

and dyed with opaque water colors to produce the bright tones found in the sand paintings. The basic colors are white, black, blue, red, yellow, and green. Each of the last four colors has three shades— dark, medium, and light—for a total of fourteen colors. The sand mandalas illustrated in this book are made from crushed white stones and marble dust dyed with opaque water colors.

Deities in the Kalachakra Sand Mandala are represented by Sanskrit syllables or dots. This detail of the southwest corner of the Mandala of Enlightened Body shows a peacock on which a lotus flower supports thirty deities, each represented by a dot.

Each of the 722 deities that resides in the Kalachakra Sand Mandala, as well as the mandala itself, is a manifestation of the principal deity Kalachakra. Each represents a different aspect of Kalachakra's enlightened qualities.

On seeing the Kalachakra Sand Mandala for the first time, people often ask, "Where are the deities?" and "Why are they represented by symbols?" It is important that full-body representations of deities be drawn exactly according to prescribed proportions or they cannot be blessed. Since spatial limitations prevent this, representative symbols are used in the mandala.

Here in the Kalachakra Sand Mandala, all deities are symbolized by either Sanskrit "seed" syllables or by dots. The exception is Kalachakra himself, who is symbolized by his own hand implement, the vajra, which is a diamond-hard or adamantine scepter, or "thunderbolt," as translated literally from the Tibetan word *dorje,* representing the indestructible mind.

Practitioners of the Kalachakra Tantra use the sand mandala as a diagram, or blueprint, to help them visualize Kalachakra's three-dimensional palace during the initiation ceremony and their daily meditation practices. It is said, and attested to by many, that by simply contemplating a mandala one can gain deep insight or inner peace, and that those who only see the mandala can establish a

The Mandala of Enlightened Great Bliss

The Mandala of Enlightened Wisdom

The Mandala of Enlightened Mind

The Mandala of Enlightened Speech

The Mandala of Enlightened Body

Gate of the Mandala of Enlightened Mind

Gate of the Mandala of Enlightened Speech

Gate of the Mandala of Enlightened Body

Side view of southern gate

Eastern entrance Mandala of Enlightened Body

Elevation of the three-dimensional palace of Kalachakra. This drawing and those on pp. 183–92 were generated with an Auto-cad computer program by Daniel Maciejczyk under the supervision of Christian Lischewski.

strong connection with the deity. This mandala also serves to purify the polluted environment and bring prosperity to the world.

The Kalachakra Sand Mandala consists of five square mandalas, one within the other, surrounded by six concentric circles. Each square mandala represents one of the five levels of Kalachakra's palace. The largest is known as the Mandala of Enlightened Body. Within it is the Mandala of Enlightened Speech. And within that is the Mandala of Enlightened Mind. These are characterized by elaborate gates at each of the cardinal directions.

Within the Mandala of Enlightened Mind, we find the Mandala of Enlightened Wisdom, and, at the very center, the Mandala of Enlightened Great Bliss. In the three-dimensional mandala, this is the

The six concentric circles surrounding the square palace of Kalachakra represent the elements earth, water, fire, wind, space, and wisdom, or bliss consciousness.

uppermost or "penthouse" level of the palace. Here, the principal deity, Kalachakra, resides with his consort, Vishvamata, and ten female deities known as shaktis.

Each of the six concentric circles surrounding the palace represents one of the six elements. The innermost circle is the element of earth, upon which the base of the palace firmly rests. Going outward from the earth are the elements of water, fire, wind, and space. When the practitioner visualizes the mandala in its three-dimensional form, these elements are imagined directly below the palace, with that of space being all-pervasive. The sixth element, wisdom, or the light of blazing consciousness, is the outermost circle, described by the Dalai Lama as "pristine awareness."

orientation

The organization of the two-dimensional Kalachakra Mandala is based on the four cardinal directions, like a map. These directions are represented according to Buddhist orientation. We Westerners are accustomed to finding north at the top of every map, and east at the right. During the Kalachakra Initiation, the two-dimensional mandala lies flat, and students always circumambulate it beginning in the east, then continue walking around it in a clockwise direction. When a painted or photographic Kalachakra Mandala is hung on a wall, the black eastern quadrant is located at the bottom, so that the viewer is facing the central face of the deity.

The colors of the four quadrants of the Kalachakra Mandala correspond to the colors of the four faces of the deity Kalachakra. If the viewer is observing the mandala from the east, he is not only facing Kalachakra at the center, he is also facing the eastern quadrant, which corresponds to Kalachakra's central black face, associated with the element of wind. To the viewer's left, the red southern quadrant is associated with the element of fire, and corresponds to Kalachakra's red face. To the viewer's right, the white northern quadrant is associated with the element of water, and corresponds to

Kalachakra's white face. The yellow western quadrant is associated with the element of earth, and corresponds to Kalachakra's yellow face, which is actually his rear face.

preparation for ritual artists

In Tibetan Buddhism, altruistic motivation and the realization of emptiness are basic to one's practice. These are essential in performing ritual arts such as constructing the sand mandala, in both sacred and cultural contexts.

The creation of ritual art is both a meditative practice and a means of transmitting the message of the Buddha, which for Buddhists is the main source of happiness and prosperity for all sentient beings. With this in mind, the monks strive for perfection of their art as a means to perfect their minds and to maintain the authenticity of the teaching.

The monks begin their work each day by reciting special prayers of purification to cultivate a state of inner peace. The prayers reinforce the artists' motivation to work for the benefit of all beings. Through a visualization process, they generate themselves as the deity Kalachakra to remain mindful of their goal of perfection as they work.

As they paint they are conscious that the quality of their work reflects upon their spiritual lineage, as vital today as it has been since the time of the Buddha. Thus, their efforts toward perfection in constructing the sand mandala ensure that everyone, from the experienced Buddhist practitioner to the first-time viewer, will receive the blessings and teachings as the Buddha presented them more than 2,500 years ago.

While constructing the sand mandala, the monks work as a team, discussing the symbolic and philosophic meaning of the details they are painting, all the while offering one another advice and support. Each strives to do his best, enhancing his own spiritual practice through the development and perfection of mental and artistic activity.

drawing the mandala

In the Kalachakra Initiation, the mandala is drawn on the second day. As with other Tibetan mandalas, the drawing and design of the Kalachakra Mandala are taken from written instructions in ancient Buddhist texts and have remained unchanged since the Buddha taught them. The steps for drawing in this chapter use Western architectural language interpreted by Stan Bryant in conjunction with the system used at Namgyal Monastery, which is based on the Kalachakra text.

The drawing begins with the geometric outline of the mandala with chalk, charcoal, or pencil on a solid wooden platform, usually

Prajnaparamita, the holy teaching of transcendental wisdom. To the left and right of the text is painted the Kalachakra mantra known as the "Power of Ten." Gilt wood covers, 346 folios, colored ink and gold, 8.75"x 27", 60 lbs., Tibet, 16th century.

painted red or blue. If the mandala base is square, the center of the base is the center of the mandala. The mandala can be made any size; its maximum diameter is determined by the shortest width of the mandala base. Some of the construction lines are drawn lightly, as they will be erased later.

STEP I: Brahman Lines

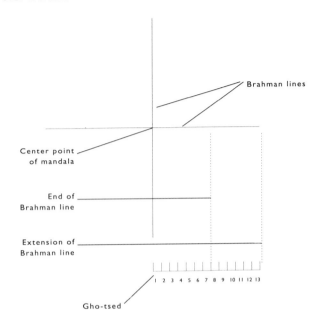

Brahman lines

Center point
of mandala

End of
Brahman line

Extension of
Brahman line

1 2 3 4 5 6 7 8 9 10 11 12 13

Gho-tsed

STEP 2: Diagonal Lines

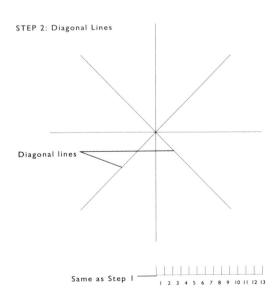

Diagonal lines

Same as Step I ———
1 2 3 4 5 6 7 8 9 10 11 12 13

Step 1: Two lines are drawn perpendicular to each other with the center of the mandala as the point of intersection, and both are extended in each direction to the desired perimeter of the mandala. These two lines are the Brahman lines. The radius of the mandala is then divided along the Brahman lines into thirteen equal parts. The monks do not do this division mathematically but by trial-and-error, folding a strip of paper until it has thirteen equal parts totaling the length of the radius. The result is a scale with thirteen equal basic units for the Mandala of Enlightened Body, the largest of the five mandalas that compose the Kalachakra Mandala. The basic unit is called a *gho-tsed* in Tibetan. In each scale there are always thirteen gho-tseds, their length varying in proportion to the size of the mandala. The gho-tsed is also referred to as the entrance measurement because its measurement is the same as the width of the entrance to the mandala.

Step 2: Two diagonal lines are drawn through the center point of the mandala, defining eight equal pie-shaped spaces radiating from the center point.

183

Step 3: There are three gho-tsed measurement scales used in drawing the Kalachakra Mandala: that of the Mandala of Enlightened Body, the Mandala of Enlightened Speech, and the Mandala of Enlightened Mind. The scale of the Mandala of Enlightened Mind is one-half that of the Mandala of Enlightened Speech, and the scale of the Mandala of Enlightened Speech is one-half that of the Mandala of Enlightened Body. Within the Mandala of Enlightened Mind are the Mandala of Enlightened Wisdom and the Mandala of Enlightened Great Bliss, both of which use the scale of the Mandala of Enlightened Mind. Each gho-tsed is divided into six equal "small parts," or *chachung* in Tibetan.

Step 4: Starting at the center of the mandala, using the scale of the Mandala of Enlightened Body, a point measured one gho-tsed unit from the center determines the root line of the Mandala of Enlightened Mind. Two units out determines the root line of the Mandala of Enlightened Speech, and four and six units out determine the root line and parapet line respectively, of the Mandala of Enlightened Body. Eight units from the center locates the point at which the Brahman line will end, which is where the innermost of the six concentric circles will later be drawn. Thirteen gho-sted units out from the center of the mandala locates the perimeter of the entire mandala.

STEP 3: Proportional relationship of three mandalas and their respective gho-tseds

Enlightened Body

Enlightened Speech

Enlightened Mind

6 chachung

chachung

chachung

STEP 4: Constructing the root lines

End of Brahman line

Extension of Brahman line and perimeter of entire mandala

Root line of Enlightened Mind

Root line of Enlightened Speech

Root line of Enlightened Body

Parapet line of Enlightened Body

Gho-tsed of Mandala of Enlightened Body

STEP 5: The Mandala of Enlightened Mind

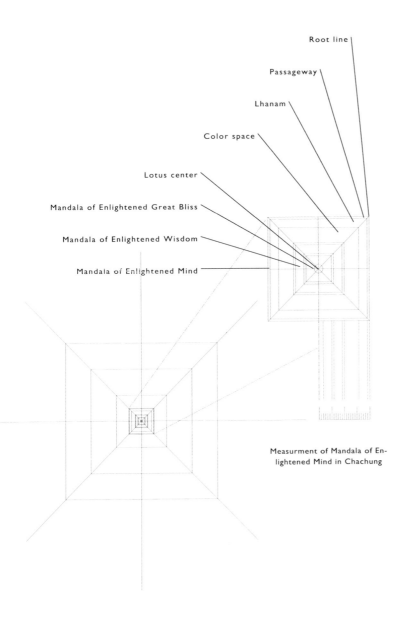

Root line

Passageway

Lhanam

Color space

Lotus center

Mandala of Enlightened Great Bliss

Mandala of Enlightened Wisdom

Mandala of Enlightened Mind

Measurment of Mandala of Enlightened Mind in Chachung

the mandala of enlightened mind

Step 5: Using the gho-tsed measurement scale of the Mandala of Enlightened Mind, a circle is drawn from the center point with a radius of 2 chachung, or "small parts"; this is the center of the lotus flower and the seat of Kalachakra and Vishvamata.

The following points are marked, measuring out in chachung from the center point along the Brahman line: 6, 7, 11, 12, 19, 23, and 24. From these points lines are drawn perpendicular to the Brahman line and end at the diagonal construction line which represent:

• at 6 chachung, the inner square of the Mandala of Enlightened Great Bliss.

• at 7 chachung, the outer line of the inner square of the Mandala of Enlightened Great Bliss.

• at 11 chachung, the inner line of the square of the Mandala of Enlightened Wisdom.

• at 12 chachung, the outer line of the square of the Mandala of Enlightened Wisdom.

• at 19 chachung, the outer line of the area called the "color space" of the Mandala of Enlightened Mind.

• at 23 chachung, the outer line of the area called the *lhanam* of the Mandala of Enlightened Mind.

• at 24 chachung from the center point, the root line, or outer line, of the passageway of the Mandala of Enlightened Mind.

Step 6: A circle is drawn with a radius of 6 chachung from the center point. This circle is divided into eight equal parts, rotated 22.5 degrees off the Brahman line, to define the eight petals of the lotus flower, which are the locations of the eight shaktis. To define the sixteen chambers that will house the eight pairs of deities and eight vases, the following points are marked along the Brahman line, in both directions out from the center: 2, 3, 6, and 7 chachung.

From these points lines are drawn perpendicular to the Brahman line 4 chachung long, located between the points 7 and 11 chachung out from the same Brahman line. Repeating this in all four directions creates the eight deity chambers, each 4 chachung wide, one in each of the four cardinal directions and one in each corner. It will also create the eight vase chambers, each 3 chachung wide and each separated from the adjacent deity chambers by pillars 1 chachung wide.

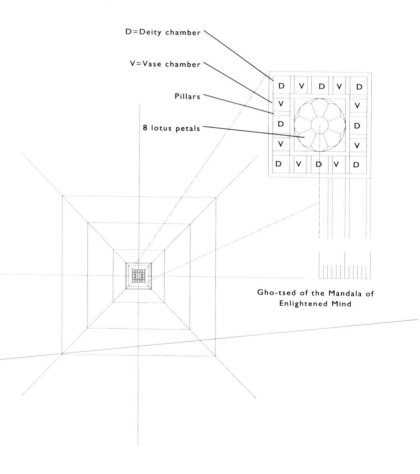

STEP 6: The Mandala of Enlightened Great Bliss and the Mandala of Enlightened Wisdom

D=Deity chamber

V=Vase chamber

Pillars

8 lotus petals

Gho-tsed of the Mandala of Enlightened Mind

entrance and parapet of the mandala of enlightened mind

Step 7: Note that the following descriptions and illustrations apply to all of the three mandalas of the Kalachakra Mandala; only the measuring scale is changed. All the measurements above are based on only one side of the entrance or Brahman line. The calculations are repeated on each side of the Brahman lines on all four sides of the mandala.

The following points are marked in chachung out from the center of the mandala along the Brahman line: $25^{1}/_{2}$, $2^{1}/_{2}$, 30, 33, $34^{1}/_{2}$, and 36.

The width of the mandala entrance is 6 chachung wide, and is determined by measuring at the root line 3 chachung on either side of, and perpendicular to, the Brahman line.

At this location, the inner line of the entrance wall is begun. A line is drawn 6 chachung out from this point parallel to the Brahman line.

A line is drawn from this point 6 chachung away from the Brahman line, parallel to the root line.

A line is drawn parallel to the Brahman line 6 chachung out. The line ends at the parapet line.

At the point $4^{1}/2$ chachung from the Brahman line and $1^{1}/2$ chachung out from the root line, a line is drawn 3 chachung out, parallel to the Brahman line.

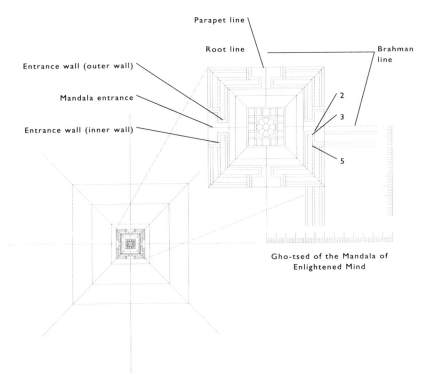

STEP 7: Entrance and parapet of the Mandala of Enlightened Mind

Parapet line

Root line

Brahman line

Entrance wall (outer wall)

Mandala entrance

Entrance wall (inner wall)

2

3

5

Gho-tsed of the Mandala of Enlightened Mind

From this point, a line is drawn parallel to the root line and two points are located at a distance of $10^1/2$ and 12 chachung from the Brahman line. From these two points, lines are drawn parallel to the Brahman line $7^1/2$ chachung out, ending at the parapet line.

Again the point is located $4^1/2$ chachung from the Brahman line and $1^1/2$ chachung out from the root line. A line is drawn parallel to the root line ending 26 chachung from the Brahman line, where it meets the diagonal construction line.

The foundation wall and pillars adjacent to the wall have now been completed.

Now all construction lines within the entrance, walls, and pillars are erased. The entrance, wall, and pillar lines are then drawn heavily, as all other major mandala lines constructed up to this point.

gate of the mandala of enlightened mind

Step 8: To construct the gate, eleven construction lines are lightly drawn parallel to and measured out from the parapet line at these points (with corresponding letter identification): 1(A), 2(B), 6(C), $6^1/2$(D), $7^1/2$(E), $10^1/2$(F), 11(G), 12(H), 14(I), $14^1/2$(J), and 16(K) chachung. Each gate has three stories and a roof. Each story has a beam, a fence, and three chambers. The construction point measure-

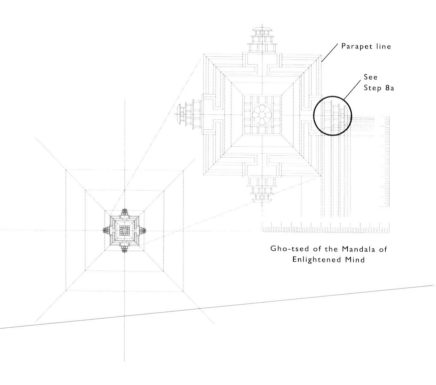

STEP 8: Gate of the Mandala of Enlightened Mind

Parapet line

See
Step 8a

Gho-tsed of the Mandala of
Enlightened Mind

STEP 8a: Gate of the Mandala of Enlightened Mind

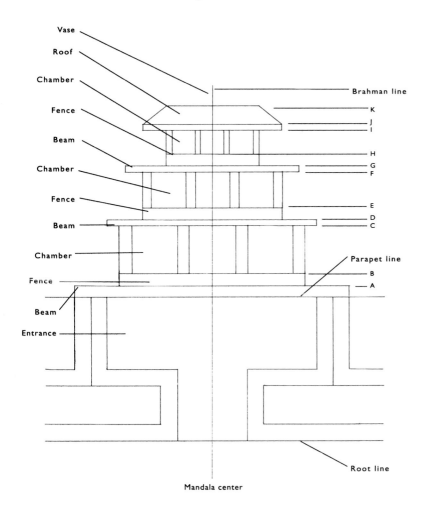

ments (dimensions between each successive point), line designations, and part names are given below.

Step 8a: The following lines are drawn parallel to and measured from the Brahman line. All of the measurements are based on only one side of the Brahman line. To complete the gate, the same measurements are duplicated on the other side of the Brahman line.

A line from the parapet line to A, 12 chachung from the Brahman line.

A line from A to C, 8 chachung from the Brahman line.

Three lines from B to C, 2, 3, and 7 chachung from the Brahman line.

A line from C to D, 9 chachung from the Brahman line.

A line from D to F, 6 chachung from the Brahman line.

Three lines from E to F, $1^1/2$, $2^1/4$, and $5^1/4$ chachung from the Brahman line.

A line from F to G, $7^1/2$ chachung from the Brahman line.

A line from G to I, 4 chachung from the extension of the Brahman line.

Three lines from H to I, 1, $1^1/2$, and $3^1/2$ chachung from the Brahman line. Note that the Brahman line ends at line H.

A line from I to J, 6 chachung from the extension of the Brahman line.

A point is located on line K, 4 chachung from the extension of the Brahman line. A line is drawn from this point to a point on line J, 6 chachung from the extension of the Brahman line.

At the intersection of the Brahman line extension and line K, a vase is drawn 2 chachung tall.

All the lines noted above are darkened. Between these darkened lines, all lines parallel to the root line are darkened. The gate to the Mandala of Enlightened Mind has now been completed.

STEP 9: Gate of the Mandala of Enlightened Speech

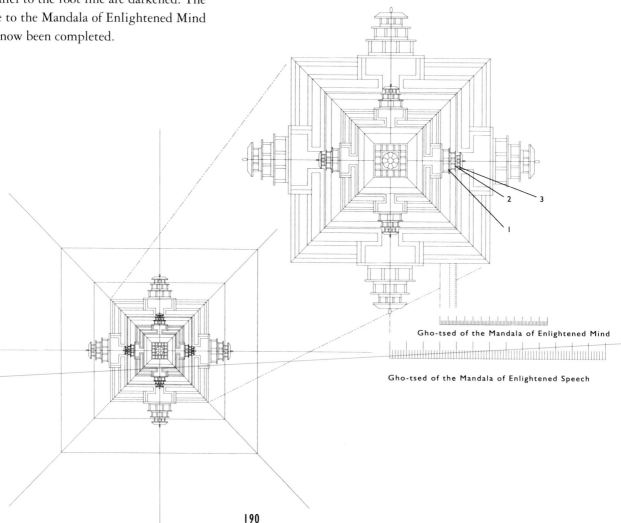

Gho-tsed of the Mandala of Enlightened Mind

Gho-tsed of the Mandala of Enlightened Speech

STEP 10: The Mandala of Enlightened Body

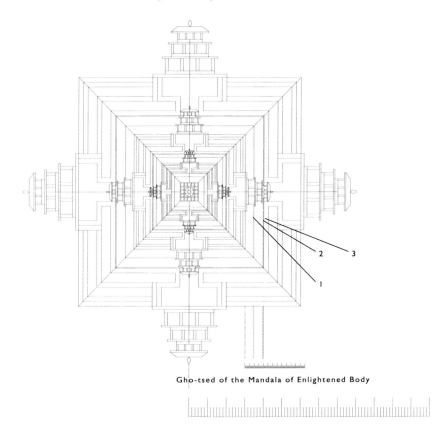

Gho-tsed of the Mandala of Enlightened Body

the mandala of enlightened speech

Step 9: Two measurement scales are used in drawing the Mandala of Enlightened Speech.

Using the gho-tsed measurement scale of the Mandala of Enlightened Mind, three construction lines are drawn parallel to and measured out from the parapet line of the Mandala of Enlightened Mind at these points: 7, 11, and 12 chachung, ending at the diagonal construction line.

Using the measuring scale of the Mandala of Enlightened Speech, the steps from the root line to the top of the mandala gate are repeated as shown in Steps 7 and 8, to complete the construction lines for the Mandala of Enlightened Speech.

the mandala of enlightened body

Step 10: As with the Mandala of Enlightened Speech, the Mandala of Enlightened Body uses two gho-tsed measurement scales.

Using the gho-tsed measurement scale of the Mandala of Enlightened Speech, three construction lines are drawn parallel to and measured from the parapet line of the Mandala of Enlightened Speech at these points: 11, 23, and 24 chachung, between the diagonal construction line and the mandala gate.

Using the measuring scale of the Mandala of Enlightened Body, the steps are repeated from the root line to the top of the mandala gate as shown in Steps 7 and 8 to complete the construction lines for the Mandala of Enlightened Body.

Step 11: Now that the drawing of the palace of Kalachakra has been completed, the six concentric circles are drawn. The innermost of these is located at the end of the Brahman line, 8 gho-tseds from the center using the scale of the Mandala of Enlightened Body. This is at the top of the fence of the highest gate of the Mandala of Enlightened Body.

The remaining circles are constructed at the following points out from the center: $8^{1}/_{2}$, $9^{1}/_{2}$, $10^{1}/_{2}$, $11^{1}/_{2}$, 12, and 13 gho-tseds.

Once the circles are drawn, all remaining construction lines are erased, including the diagonal lines, except for the segment within the color space. All the lines are enhanced with ink. This completes the drawing of the Kalachakra Mandala.

STEP 11: The Six Concentric Circles of the Kalachakra Mandala

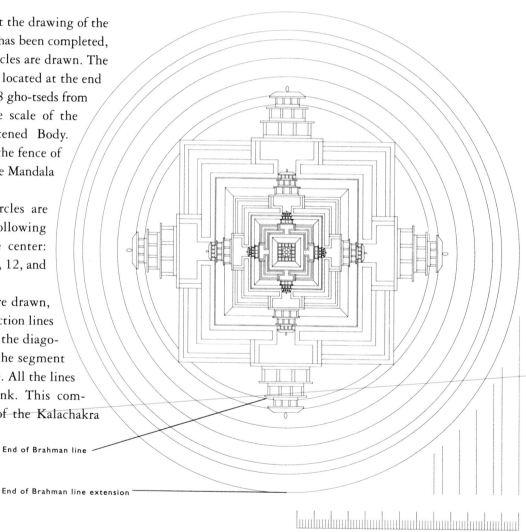

End of Brahman line

End of Brahman line extension

Gho-tsed of the Mandala of Enlightened Body

painting the mandala

During the initiation, four monks begin applying the sand to the Kalachakra Mandala. They are joined by four more monks as the mandala grows larger. The drawing of the mandala is completed on the second day and from the third to the eighth day the monks apply the sand for a total of seven days' work. When the mandala is presented as a cultural offering in a museum or gallery, the drawing

is completed before the exhibition is open to the public. The painting with sand by four monks takes from three to five weeks, depending on the schedules of both the facility and the monks.

The Kalachakra Sand Mandalas depicted in this book measure 6^{1}/$_{2}$ feet in diameter. No matter what the mandala size, the work of applying sand is always started at the very center and progresses out-

With fixed concentration, Namgyal monks seated on the mandala base apply sand to the eastern and western gates of the Mandala of Enlightened Mind.

ward. In the early stages of painting, while the sand mandala is smaller than three feet in diameter, the monks sit on the outer part of the unpainted mandala base, always facing the center.

In the tantric teachings, it is said that if one steps over or on or sits on a ritual instrument, image of the deity, scripture, or mandala, it is not only a sign of disrespect but is also equivalent to the breaking of one's vow. To avoid such an incident, the monks recite prayers each day, visualizing the lines of the drawing of the mandala as being lifted above the mandala base and remaining suspended above it until the workday is completed. When the mandala is about halfway completed, the monks then stand on the floor, bending forward over the base in order to apply sand.

Traditionally, one monk is assigned to each of the four quadrants. At the point where the monks stand to apply the sand, an assistant joins each of the four. Working cooperatively, the assistants help by filling in areas of color while the primary four monks outline the other details with sand.

The monks memorize each detail of the mandala as part of Namgyal Monastery's training program. It is important to note that the mandala is explicitly based on the scriptural text, and that there is no creative invention along the way. Any creative inspiration is manifested within the perfecting of skill.

At the end of each work session, the monks dedicate any artistic or spiritual merit accumulated from this activity to the benefit of others. This practice prevails in the execution of all ritual arts.

There is a good reason for the extreme degree of care and attention that the monks put into their work: they are actually imparting the Buddha's teachings. Since the mandala contains instructions by the Buddha for attaining enlightenment, the purity of their motivation and the perfection of their work allows viewers the maximum benefit.

Each detail in all four quadrants of the mandala faces the center, so that it is facing Kalachakra. Thus, from the perspective of both the monks and the viewers standing around the mandala, the details in the quadrant closest to the viewer appear upside down, while those in the most distant quadrant appear right-side up.

previous page: The Kalachakra Sand Mandala at the American Museum of Natural History, 1988.

left: Namgyal monk Lobsang Samten applies sand to the Kalachakra Mandala. above: Shinga, or wooden scraper, used to adjust sand.

overleaf, left: The four faces of Kalachakra, which correspond to the four quadrants of the Kalachakra Mandala:

black = wind / East, red = fire / South, yellow = earth / West, and white (rear face) = water / North.

In the foreground is the yellow face of Vishvamata.

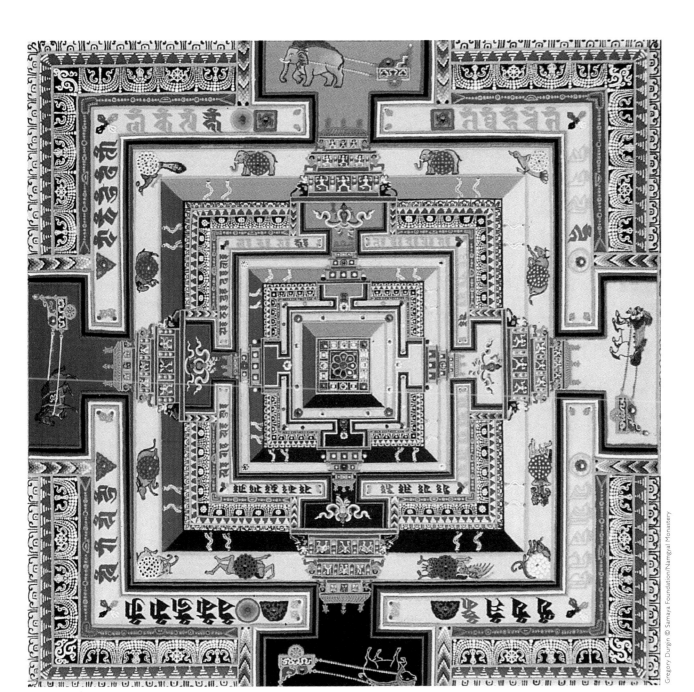

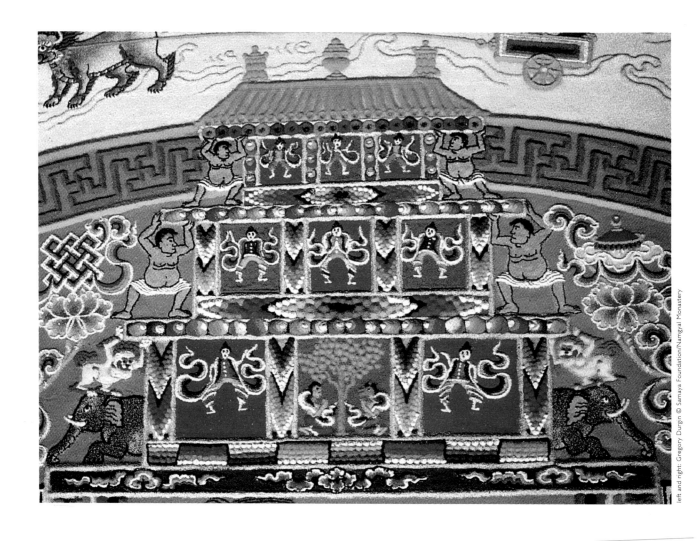

above: The Western Gate of the Mandala of Enlightened Body.

right: Detail of the Western Gate. Male and female "probable humans" seated in devotion beneath a sweet-smelling yellow tree.

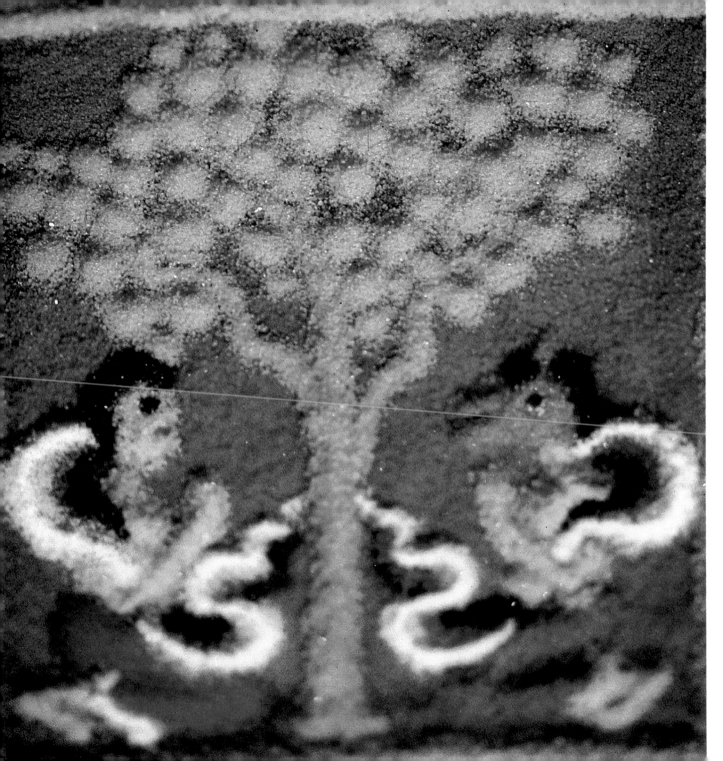

Seven animals at each entrance of the Mandala of Enlightened Body draw a chariot with a green lotus flower which serves as a cushion for male and female wrathful protective deities represented by a pair of colored dots.

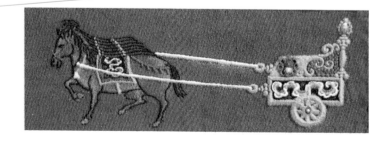

top: In the West, seven elephants against a yellow background. left: In the South, seven horses against a red background.
middle: In the East, seven boars against a black background. right: In the North, seven snow lions against a white background.

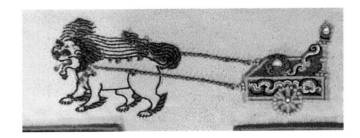

left: Bowl of fruit offering.

above: Offering garden surrounding the eastern gate of the Mandala of the Enlightened Body of the Kalachakra Mandala.

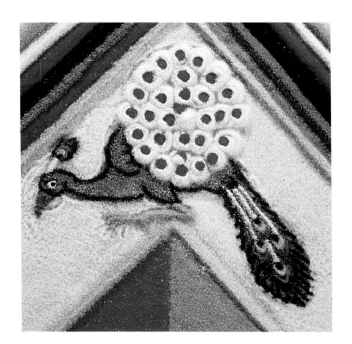

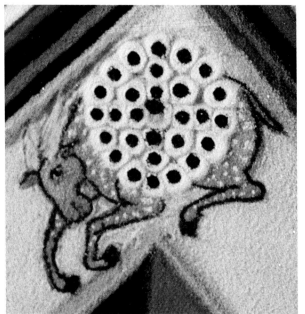

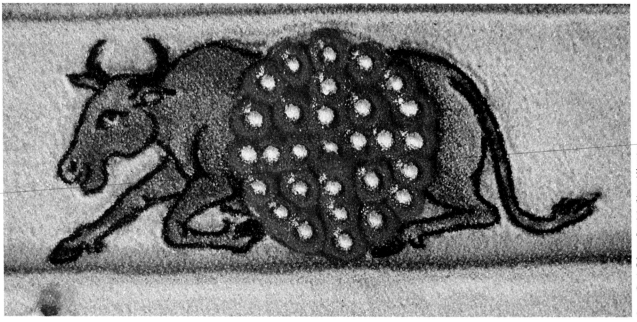

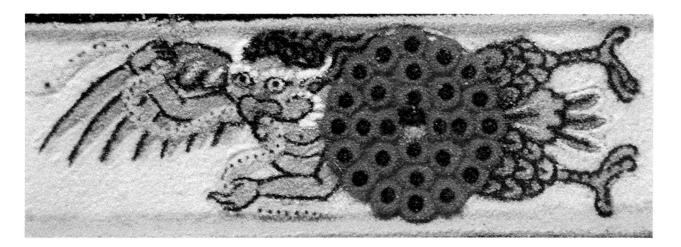

Twelve animals act as mounts for 360 deities symbolized by dots of colored sand, representing the calendar according to the Kalachakra system. The five animals pictured here are: the peacock, ram, Garuda, bull, and mouse.

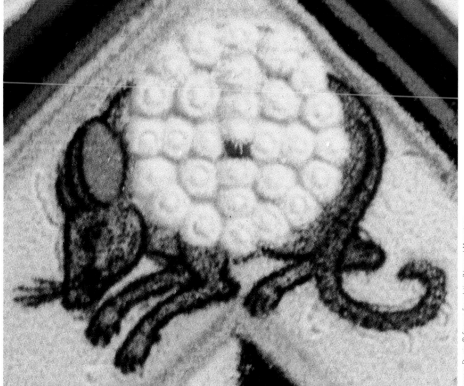

left: Detail of the eastern gate of the Mandala of
Enlightened Body and the adjacent offering garden.
above: Detail of offering garden of the
eastern gate; A blue sword with red flames,
used to cut through ignorance.

above: Offering goddess offers her dance at the northern gate of the Mandala of Enlightened Speech.

right: Wheel symbolizing cemetery grounds where the elements of fire and wind meet.

The wheel serves as a cushion for two protective deities represented by dots at the center.

left: The king of the bird kingdom, Khading Anila, in the East. above: The snow lion Senge Kangpa Gyepa in the West. These two mythical animals, depicted in the water element, are actually visualized above and below the mandala.

Dismantling the Kalachakra Sand Mandala. left: Here, the Dalai Lama cuts the energy of the mandala at the Kalachakra Initiation at Madison Square Garden, New York City, 1991. above: Namgyal monks, led by Tenzin Yignyen, sweep up sand at the Natural History Museum of Los Angeles County, 1989.

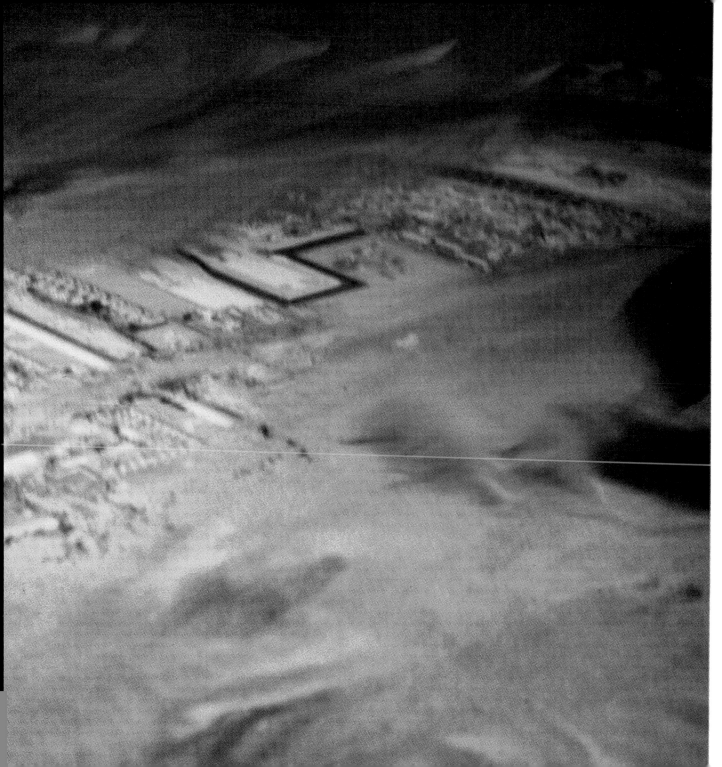

Generally, each monk keeps to his quadrant while painting the square palace. When they are painting the outer six concentric circles, they work in tandem, moving all around the mandala. They wait until an entire cyclic phase or layer is completed before moving outward together. For example, they complete the Mandala of Enlightened Mind before starting the Mandala of Enlightened Speech. This insures that balance is maintained, and that no quadrant of the mandala grows faster than another.

In the tradition of Namgyal Monastery, sand is applied through the end of a long, narrow metal funnel, or *chakpu* in Tibetan. Each chakpu is part of a pair; one is used to rasp a corrugated metal strip attached to the other, which is filled with sand. This rasping vibrates the chakpu, causing the sand to be released through the smaller opening.

There are different sizes of chakpus. Those with larger openings, through which sand flows liberally, are used for filling in background space and making thick walls and borders. Other chakpus with smaller openings for the elaboration of fine details, release sand grains in more delicate streams. The flow of sand is also controlled by the speed and pressure used in rasping. Slow, soft rasping causes the sand to trickle out, even just a few grains at a time, while harder, faster rasping causes it to pour out in a steady stream. Mastery of this technique takes great patience and diligent practice.

The chakpu is not grasped; rather, it lies flat in the open palm of the left hand (assuming one is right-handed), with the thumb placed on top to balance it. The rasping side is facing up. The hand itself is supported by the mandala base, which allows for ease of lateral movement from the wrist. The right hand holds the second chakpu, which does the actual rasping. The monks interpret the sound of the hollow metal chakpus being rubbed together as an expression of the Buddhist concept of emptiness, or the interdependence of phenomena.

The two chakpus symbolize wisdom and compassion; their action symbolizes the Buddhist practice. The result of their action, the completed mandala, symbolizes the enlightened state.

A monk uses an elongated metal funnel called a *chakpu*, which is filled with colored sand. It rests in the palm of his left hand, secured by his thumb and guided by the movement of his wrist. In his right hand, he holds a second chakpu, used to rasp the corrugated surface of the first. This causes a vibration that releases the fine grains of sand from the tip of the instrument.

Traditionally, in ancient times, the Indian Buddhist sand painters used their fingers to apply sand, as monks in some Tibetan monasteries do today. The chakpus used by the Namgyal monks were invented in the 18th century by the Tibetans.

Special wooden scrapers, known in Tibetan as *shinga,* usually measuring one to four inches wide, are used to adjust and straighten lines of sand into narrow, raised walls and other demarcations. The shinga are also used to remove excess sand.

Within Namgyal Monastery, before a monk is permitted to work on constructing a sand mandala he must undergo at least three years of technical artistic training and memorization, learning how to draw all the various symbols and studying related philosophical concepts.

Because of the monks' level of concentration and the high degree of cooperation among them as they work, mistakes or accidents are rare. But occasional errors made within the mandalas may be discov-

Above: The *shinga,* a wooden scraper, is used to straighten the edges of the sand and to remove stray grains from the work area.

Right: Young monks learn to draw the Kalachakra Sand Mandala in a class at Namgyal Monastery in Dharamsala, India.

ered too late for correction by the wooden scrapers. In such a case, a piece of cloth is placed over the large end of a chakpu, and a monk will gently suck up the sand of the mistake into the chakpu.

Viewers of the sand mandala often ask, "What keeps the sand in place?" The answer is that the layers of sand simply rest one upon the other. Gravity is the only adhesive used.

entering the mandala

During the Kalachakra Initiation, the ritual master introduces the students to the deities. His description of the mandala begins at the outermost concentric circle, moving inward toward the center of the innermost mandala, wherein he describes the principal deity Kalachakra and his consort Vishvamata.

We begin our description of the mandala in the center and move gradually outward, the way it is painted. Every motif depicted in the mandala has symbolic meaning and can be interpreted on various levels.

the mandala of enlightened great bliss

The Mandala of Enlightened Great Bliss represents the transcendent experience of enlightened awareness. To the practitioner, this is understood as the consummate union of Kalachakra and Vishvamata; that is, the union of wisdom and compassion.

1. The mandala represents the fifth and uppermost level of the five-story, three-dimensional palace of Kalachakra.

2. A solitary monk sitting in the eastern quadrant of the mandala begins by painting the outline of a small circle at the center, which is the center of the lotus flower. He paints the outline of the eight petals with light blue sand.

3. In the center of the lotus flower, five layers of colored sand are painted, one on top of the other. At the bottom is green, then white, red, blue, and finally yellow, which is the only one we can see. They symbolize, respectively, a lotus flower, the moon, the sun, Rahu, and Kalagni (lunar nodes in the Tibetan cosmology). These layers serve as cushions for the central deities, Kalachakra and Vishvamata, and together they represent the central themes of Buddhist practice: renunciation, bodhicitta, and the realization of emptiness.

4. The lotus flower represents renunciation, the letting go of all attachments. Just as the lotus grows in the muddy water but is not

The first step in painting the Kalachakra Sand Mandala: a monk seated in the eastern quadrant outlines the central lotus flower.

defiled by the muck, renunciation, which is born of the suffering of cyclic existence, bears none of the qualities of the suffering. The lotus flower also represents the pure nature of the mind.

5. The next layer is the moon, symbolizing the pure nature of bodhicitta which cools the disturbing emotions and their results.

6. Directly on top of the moon layer is that of the sun. The sun symbolizes the realization of emptiness, which eliminates one's delusion or ignorance, a primary cause of suffering. Thus, it represents the fire of enlightened awareness which consumes all confusion and distortion caused by human ignorance.

7. On top of the sun layer is a Rahu disc, which symbolizes the wisdom unique to the Kalachakra Tantra, that of immutable bliss.

8. The last and topmost layer, the Kalagni, totally covers the layers just painted. It symbolizes the special method of the Kalachakra Tantra, empty form.

9. Next, a blue vajra is painted on top of the raised mound of layered sand. It represents the mandala's principal deity, Kalachakra. The vajra itself is Kalachakra's main ritual implement, symbolizing the indestructible mind of the Buddha, which has the ability to cut through illusion. The blue color of the vajra, symbolizing Kalachakra's blue body, represents immutability; Kalachakra dwells in an unending state of being, far beyond any kind of degeneration, impairment, or exhaustion.

10. The monk, still sitting in the eastern quadrant, then paints a yellow orange dot to the right of the blue vajra. This dot represents Kalachakra's consort, Vishvamata. Yellow is associated with the earth element and with the fulfillment of one's potential through spiritual practice.

Although we see only the symbols of Kalachakra and Vishvamata, residing in that same space and inseparable from them are two male deities, Akshobhya and Vajrasattva, embraced by their consorts Prajnaparamita and Vajradhatvishvari, respectively. In all, six deities reside in the center of the lotus flower.

The green color to the left of the blue vajra represents Kalachakra's cloak.

11. Like all of the deities represented in pairs throughout the mandala, Kalachakra lives in perpetual embrace with his consort. Their position symbolizes the simultaneous union of compassion and wisdom, which in the Kalachakra Tantra refers to empty form and immutable bliss, the state of enlightenment.

12. The eight petals surrounding the center of the lotus flower are filled with green sand. The petals serve as seats for the eight shaktis. They (with Prajnaparamita and Vajradhatvishvari, who reside at the center with the four other deities) represent the spiritual powers which are the purification of ten winds and represent, in turn, the ten perfections. Each shakti is represented by a symbol—in this case, a dot of color for the seed of the deity's essence or vitality. The term for this in Tibetan is *tigley* and in Sanskrit *bindu,* which translates into English literally as "drop." The colors of the dots correspond to the quadrants of the mandala in which the shaktis reside.

13. The square that surrounds the eight-petaled lotus is now filled with blue sand. This color reflects that of the central vajra or Kalachakra.

14. In each corner of the central square, colored dots representing Kalachakra's

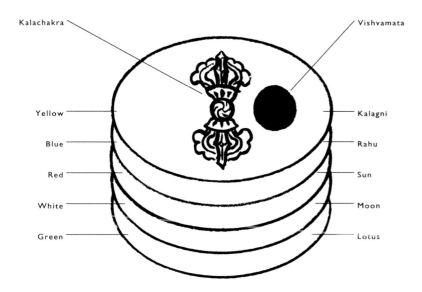

Kalachakra

Vishvamata

Yellow

Blue

Red

White

Green

Kalagni

Rahu

Sun

Moon

Lotus

Cross section of the five layers of sand at the center of the lotus flower of the Mandala of Enlightened Great Bliss.

qualities of body, speech, mind, and wisdom consciousness are placed upon the blue background. The white dot (northeast corner), symbolizing a white conch shell, represents his body; the red dot (in the southwest corner), symbolizing a wooden ritual gong sounding the dharma, represents his speech; the black dot (in the southeast corner), symbolizing a precious black jewel, represents Kalachakra's mind; and the yellow dot (northwest corner), symbolizing a wish-fulfilling tree, represents his wisdom consciousness.

15. A sky-blue border of sand forms a square surrounding the eight-petaled lotus flower. A second sky-blue border surrounds the first. The space between these borders is filled with a chain or rosary of black vajras which represents beams supporting the ceiling of the Mandala of Enlightened Great Bliss. These decorated beams mark the outside boundary of this innermost mandala. It is inhabited by a total of fourteen deities, although only ten are visible here.

the mandala of enlightened wisdom

The Mandala of Enlightened Wisdom represents the subtle mind.

1. The mandala represents the fourth level (second to the top) of the five-story palace of Kalachakra.

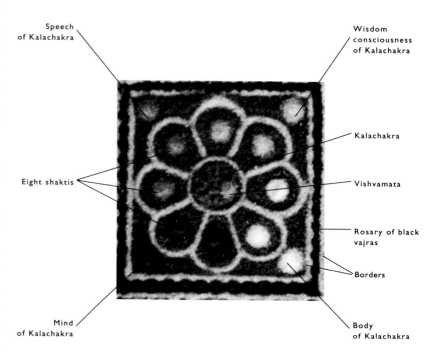

Speech of Kalachakra

Wisdom consciousness of Kalachakra

Kalachakra

Eight shaktis

Vishvamata

Rosary of black vajras

Borders

Mind of Kalachakra

Body of Kalachakra

The Mandala of Enlightened Great Bliss.

2. It contains sixteen black pillars, four in each direction. These pillars, symbolic of the sixteen different kinds of emptiness, are decorated with dots of sand which represent black swords (east), red jewels (south), yellow dharma wheels (west), and white lotus flowers (north).

Surrounding these sixteen pillars are two light blue borders like those surrounding the previous mandala. Between these borders is a chain or rosary of green vajras symbolizing the indestructible nature of the enlightened mind. The spaces between these pillars form sixteen chambers.

3. Four of the sixteen chambers are located in the four cardinal directions and four in the four corners. Each houses an eight-petaled lotus flower, which acts as a cushion for a pair of male and female peaceful deities represented by dots of colored sand. The lotus flowers in the four cardinal direction chambers are white and those in the four corner chambers are red. The sixteen deities residing in this mandala are called the "Ones Gone Thus."

4. The remaining eight chambers located between the chambers of the eight pairs of deities house eight white vases, each between two white lotus flowers. The vases are filled with a nectar of purified substances of the human body, such as blood and bone marrow. Along with the

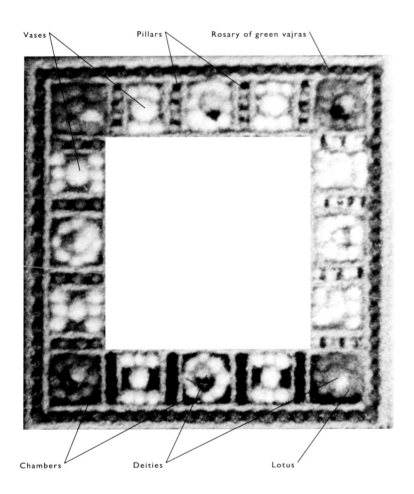

Vases   Pillars   Rosary of green vajras

Chambers   Deities   Lotus

The Mandala of Enlightened Wisdom.

two vases located at the eastern and western entrances to the Mandala of Enlightened Mind, they are the vases of the ten directions. They represent the power of spiritual transformation.

the mandala of enlightened mind

The Mandala of Enlightened Mind represents the coarse mind, which perceives relative truth and absolute truth as one. It houses 70 deities.

1. The Mandala of Enlightened Mind represents the third level of the five-story palace of Kalachakra.

2. The area surrounding the light blue border of the Mandala of Enlightened Wisdom, which appears as four trapezoids, is called the "color space," because the colors of this area correspond to the colors of their respective directions.

3. The white square surrounding the color space is called the *lhanam,* which, in Tibetan, means the actual place where the deity resides. This lhanam is home to twelve pairs of deities, represented by colored dots, residing on four red lotus cushions at each corner and eight white lotus cushions between the four corners. These twenty-four deities are bodhisattvas.

4. The area immediately beyond the border surrounding the white lhanam is a very narrow passageway in the colors of the four directions.

5. Surrounding the passageway are the foundation walls, composed of three parallel lines. From the inside out, white, red, and black give shape to the four entrances. These three foundation walls represent the three vehicles or paths: the Theravada, the Mahayana, and the Vajrayana.

6. At each of the four entrances, a pair of wrathful deities resides on a red or white lotus flower. At the eastern entrance, an additional pair of wrathful deities sits on a second lotus. These ten wrathful deities act as protectors of the entrances to the Mandala of Enlightened Mind.

Opposite: The Mandala of Enlightened Mind, inside of which are the Mandala of Enlightened Wisdom and the Mandala of Enlightened Great Bliss.

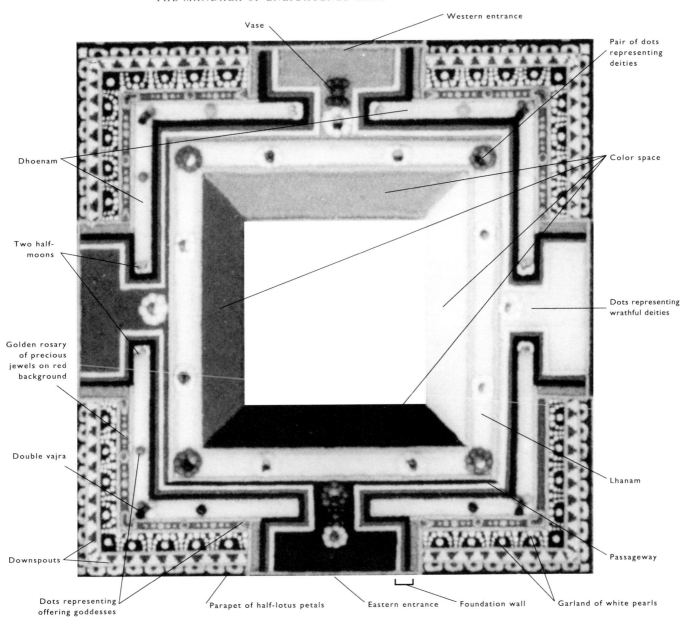

Western entrance

Vase

Pair of dots representing deities

Dhoenam

Color space

Two half-moons

Dots representing wrathful deities

Golden rosary of precious jewels on red background

Double vajra

Lhanam

Downspouts

Passageway

Dots representing offering goddesses

Parapet of half-lotus petals

Eastern entrance

Foundation wall

Garland of white pearls

203

Also at the eastern entrance, a green vase is found between two lotuses, and in the western entrance is a blue vase. As mentioned above, these two vases, plus the eight vases in the Mandala of Enlightened Wisdom, are the vases of the ten directions.

7. The white areas just beyond the foundation walls are known in Tibetan as the *dhoenam*. Two offering goddesses represented by dots reside on each of the four L-shaped white dhoenam, which are at the exterior of the foundation walls. These eight offering goddesses, plus four located in the entrance of the Mandala of Enlightened Speech, are part of a group of six pairs of offering goddesses.

8. At each of the four corners of the dhoenam there is a multicolored double vajra, which symbolizes the four means of a bodhisattva to gather students. They are:

- giving whatever is necessary
- speaking pleasantly
- speaking in accordance with the doctrine
- practicing the doctrine

The colors of the four points of the vajras and their centers represent the five Buddha families—Akshobhya (blue), Vairocana (white), Ratnasambhava (yellow), Amitabha (red), and Amoghasiddhi (green). The vajras also represent stability.

At either end of the dhoenam, situated next to the entrance, are two white half-moons connected to yellow half-vajras, adorned with a red jewel where they meet. These symbolize the Four Noble Truths.

9. The red area outside the dhoenam is decorated with a golden rosary of precious jewels. This area symbolizes wisdom, and the jewels represent method.

10. The next three areas beyond this red area contain architectural decorations and offerings. In the first, on a black background, hang garlands and half-garlands of white pearls. They symbolize the particular qualities of the Buddha which are not shared by others. These pearls spill out of the mouths of decorative sea monsters, which are represented by a dot of white sand in each corner.

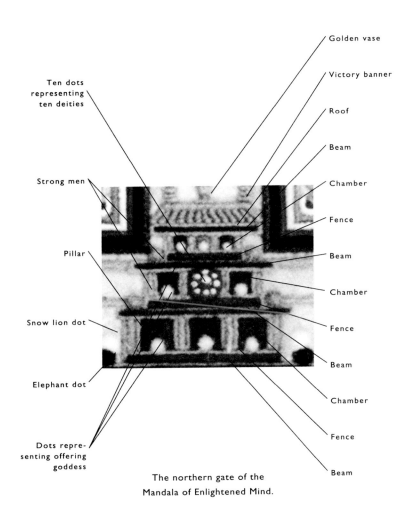

Golden vase

Victory banner

Roof

Beam

Chamber

Fence

Beam

Chamber

Fence

Beam

Chamber

Fence

Beam

Ten dots representing ten deities

Strong men

Pillar

Snow lion dot

Elephant dot

Dots representing offering goddess

The northern gate of the
Mandala of Enlightened Mind.

At the ends of the garlands are offerings, represented by dots, symbolizing mirrors, bells, yak-tail fans, half-moons, flowers, etc. And inside the garlands are many more offerings represented by white dots. The mirrors represent the emptiness of all phenomena. The bells ring the sound of emptiness. The yak-tail fans remove the heat of suffering.

Moving outward, the blue area representing the sky contains white lines forming triangles, which represent downspouts that release rainwater from the palace roof.

The last and outermost detail of the Mandala of Enlightened Mind is a white parapet with the design of a half-lotus petal outlined in black. The parapet symbolizes protection from afflictive emotions and fortifies one against obstruction on the path. There are yellow pillars to the left and right of each entrance.

the gates of the mandala of enlightened mind

1. There are four gates in the Mandala of Enlightened Mind. Each is made up of eleven levels which appear as three stories. Each story has a beam, fence, and chambers, totaling nine levels, all crowned by a beam supporting the roof and the roof itself. These eleven levels represent the levels leading toward the realization of Buddhahood. Each story has three cham-

bers, for a total of nine chambers separated by yellow pillars. Each gate has a golden roof with a golden vase, flanked by golden victory banners.

2. In each of the nine chambers of each gate, a playful offering goddess makes offerings to Kalachakra. Here, they are represented by white dots.

3. The gray and white dots seen at either side of the lower level of each gate symbolize elephants and snow lions, which help to support the gate. The elephant represents empty form and the snow lion represents immutable bliss. On either side of the two upper levels, light flesh-colored dots represent strong men, who help to support the upper beams. The strong men at the lower of these two levels represent the union of empty form and immutable bliss which is dependent on ongoing learning; the strong men at the upper level represent the union of empty form and immutable bliss that requires no more learning; that is, the state of Buddhahood, or enlightenment.

4. On either side of the gate, near the corners of the parapet of the Mandala of Enlightened Mind, are two banners attached to golden poles set in golden vases, with a half-moon and a half-vajra at the top of each. These banners (here depicted as white) appear superimposed over the color space of the Mandala of Enlightened Speech. They wave from the roof of the palace and symbolize conquest over all afflictions that cause suffering.

the mandala of enlightened speech

The Mandala of Enlightened Speech represents the pure qualities of the Buddha's speech. This quality of enlightened speech enabled Shakyamuni Buddha to teach both sutras and tantras, so that disciples of all levels and dispositions could attain realization. The speech of the Buddha is considered most supreme, because it is through speech that he was able to teach and liberate sentient beings. This mandala houses 116 deities.

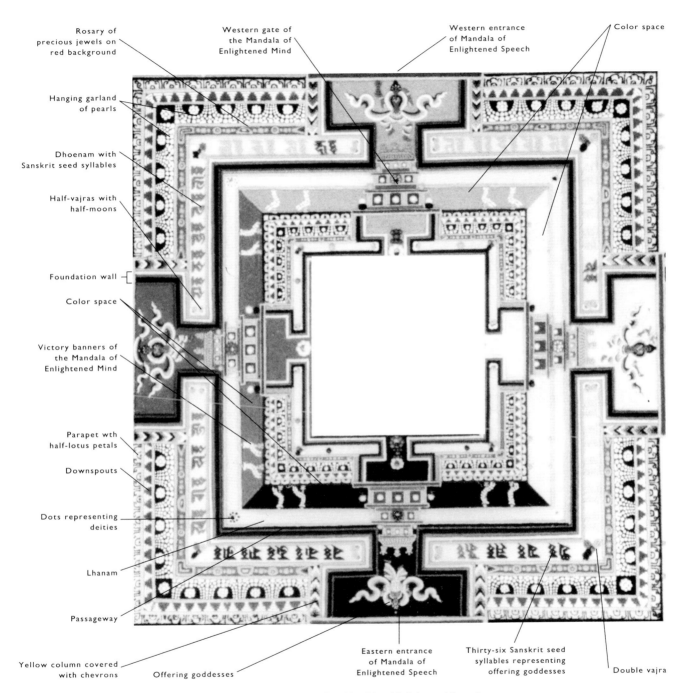

Rosary of precious jewels on red background

Western gate of the Mandala of Enlightened Mind

Western entrance of Mandala of Enlightened Speech

Color space

Hanging garland of pearls

Dhoenam with Sanskrit seed syllables

Half-vajras with half-moons

Foundation wall

Color space

Victory banners of the Mandala of Enlightened Mind

Parapet wth half-lotus petals

Downspouts

Dots representing deities

Lhanam

Passageway

Yellow column covered with chevrons

Offering goddesses

Eastern entrance of Mandala of Enlightened Speech

Thirty-six Sanskrit seed syllables representing offering goddesses

Double vajra

The Mandala of Enlightened Speech.

207

1. The Mandala of Enlightened Speech is the second level of the five-story palace of Kalachakra.

2. The color space, the white lhanam, and the narrow passageway of the Mandala of Enlightened Speech are the same as those of the Mandala of Enlightened Mind.

3. This lhanam contains 80 deities. It has eight lotus flowers, each with ten deities, represented by colored dots. A female deity resides on each of the eight petals, and at the center of each are a male and female deity in union. The lotuses at the four corners are white. At the cardinal directions, superimposed on the middle chamber of each of the four gates, is a red lotus. These 80 deities are called *yoginis* of the Mandala of Enlightened Speech.

4. Five parallel colored lines beyond the narrow passageway serve as foundation walls of this mandala, giving structure to the four entrances. These walls represent the five cognitive faculties: faith, effort, mindfulness, concentration, and wisdom. The colors of the walls (green, black, red, white, yellow) are symbolic of the five wisdoms of the Buddha, which correspond to the five Buddha families. They are as follows:

- Vairocana: Wisdom of Ultimate Reality = Purification of Ignorance
- Akshobhya: Mirror-like Wisdom = Purification of Anger/Hatred
- Ratnasambhava: Wisdom of Equanimity = Purification of Pride
- Amitabha: Wisdom of Discrimination = Purification of Attachment/Desire
- Amoghasiddhi: Wisdom of Accomplishment = Purification of Jealousy

5. In each of the four entrances of the Mandala of Enlightened Speech are colorful offering goddesses, appearing to be directly above the gates of the Mandala of Enlightened Mind. They aren't shown here, due to spatial limitations, although they are actually two of the six pairs of offering goddesses, the first four of which were

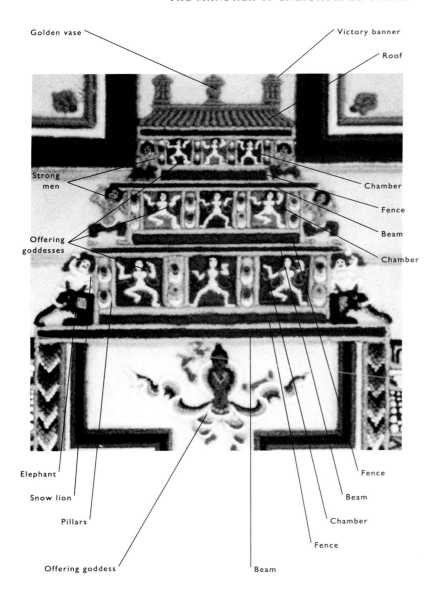

Golden vase

Victory banner

Roof

Strong men

Chamber

Fence

Beam

Offering goddesses

Chamber

Elephant

Snow lion

Pillars

Fence

Beam

Chamber

Fence

Offering goddess

Beam

The northern gate of the Mandala of Enlightened Speech. In the entrance, below the gate with the white background, is an offering goddess.

seen in the dhoenam of the Mandala of the Enlightened Mind (p. 204, Step 7).

6. The four white, L-shaped dhoenams (four Ls) found just beyond the five colored walls are home to 36 *dhoema,* or offering goddesses, represented here by Sanskrit seed syllables. The color of each syllable is the color associated with the goddess.

Just as in the Mandala of Enlightened Mind, double vajras appear in the corners of the dhoenam, and half-vajras with half-moons, each adorned with a red jewel, appear on either side of the entrance.

7. Beyond the dhoenam, the red area decorated with golden jewels, the black background areas containing garlands of white pearls, the blue area with white triangular-shaped downspouts, the parapets, and the victory banners are exactly the same as those of the Mandala of Enlightened Mind.

8. At either side of the entrance is a decoration of multicolored chevrons hung over the columns.

the gate of the mandala of enlightened speech

The four gates of the Mandala of Enlightened Speech are the same as the gates of the Mandala of Enlightened Mind, except the pillars are decorated with multicolored sand. The offering goddesses, elephants, snow lions, and strong men are represented in their natural forms rather than by dots.

Within the square chambers of the gates, offering goddesses make offerings, including the joy of their music and dance, to all deities of the mandala.

At the top of each gate is a golden roof with a golden vase flanked by golden victory banners.

the mandala of enlightened body

1. The outermost mandala, the Mandala of Enlightened Body, is the bottom level of the five-story palace of Kalachakra. Here are housed 536 deities, including the 108 deities depicted in the cemetery grounds.

The three areas surrounding the parapet of the Mandala of Enlightened Speech are, moving outward, color space, lhanam, and the narrow passageway of the Mandala of Enlightened Body.

2. 360 deities reside in this lhanam. Here we see 12 animals, each carrying one of the 12 months of the year.

Each animal bears a lotus with 28 petals. Each of the 28 petals supports a deity representing one day of the 28-day lunar cycle, and a pair of deities in union, symbolizing the full moon and the new moon, resides at the lotus center. Thus, every figure bears a lotus flower which serves as a cushion to 30 deities representing the 30 days of the lunar month, totaling the 360 days of the lunar year.

Note that the lotus flowers in the corners of the lhanam are white and that the more central lotuses are red. These colors are the opposite of, and thus complement and balance, those found in the lhanam in the Mandala of Enlightened Mind. In Tibetan astrology, red is symbolic of the energy of the sun and white is symbolic of the energy of the moon. Red is identified with the female essence and white with the male essence.

3. The five colored lines surrounding the narrow passageway of the Mandala of Enlightened Body are the same as those of the Mandala of Enlightened Speech, which represent the walls. Here the five

walls symbolize the five powers: faith, effort, memory, concentration, and wisdom.

4. The four white, L-shaped dhoenam found just beyond the five colored walls are home to 36 offering goddesses, represented by Sanskrit seed syllables. The colors of the syllables correspond to the colors of their respective directions, with the exception of one syllable in each direction. East and north each have a single blue syllable, and the south and west a single green syllable. All of these goddesses are known as *Chir-Doema* in Tibetan.

5. Just as in the Mandalas of Enlightened Mind and Enlightened Speech, in each corner of the dhoenam of the Mandala of Enlightened Body double vajras are painted in four colors which correspond to each of the four directions. Half-vajras with half-moons, each adorned with a red jewel, appear on either side of the entrances.

6. In the dhoenam and in the space between the Sanskrit seed syllables and the half-vajras are geometric shapes symbolizing the six constituencies, which are the five elements (water, earth, fire, air, and space) plus the wisdom element. On either side of the eastern entrance, two gray crescent shapes symbolize the element of wind, and to the left of the entrance a green circle symbolizes space. On either side of the southern entrance, two red triangles symbolize fire. On either side of the western entrance, two yellow squares decorated with the green swastika (a symbol of stability) symbolize earth, and on the right side of the entrance a blue circle symbolizes the wisdom element. Two white circles on either side of the northern entrance are symbolic of the element of water.

These four geometric shapes represent the types of tasks performed by the Buddha and the protective deities:

- pacifying (circle)
- increasing (square)
- empowering (bow)
- enforcing (triangle)

The geometric shapes in all four directions bear green lotus flowers at their centers. These each act as a cushion for a male *naga* deity

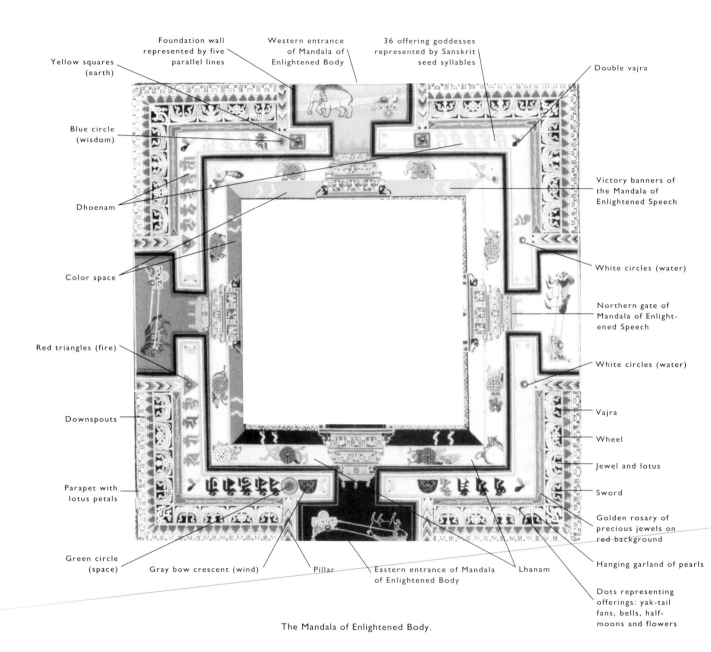

Yellow squares (earth)

Blue circle (wisdom)

Dhoenam

Color space

Red triangles (fire)

Downspouts

Parapet with lotus petals

Green circle (space)

Gray bow crescent (wind)

Pillar

Foundation wall represented by five parallel lines

Western entrance of Mandala of Enlightened Body

36 offering goddesses represented by Sanskrit seed syllables

Double vajra

Victory banners of the Mandala of Enlightened Speech

White circles (water)

Northern gate of Mandala of Enlightened Speech

White circles (water)

Vajra

Wheel

Jewel and lotus

Sword

Golden rosary of precious jewels on red background

Hanging garland of pearls

Eastern entrance of Mandala of Enlightened Body

Lhanam

Dots representing offerings: yak-tail fans, bells, half-moons and flowers

The Mandala of Enlightened Body.

Opposite: Detail of the Mandala of Enlightened Body showing the twelve animals residing in the lhanam. They serve as mounts for the 360 deities that represent the calendar according to the Kalachakra system.

212

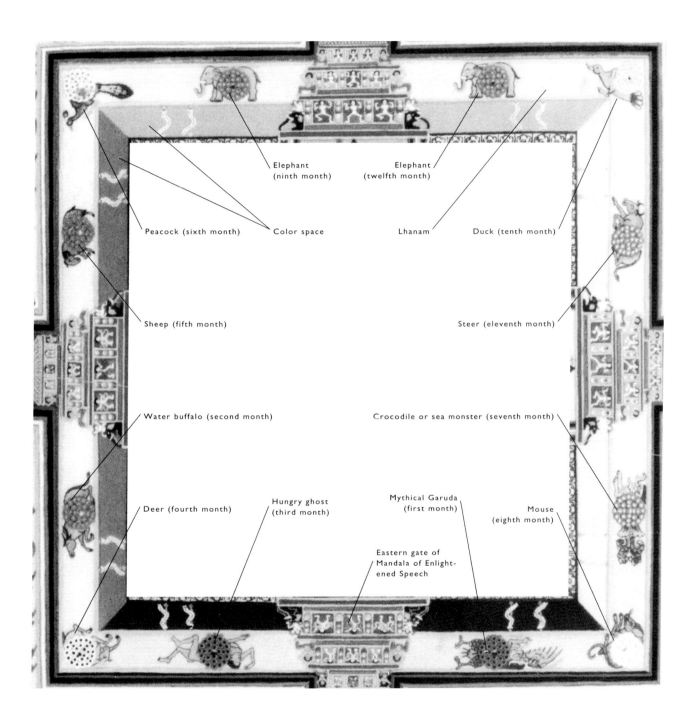

Elephant
(ninth month)

Elephant
(twelfth month)

Peacock (sixth month)

Color space

Lhanam

Duck (tenth month)

Sheep (fifth month)

Steer (eleventh month)

Water buffalo (second month)

Crocodile or sea monster (seventh month)

Deer (fourth month)

Hungry ghost
(third month)

Mythical Garuda
(first month)

Mouse
(eighth month)

Eastern gate of
Mandala of Enlight-
ened Speech

(associated with water and jewel treasures) in embrace with a very fierce female deity. The green circle in the east and the blue circle in the west have no lotus flowers, although the deities are the same as described above. There are 20 deities residing in these elements.

7. Just as in the Mandala of Enlightened Speech, the red area beyond the dhoenam is decorated with a golden rosary of precious jewels. The red area symbolizes wisdom, and the jewels represent method. The different shapes of the jewels represent the four tasks performed by the Buddha, described above.

8. The next three areas beyond this red area contain offerings to Kalachakra, which are the same as those of the Mandala of Enlightened Mind and the Mandala of Enlightened Speech, but with more space available, they are portrayed with different and even more elaborate designs. In the first, a black background is filled with hanging garlands and half-garlands of white pearls, symbolizing the qualities particular to the Buddha which are not shared by others. These pearls spill out of the mouths of decorative designs meant to evoke sea monsters. At the ends of the garlands are offerings symbolized by dots representing mirrors, yak-tail fans, bells, half-moons, and flowers. Inside the garland of white pearls are offerings symbolizing the implements of the five Buddha families, which are:

- a vajra (Akshobhya)
- a wheel (Vairocana)
- a jewel (Ratnasambhava)
- a lotus placed under the jewel (Amitabha)
- a sword (Amoghasiddhi)

These symbols are repeated two times in each direction.

The choice of what kind of offerings to depict here is left to the artists, as an expression of their devotion to Kalachakra. They can draw symbols representing the five sensual objects, eight auspicious signs, eight auspicious substances, seven precious royal emblems, or (as here) the five Buddha families.

If the mandala were viewed as a three-dimensional structure, we would see that the garlands described above adorn part of the upper wall of the palace. In the three-dimensional mandala, viewed from

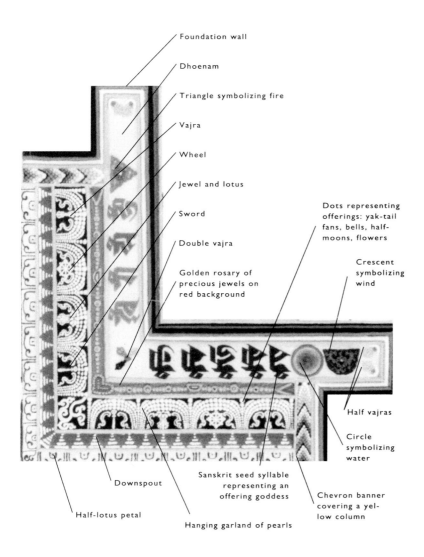

Foundation wall

Dhoenam

Triangle symbolizing fire

Vajra

Wheel

Jewel and lotus

Sword

Double vajra

Golden rosary of
precious jewels on
red background

Dots representing
offerings: yak-tail
fans, bells, half-
moons, flowers

Crescent
symbolizing
wind

Half vajras

Circle
symbolizing
water

Downspout

Sanskrit seed syllable
representing an
offering goddess

Chevron banner
covering a yel-
low column

Half-lotus petal

Hanging garland of pearls

Foundation and wall detail of the northeast corner
of the Mandala of Enlightened Body.

the exterior, most of the details are obscured; these garlands are among the few design details that are clearly visible in the photograph of the three-dimensional mandala shown on pages 58 and 218.

The next area, painted blue and representing the sky, contains triangular shapes representing downspouts, which release rainwater from the palace roof.

The last and outermost detail of the Mandala of Enlightened Body is a white parapet with a half-lotus petal design outlined in black. The parapet symbolizes protection from afflictive emotions as well as fortification against obstructions on the path.

On either side near the corners of the parapet are two victory banners, depicted here as white, although they are often golden. The banners are set in golden vases and supported by golden poles, with a half-moon and half-vajra on top of each.

entrances of the mandala of enlightened body

At the entrance of each of the four quadrants of the Mandala of Enlightened Body, seven animals pull wheeled chariots, each carrying a pair of wrathful protective deities on a green lotus flower cushion. These chariots are symbolic of attaining the realization of emptiness mind through the conceptual mind.

In the east, seven boars draw a chariot against a black background associated with the element of wind.

In the south, seven horses draw a chariot against a red background associated with the element of fire.

In the west, seven elephants draw a chariot against a yellow background associated with the element of earth.

In the north, seven snow lions draw a chariot against a white background associated with the element of water. In Tibetan astrology, the 28 animals that draw the chariots represent 28 constellations.

At either side of the entrance is a decoration of multicolored chevrons hung over the columns.

Due to spatial limitations, two additional pairs of deities included in this group are located in the second innermost concentric circle (the water element) surrounding the square palace, sitting on carts pulled by mythical animals (see pages 224 and 225).

Detail: Two victory banners of the Mandala of Enlightened Body in the garden of the western quadrant.

the gates of the mandala of enlightened body

1. Each gate has three stories separated by blue beams and red fences. Each story has three chambers, for a total of nine, separated by yellow pillars. Each gate has a golden roof with a golden vase flanked by golden victory banners. Starting at the bottom and working toward the top, each gate consists of eleven levels which include the beam, fence, and chamber (three times), the beam supporting the roof, and the roof itself.

2. The four gates here are essentially the same as those in the Mandala of Enlightened Mind and the Mandala of Enlightened Speech. Here the blue beams, red fences, and yellow pillars are decorated in beautiful multicolored designs.

3. On the lower level, on either side of the gates, elephants bear snow lions on their backs and together they support the beams. Additional support on the upper levels is provided by strong men. The nine chambers of each gate are the same as in the other two mandalas, with the exception of the middle chamber of the lower level.

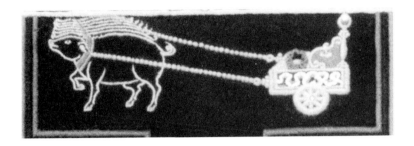

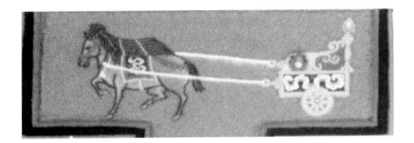

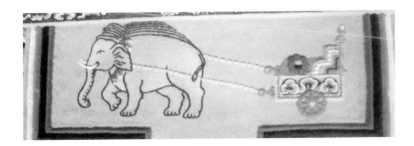

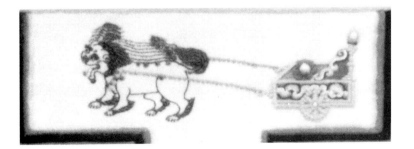

In each of the entrances of the Mandala of Enlightened Body are seven animals pulling a chariot that holds two protective deities.

Top to bottom: The seven boars in the eastern quadrant.

The seven horses in the southern quadrant.

The seven elephants in the western quadrant.

The seven snow lions in the northern quadrant.

4. In the middle chamber of the lower level of the eastern gate, two deer focus their attention on a dharma wheel. They symbolize the instruction that the practitioner must completely concentrate on the teachings without distraction. In this mandala, the black dharma wheel symbolizes the Kalachakra teaching and the two deer symbolize the practitioner's generation and completion stages of the Kalachakra Tantra.

In the middle chamber of the lower level of the southern gate, we see a red offering vase, a conch shell, and a lotus flower. The vase filled with nectar symbolizes the doctrine that, to practice the Buddha's teaching, one must first fill one's mind with the knowledge—which is the nectar—of the Buddha dharma.

The standing conch-shell horn symbolizes the responsibility of the practitioner not to possess the teachings but to pass them on to others. The lotus flower symbolizes spreading the teachings with the motivation to benefit others.

In the middle chamber of the lower level of the western gate, a yellow tree fulfills all aspirations. The tree is also a symbol of the state of enlightenment. At the base of the tree, male and female "probable humans" (a class of beings included within the realm of the gods of desire) are kneeling, showing respect and devotion to the tree. The male has the face of a horse, one of the traditional ways of representing this type of spirit, which is attached to fragrance. Just as the sweet-smelling tree attracts the spirits, here it symbolizes the qualities and knowledge of Kalachakra that attract practitioners.

In the middle chamber of the lower level of the northern gate are a hanging drum, a club, and a hammer. The sound of the drum symbolizes the awakening of sentient beings from the sleep of ignorance. The club represents a support, symbolizing bodhicitta, necessary for the attainment of enlightenment. The hammer symbolizes the realization of emptiness, which removes major obstacles on the path.

5. As in the other two mandalas, on top of the roof of each gate, golden victory banners flank the central, golden offering vases.

A gate of the Mandala of Enlightened Mind, a detail from the three-dimensional Kalachakra Mandala in the Potala Palace in Lhasa, Tibet. 12' in diameter, 18th century.

218

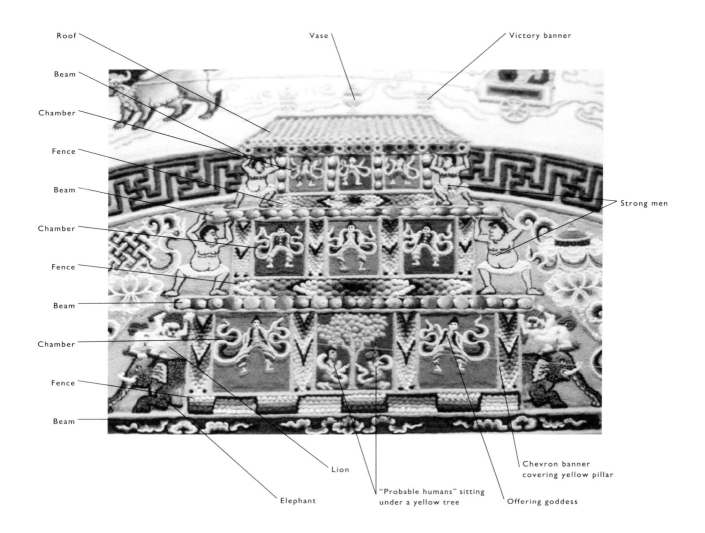

Roof

Beam

Chamber

Fence

Beam

Chamber

Fence

Beam

Chamber

Fence

Beam

Vase

Victory banner

Strong men

Lion

Elephant

"Probable humans" sitting
under a yellow tree

Offering goddess

Chevron banner
covering yellow pillar

The western gate of the Mandala of
Enlightened Body.

offering gardens

On each side of the square palace are four crescent-shaped gardens, formed by the innermost circle surrounding the palace. The background color of each garden—black, red, yellow, or white—corresponds to its respective direction. At the center of each garden is a gate as described above. Each crescent contains various offerings to the deities.

On both sides of the gates are golden vases from which grow vines with green leaves. Lotus flowers growing from these vines serve as supports for the offerings, and symbolize the act of making offerings with the selfless intention to benefit others. Each offering is wrapped in flowing cloths, similar to the way gifts are wrapped in decorative paper.

The purpose of these offerings is to generate great bliss in the deities, which results in the accumulation of merit for the practitioner and the increase of prosperity for all sentient beings.

Here the artists have the liberty to paint their own choice of offerings, usually representing the five senses, the five Buddha families, or teachings of the Buddha. Quadrant by quadrant, the Namgyal monks have painted the following offerings:

1. In the offering garden of the eastern quadrant, as viewed from the west, against a black background, are found from left to right, a red lotus flower representing a garland for the deity, an orange vajra representing indestructible mind, and a standing conch shell representing the melodious sound of the Buddha's speech. To the right of the gate are a white wheel representing the teachings of the Buddha; a multicolored jewel representing the wish-fulfilling gem; and a blue sword with red flames, used to cut through ignorance. The lotus, vajra, wheel, jewel, and sword represent the implements of the five Buddha families.

2. In the southern quadrant, as viewed from the north, against a red background, is seen from left to right, first a golden wheel representing the teachings of the Buddha. The next five offerings rep-

resent the objects of the five senses: a bowl of fruit for taste; a pair of cymbals for sound; to the right of the gate, a mirror for sight, which is form; and a conch shell on its side (filled with fragrant nectar) for smell. A cloth scarf representing touch is found surrounding all the various offerings. Continuing to the right, a standing (sounding) conch shell represents the melodious sound of the Buddha's teachings.

3. In the western quadrant, as viewed from the east, against a yellow background, moving from left to right, offerings include the eight auspicious signs: first, a yellow dharma wheel represents the teachings of the Buddha; second, a victory banner symbolizes the victory of body, speech, and mind over obstacles and negativities; third, an eternal knot symbolizes the interdependence of all phenomena, and the union of wisdom and compassion; fourth, a precious umbrella symbolizes protection from the heat of illness, obstacles, and harmful forces; fifth, two golden fish with colored fins

Opposite: The lower middle chambers of the four gates of the Mandala of Enlightened Body. From top are the eastern gate, southern gate, western gate, and northern gate.

Below: Two fish in the yellow offering garden of the western quadrant, together representing fearlessness, freedom, and spontaneity.

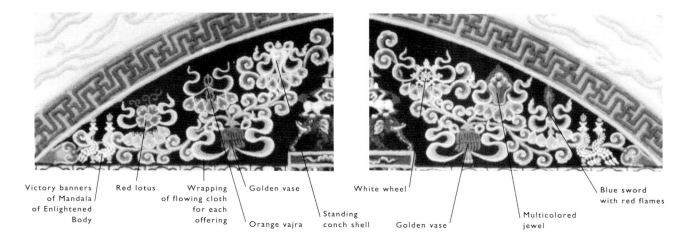

Victory banners of Mandala of Enlightened Body

Red lotus

Wrapping of flowing cloth for each offering

Golden vase

Orange vajra

Standing conch shell

White wheel

Golden vase

Multicolored jewel

Blue sword with red flames

Black offering garden of the eastern quadrant.

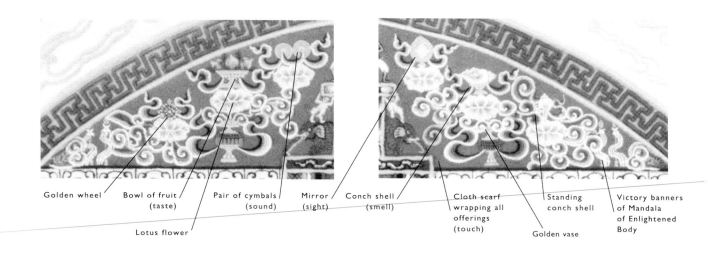

Golden wheel

Bowl of fruit (taste)

Lotus flower

Pair of cymbals (sound)

Mirror (sight)

Conch shell (smell)

Cloth scarf wrapping all offerings (touch)

Golden vase

Standing conch shell

Victory banners of Mandala of Enlightened Body

Red offering garden of the southern quadrant.

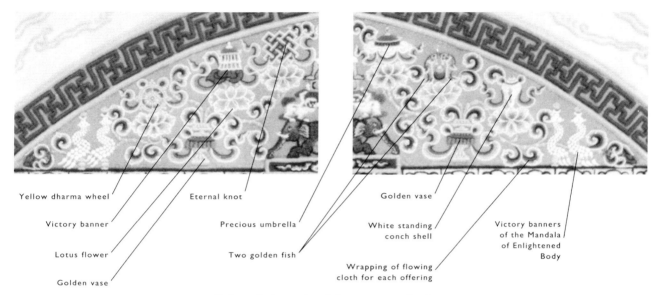

Yellow dharma wheel

Victory banner

Lotus flower

Golden vase

Eternal knot

Precious umbrella

Two golden fish

Golden vase

White standing
conch shell

Wrapping of flowing
cloth for each offering

Victory banners
of the Mandala
of Enlightened
Body

Yellow offering garden of the western quadrant.

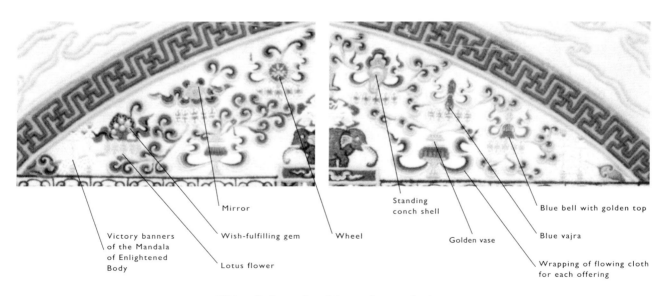

Mirror

Wish-fulfilling gem

Wheel

Victory banners
of the Mandala
of Enlightened
Body

Lotus flower

Standing
conch shell

Golden vase

Blue bell with golden top

Blue vajra

Wrapping of flowing cloth
for each offering

White offering garden of the northern quadrant.

together represent fearlessness, freedom, and spontaneity; sixth, a white standing conch shell sounds the call that urges practitioners to work for the welfare of others; seventh, lotus flowers that appear beneath all the offerings symbolize purification and the blossoming of wholesome deeds; and eighth, the golden treasure vases described in the eastern quadrant here also symbolize long life, wealth, prosperity, and all the benefits of liberation from suffering.

4. In the northern quadrant, as viewed from the south, against a white background, moving from left to right, are found a wish-fulfilling gem, a mirror, and a wheel. On the right side are a standing conch shell, a blue vajra (the description of these symbols is the same as in the other quadrants), and a blue bell with a golden top, which symbolizes wisdom and the sound of emptiness. Each of these offerings rests on a lotus flower.

the six outer circles

The six concentric circles that surround Kalachakra's square palace represent the six constituencies. From the innermost to the outermost circle are the elements of earth, water, fire, wind, space, and wisdom.

1. The first, innermost circle of earth is characterized by the earth color, yellow. An unbroken chain of green swastikas, or interlocking crosses, represents the earth's stability. In addition, the earth circle bears two symbols of the cosmos: a rising full moon in the northeast and a setting sun in the southwest.

2. The water circle is white, containing continuous, wavelike ripples and two mythical animals, each pulling a chariot. In the east we see Khading Anila, king of the bird kingdom; in the west is Senge Kangpa Gyepa, otherwise known as the "eight-legged lion." Each chariot bears a lotus flower upon which sits a pair of wrathful, protective deities in the form of dots. These deities are located above and below the square palace of Kalachakra in the three-dimensional mandala.

These two pairs of wrathful dieties are added to the four pairs that we have already seen (pages 216 and 217) in the four entrances

Setting sun

Earth

Water

Fire

Wind

Space

Wisdom

Senge Kangpa
Gyepa (lion)

Sanskrit seed syllables
representing
protective deities

Wheel with pair of
protective deities

Cart with
pair of
deities

The six circles representing the elements at the
southwest corner of the Mandala of Enlightened
Body, including the setting sun on the innermost
circle (the earth element) and the mythical
animal Senge Kangpa Gyepa.

of the Mandala of Enlightened Body, resting on chariots pulled by seven animals. Together, they make a total of six pairs of deities who protect the six directions (four, plus above and below) of Kalachakra's palace.

Above: Chain of swastikas representing stability in the circle of the earth element, which is surrounded by the water element.

Right: 88 Sanskrit seed syllables representing elemental spirits are found on the pink and gray circles (fire and wind) known as the cemetery grounds.

3 and 4. Beyond the water circle, the pink and gray circles represent the elements of fire and wind respectively. This whole area is known as the cemetery grounds. The ten wheels include one red wheel in each of the four cardinal directions, one white wheel at each corner, and an additional two red wheels, one each in the east and west. Seated on a lotus flower at the center of each wheel is a fierce female deity embraced by a male naga deity, each represented by a dot of sand. In each great cemetery are eleven Sanskrit seed syllables. These 88 seed syllables represent the main elemental spirits among millions.

5. Surrounding the dark gray wind circle is the green circle representing the element of space, which has an interlinking fence of golden vajras. The vajras depicted here have five points and are joined by a golden decorative design. This protective circle of vajras prevents evil spirits from harming the practitioner.

6. The outermost circle is also known as the "great protective circle," as well as the mountain of flames, circle of wisdom, or blazing light. Symbolic of the wisdom element, it has a design of 32 alternating sections of shaded colors. The red and yellow are drawn as

fire, whereas the blue and green are drawn as leaves. These four colors, plus white used in shading, represent the rays of the Buddha's five wisdoms in the form of a rainbow.

There is no border surrounding the great protective circle, illustrating that there are no limitations for the deeds of the Buddha and that his great compassion for all beings is extended with complete equanimity.

In the three-dimensional mandala, five of the six circles are actually layers upon which the mandala rests. The uppermost layer is earth, as the palace itself is constructed directly upon the earth. Below the earth is water, fire, wind, and finally space, which is all-pervasive. The sixth circle, the wisdom element, forms a protective sphere which totally surrounds the palace of Kalachakra.

This completes our description of the Kalachakra Sand Mandala.

Above left: The fifth and sixth circles, representing the space and wisdom elements. The space element displays a chain of vajras, while wisdom is represented by a symbolic "mountain of flames."

Above: Two red wheels, each serving as a cushion for two fierce deities, located in the southern quadrant, in the circles representing the fire and wind elements.

dismantling
ceremony

At the conclusion of the Kalachakra Initiation on the twelfth day, the sand mandala is dismantled. During its presentation as a cultural offering, the mandala is usually dismantled on the final day of the exhibition as follows:

1. Prayers request that the deities return to their sacred abodes. Once the monk presiding over the prayers is satisfied that the deities have left, the dismantling process is begun.

2. All 722 deities symbolized by the colored sands are skillfully picked up one by one by the head monk as he recites the Kalachakra mantra, and the sand is placed in an urn.

Deities that are singular or in a yab-yum grouping are picked up individually. For larger groupings of deities, the monk simply passes his fingers through them, picking up a portion of each. The disman-

tling is done in the reverse order of the making of the sand mandala. The head monk begins at the perimeter, picks up the protective deities in the cemetery grounds, and works clockwise toward the

Opposite and above: Namgyal monks dismantle the Kalachakra Sand Mandala. The American Museum of Natural History, August 1988.

center until he removes Kalachakra and Vishvamata.

3. Next, starting at the outer edge of the eastern quadrant, the head monk cuts through the mandala along the Brahman lines with a vajra, thus cutting the energy of the mandala. This is repeated, in turn, in the southern, western, and northern quadrants, and along the diagonal lines in the southeast, southwest, northwest, and northeast corners.

4. Standing in each of the four directions, the monks sweep the remaining sand into the center of the mandala. It is then placed in a specially prepared urn.

The monks carry the sand to the river or ocean, in a procession. At the water, the monks sit on a carpet with the vase containing the sand in front of them. In their prayers, they request that the protective spirits of the water accept the consecrated sand for the benefit of all beings. They visualize the aquatic life blessed by the essence of the sand. When the purified water rises from the ocean to the clouds and falls from the clouds as rain on the land, it purifies the environment and all its inhabitants. The monks then pour the sand into the water, saving some to give to those assembled in celebration of the event. Each person receives a small amount of blessed sand, which he is instructed to take home and place in a body of water or around the foundation of a house for protection.

Carrying the consecrated sand through the streets of New York City, 1988.

The dismantling of the sand mandala may be interpreted as a lesson in nonattachment, a letting go of the "self-mind." The ceremony reflects the Buddhists' recognition of the impermanence and transitory nature of all aspects of life. The monks believe that the dismantling of the mandala is the most effective means of preserving it.

Offering sand to the Hudson River through a tube that protects it from the wind.

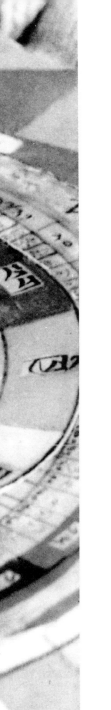

World View According *to* Kalachakra

the universe of the mandala

THE COSMOLOGY REPRESENTED by the colored sand of the mandala is the complex view of the world according to Kalachakra. When the Dalai Lama states that Kalachakra is a vehicle for world peace, he is not speaking lightly. Behind his statement is not only a suggestion that world leaders come together with renewed motivation to work for the benefit of all beings and the survival of the planet, but also an implicit plea for a new understanding based on a larger, more comprehensive vision of the world. The depth of information contained in the Kalachakra Sand

Tibetan astrological calculations.

Mandala gives us an insight into the vast network of interrelationships comprising the physical, mental, and spiritual worlds in which we exist simultaneously.

There are three cycles of time outlined by the three main chapters in the basic Kalachakra text: outer, inner, and alternate. The outer concerns the astrological calculation of planetary movement and the geography of the universe.

The inner Kalachakra focuses on the functioning of the human body and mind. The body is dealt with at length, beginning with conception and embryonic growth, and proceeding through a physical analysis of the person in Buddhist terms. The mind is treated less extensively.

The third chapter, which is known by several names in English—other, secret, and alternate—has three subdivisions. The first is the Kalachakra Initiation itself. Second is the *sadhana,* or generation stage, which includes specific instruction on the meditation practice. And lastly there is the completion stage, or the way to attain realization.

All three cycles are interrelated, as the cosmos and the human physiology and mind are reflections of one another. At the Tibetan Medical and Astrological Institute in Dharamsala, India, it is considered essential for learning anatomy and the functions of the inner elements to know the outer Kalachakra, including the planetary movements and their effects. And advanced meditation practice is dependent upon the physical condition of the internal organs, as well as the influences exerted upon them by the heavenly bodies.

history of astrology in tibet

The first astrological theory and practice came to Tibet from China in the 7th century, when Princess Wencheng, the Chinese wife of Tibetan King Songtsen Gampo, brought with her various astrological and medical texts. A second wave of Chinese influence began in the 10th century. The Chinese system introduced what are known as "element" calculations, using a sixty-year, element-animal calendar cycle.

The Kalachakra system was brought to Tibet in the 11th century. It was blended with the Chinese influences by early masters of the Sakya and Kagyu traditions of Tibetan Buddhism. The first year of the Kalachakra sixty-year cycle is 1027, when the Kalachakra Tantra reappeared from Shambala.

One of the important Sakya scholars of astrological studies was Chogyel Pagpa Lama, who was the tutor of the Mongol ruler of China, Kublai Khan, in the 13th century. It was through his influence that the Kalachakra calendar became the official calendar of Tibet. Since the Chinese system was in use when the Kalachakra system appeared, the animals and elements associated with the Chinese years were adopted for naming the years in the Tibetan calendar although the two systems have different starting years.

Two different astrological systems based on the Kalachakra Tantra evolved in Tibet. The first is the Tsurpu lineage, developed by the Third Karmapa, also in the 13th century, which is followed by the Kagyu school to this day. The second, the Phugpa lineage, was developed in the 15th century and is followed by the Nyingma, Sakya, and Gelug schools. Both systems include broad Chinese-style calculations, known as the "yellow" system.

In the 17th century, during the reign of the Great Fifth Dalai Lama, Indian-style mathematics was developed and applied to Tibetan astrology. Another influence from India is derived from the Svarodaya (Arising from the Vowels) Tantra, in which a correlation is made between the vowels of the Sanskrit alphabet and the dates of the lunar month. By comparing the initial vowel of a person's name with that of a date, predictions can be made. This system is chiefly used for personal horoscopes.

The current astrological system, compiled in the 1980s at the Tibetan Medical and Astrological Institute, comes from these traditions, and continues to use the basic Kalachakra calendar.

It has been pointed out that the positions of the planets as calculated by the Kalachakra system do not always correspond to the calculations of Western science. However, the Tibetan system was never intended for use in navigation or for sending a rocket to the

A personal horoscope prepared for the author at the Tibetan Medical and Astrological Institute in Dharamsala, India.

moon. The practical application of the Kalachakra cosmology is to help clear away obstacles to health and happiness on our Karmic path to spiritual realization.

the tibetan horoscope

Astrological influences are seen to exist, not as independent heavenly bodies guiding us from afar, but in close relationship to the individual's consciousness, as a reflection of or in correspondence to one's behavior and inner life. Therefore, calculations can tell what conse-

Above and opposite: Tibetan calendar.
Two pages from the *Astrological Handbook*,
10"x3", Tibet, 17th century.

quences may occur if corrective actions are not taken. This is all part of the elaborate system of cause and effect (*karma*) that composes the universe.

Tibetan horoscopes are less concerned with planetary influences on character and personality than Western horoscopes. What is presented is a picture of how one's life may unfold. There are a number of ways of calculating the life span; the original Kalachakra text used the maximum of 108 years. The possible life spans calculated by various techniques indicate the variety of karmic seeds which could ripen if nourished in particular ways.

The life span in a Tibetan horoscope is divided into nine periods, each ruled by one of the heavenly bodies. Since the teachings of the Kalachakra Tantra emphasize the precious nature of a human birth, the horoscope may inspire a person to find appropriate ways of overcoming unfavorable conditions by developing the most beneficial karmic seeds. Whenever an unpleasant development is predicted, a remedy is offered. For instance, it may be suggested that one give alms to the poor, or save a life, or sanctify a portrait of one's spiritual teacher. This advice is intended to help put one back on the path of altruistic motivation.

Another benefit of seeing the ups and downs of a lifetime charted in a complete picture is that it offers insight into suffering. A person with an overview of his or her own suffering may better understand

the suffering of others, and therefore develop the all-important karmic seed of compassion.

the tibetan calendar

Each year of the sixty-year cycle calendar system is ruled by one of twelve animals: the rat, steer, tiger, hare, dragon, snake, horse, sheep, monkey, bird, dog, and pig. These animal signs each correlate in turn to one of the five elements: wood, fire, earth, metal, and water. The element calculations are used for the purpose of making personal predictions, such as the prognosis and timing of illnesses, obstacles, marriage, births, and finally the timing of a person's death, including indications of which ceremonies to perform.

The Kalachakra Sand Mandala may be read as a calendar, with the twelve animals in the Mandala of Enlightened Body representing the succession of the twelve months. They each carry on their back a lotus flower hosting thirty deities, including the two in the center which represent the full moon and the new moon.

There are many complex aspects to the Kalachakra calendar, including solar days and lunar days. These refer to the precise amount of time it takes for either the sun or the moon to travel one specific unit of an entire cycle. The solar days last from dawn to dawn and are numbered by the dates of the month. Lunar days, named by the days of the week, are based on the phases of the moon and are of unequal lengths. Lunar and solar days do not correspond exactly to one

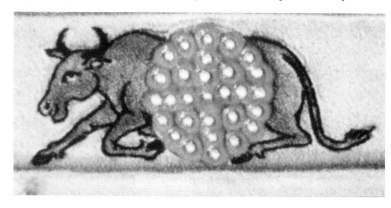

Right: The steer, one of the twelve animals in the Kalachakra Sand Mandala representing the succession of the months of the year, located in the northwest area of the Mandala of Enlightened Body.

Opposite: Model of the universe according to the Kalachakra teachings. At the center is Mt. Meru, surrounded by symbols of the four continents. The orbiting circles represent the movement of the planets and the months of the year. The twelve animals ruling the sixty-year calendar cycle surround twenty-seven constellations.

another. This makes for days with double dates, omitted dates, and an occasional additional month, which is something like our leap-year day.

A feature shared with the ancient Greek system is the naming of the days of the week after the planets—Sunday for the sun, Monday for the moon, and so on. In fact the Tibetan word for "weekday" (*gza*) is the same as that for "planet."

There are many days in the Tibetan calendar that are recognized, for various reasons, as being auspicious or inauspicious. Astrology is popularly used among Tibetans to determine dates for weddings, journeys, and business ventures. Physicians use astrology to determine the best date for administering medical treatments. And it is important for setting dates to make offerings and observe other religious rituals. For instance, it is considered auspicious to begin a

spiritual practice as the moon is waxing, so that the benefit of the practice will also expand. The Kalachakra Initiation is always given on the day of a full moon.

An astrological prediction is used much like a weather forecast. It determines the likelihood of things going well or badly according to the cosmological factors, which can always be influenced by one's behavior.

the external kalachakra

The Kalachakra system of astrology, sometimes known as the "stellar calculations" or "star studies," and the Greek (or Western) systems share a common pan-Indian source, so there are similarities. As in Western astrology, the Tibetan zodiac is divided into twelve signs and twelve related houses. The signs bear the same names as those in the modern West (Aries, Taurus, Gemini, and so on), but they are referred to as houses (*khyim*).

What we in the West would call the houses—those areas of the sky or horoscope that denote the various "departments" of a person's life (the physical body, personal finances, siblings and relatives, etc.)—are known as periods (*dus-shyor*) and carry slightly different meanings.

As in the modern Western system, ten "planets" are used, but in this system only seven of these are heavenly bodies. (This was true as well in the older Western system, before the age of the telescope.) These are the seven visible bodies of the sun, the moon, Mercury, Venus, Mars, Jupiter, and Saturn. The remaining three in the Tibetan system are the comet, and Rahu and Kalagni (known as *Ketu* in the Hindu system), which are the north node and the south node of the moon. The comet is not used in horoscopes, but the lunar nodes are important for predicting solar and lunar eclipses.

the internal kalachakra

In the internal Kalachakra, emphasis is placed on the functioning of the human body and of the coarse and subtle minds. It is very im-

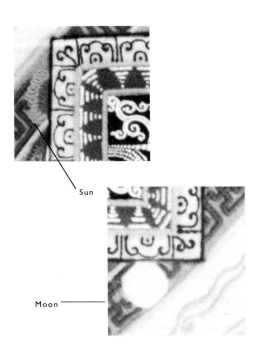

The sun and the moon depicted in the Kalachakra Sand Mandala.

Top: The sun sets in the earth element at the southwest corner of the Mandala of Enlightened Body.

Bottom: The full moon rises at the northeast corner.

portant for the Kalachakra student to know that the internal winds are in motion, just as the planets are. That is why meditators who study the Kalachakra Tantra first learn the external Kalachakra, which details the movements of the sun and the moon.

The importance of these two heavenly bodies in the tantra is underlined by their representation by the principal deities themselves, Kalachakra (the moon) and Vishvamata (the sun). The purpose of the practice of Kalachakra—to achieve the purified mind of the deity—requires harmonizing one's inner being with the structure of the cosmos.

In the internal Kalachakra, the sun and the moon correspond to, or "rule," the right and left channels of the body. It is necessary to know the solar and lunar days and how to calculate them to work effectively with the variable sun and moon energies, which affect the passage of the winds through the body.

In Kalachakra, as in other Buddhist tantras and in Tibetan medicine, the subtle energy of the body is pushed by the "winds" through the "channels" of the human anatomy. Although these concepts do not have direct correlations in Western medicine, Tibetan doctors tell us the channels are part of the nervous system.

The channels are divided into left (*kyangma*) and right (*roma*), which correspond to the polarities of male and female. The sun is female in the Tibetan astrological system and the moon is male. The pull between these polarities is the reason we experience various inclinations of energy, mood, and mind/body balance. The central channel (*ooma*) is the pathway of balance and stability.

The deity Kalachakra has three colored necks representing the three wind channels. The right (*roma*) is red and influenced by the sun, and is the channel through which the sun-wind passes. The white neck is the left channel (*kyangma*); it is influenced by the moon, being the channel through which the moon-wind passes. The blue neck (*ooma*) is the neutral and central channel.

The neutral wind corresponds to the lunar nodes, Rahu and Kalagni. This neutral wind (known as "Rahu's wind"), which is not always present, is important for the practitioner of the generation and completion stages of Kalachakra, and generally it is experienced only by accomplished meditators. The Kalachakra Initiation provides the empowerment for binding together the winds of the left and right channels into the central channel.

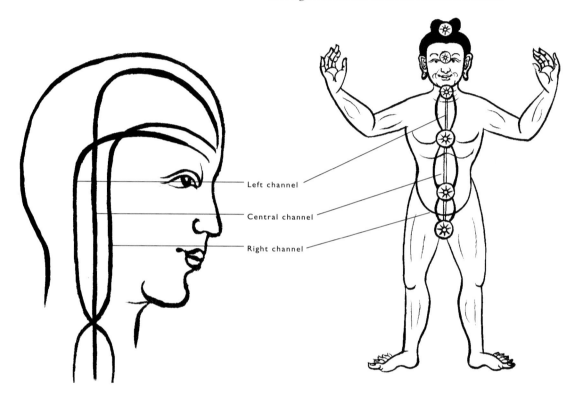

Left channel

Central channel

Right channel

The left, right, and central channels of the body. The goal of the Kalachakra practitioner is to bind together the winds of the left and right channels into the central channel, thereby stilling the ever-changing mind.

The practitioner who achieves control of the ever-moving winds inside the body, and especially the subtle winds, can still the ever-changing mind. But until that time, the winds act to agitate consciousness.

The axis and balance of the regenerative fluids are also influenced by the planetary movements. For instance, the red, female regenerative fluid, or blood, is influenced by the sun; the white, regenerative fluid of the male, or semen, is influenced by the moon.

the alternate kalachakra

When we speak of the alternate Kalachakra, we are referring to the stages of initiation, generation, and completion. The chapter in this book on the Kalachakra Initiation provides more detail about the first of these three stages.

The alternate Kalachakra is the path of transformation. The initiation forms the basis for developing the actual practice. During the generation stage, the practitioner develops a clear visualization of himself or herself as the deity Kalachakra, including his abode, the Kalachakra Mandala. This can only be done sketchily at first, but the meditator continues until he or she is able to maintain a precise visualization of the entire mandala with all its details in a space the size of a pea, for as long as desired.

The generation stage, in turn, provides the basis for the completion stage. Once the meditator can maintain the self-visualization as the deity in the mandala, he or she employs advanced techniques to actualize the visualization. This requires manipulation of the physiological processes, including control of the winds and the endocrine system. The realization of the state of mind produced by this practice is not yet the end of the path; it is but the first of twelve stages leading to the ultimate goal of enlightenment, or Buddhahood.

It is clear why the Kalachakra is among the highest levels of tantra, requiring dedicated practice based on firm motivation. This is why the Dalai Lama says that before we can bring about world peace, we must work to attain our own inner peace, always motivated by the desire to benefit all sentient beings. We must go beyond the illusion that we are each a single entity at the mercy of powerful physical forces. The Kalachakra Tantra makes it clear that attaining enlightenment affects not only our inner being but our bodies, the stars, and ultimately, our entire cosmos as well. We are all an integral part of the interrelated universe, the Mandala of Kalachakra. Our very survival depends on our awakening to this truth.

Transformation *of* Consciousness

"IN THE SEVENTEEN YEARS I've been here," said Dr. Malcolm Arth, director of the Department of Education of the American Museum of Natural History, of the Kalachakra Sand Mandala exhibition and demonstration, "I don't think I've ever experienced this combination of quiet and intensity among the public. The average museum visitor spends about ten seconds before a work of art, but for *this* exhibit, time is measured in minutes, sometimes *hours*. Even the youngsters, who come into the

Namgyal monk Lobsang Samten helps a child try her hand at sand painting at the American Museum of Natural History. New York City, 1988.

245

museum and run around as if it were a playground—these same youngsters walk into this space, and something happens to them. They're transformed."

Even the Dalai Lama has asked why people who are not Buddhist practitioners have such a strong response to the experience of seeing the Kalachakra Sand Mandala. After all, here is an aspect of one of the most advanced, most complex of all the Indo-Tibetan Buddhist tantras, a tradition that is bafflingly "foreign" to most Westerners and extremely difficult to understand even for longtime Buddhist students. Yet thousands of people every day would wander in from the hot summer sun and become transfixed, speaking only in whispers if at all, for long periods of time.

The reactions of the museum visitors demonstrated that something was being communicated. One man noticed, "You do feel calm—I mean *everybody*. Nobody's talking loud. There's no screaming and yelling. The kids are behaving themselves. It's amazing." Another said, "It has a very peaceful, very calming, and very centering impact on me, especially if I've come from the subway. To come here makes my day."

The real impact of the exhibition seemed to come from the firsthand experience of observing the extremely intense concentration of the monks at work. People were most impressed by the meditative focus they maintained as they applied the sand through the funnels. One man, who likened their work to pastry decoration, was, like many others, impressed at how they could work undisturbed with so many people moving around them and asking questions.

Perhaps some of the bystanders were able to sense the grace and ease in the monks' use of their bodies—the careful positioning of their elbows, the way in which their hands and wrists were held—that expresses the inner calm of their contemplative focus. But most of the onlookers were not consciously aware of the monks' arduous preparation through years of meditation and prayers for the purification of their motivation.

A Namgyal monk in deep concentration.

transformation of consciousness

What is it that makes the Kalachakra Sand Mandala, a highly refined and esoteric ritual art form, so powerful a vehicle for communicating with contemporary Westerners? The training the monks receive is not merely drawing lessons, nor is their meditation only the kind that clears and pacifies the mind. Their mental activity is focused on developing their altruistic intention to be of benefit to all sentient beings. They generate this in their prayers at the beginning of the day, and it continues like the steady hum of a generator.

This causes an almost electric kind of energy to pervade the room. What surprised the Dalai Lama, museum administrators, and the media was the discernible effect of this force on a random Western audience in a secular setting.

Through the careful, mindful way in which he uses his body; through the continuous monitoring of the content of his speech and mind; and through meticulous attention to every detail of the iconography, the ritual artist becomes a selfless conduit for that which is so much larger than himself. And when these artisan monks construct the canonically specific images of the sand mandala or other ritual art forms, they inevitably meditate on their symbolic meanings, which thus have the potential to become an integral part of their beings. This heightened attention to detail and awareness of implicit meaning is repeated over and over again wherever Buddhists practice ritual art, whether for a Tibetan religious ceremony or in an American museum.

Attention to detail is critical throughout the entire construction of the sand mandala. If the artists are off by even a fraction of an inch, the whole mandala will reflect this, so an attempt at perfection must be made from the very beginning of the laborious process. Inevitably the monks are at the same time contemplating, both individually and in their conversation as a group, the philosophical significance of every detail, color, and placement. There is a vast, complex logic inherent in the design of the Kalachakra Sand Mandala, and the monks

A monk practices the ritual art of making *torma* offerings.

develop an understanding of its meaning over many years, both conceptually and experientially, in the act of constructing it. Each time they participate in a sand mandala they are expanding their knowledge of the tantra as well as their peace of mind.

If we make a mistake while painting in oil or gouache, we can simply paint over it. Of course there are ways of correcting some of the mistakes possible in a sand painting, but the sand is as delicate and fragile a material as could be worked with, and even a slight deviation can cause damage difficult to correct. As in all things, the monks themselves differ in their ability to do sand painting. There are some who draw or apply sand with exceptional grace, which is not only a reflection of their training but also of their state of mind.

By learning to focus on that which is beneficial to others as an integral part of the ritual, the monks enter into a state of union with the deity. Western athletes, whose performance is dependent on deep concentration, refer to such a state as "entering the zone." This is also what is meant by "the dancer becomes the dance." The intense degree of training and physical preparedness required of their discipline brings them to a level of consciousness where they are no longer distracted by personal preoccupations but are instead completely prepared for spontaneous and appropriate action.

The difference, however, between the Western athlete or dancer and the Tibetan ritual artist is in the motivation. Instead of doing his best in order to gain praise or win a competition, the ritual artist is motivated to make a perfect sand mandala as a means of achieving liberation from suffering for all sentient beings.

the art of simply being present

The curious analogy that can be made between ritual artists and athletes is evident in that both are completely compelling to behold in action. What accounts for this magnetism is their total lack of self-consciousness, something we rarely see in our society. When we see someone who is fully engaged in what he or she is doing, without falseness, without scattered or needless activity, and without preoccupation with appearance, we respond to it.

Greg Louganis: an athlete in "the zone." He visualizes the perfect dive before executing it.

In Zen terminology this is called "being present." It is the kind of attunement that always coexists with a heightened state of attention and that has been, for millennia, consciously developed in the ritual arts of cultures throughout the world.

"The dancer becomes the dance." Laura Dean
Musicians and Dancers in performance.

Another way of speaking about it is as a state of surrender. One gives oneself over; one is totally there. This is an experience that can be found equally in the person who is creating the art or the performance and in the person who is viewing, hearing, or receiving it. Children as well as adults noticeably surrendered themselves to the experience of the sand mandala in the museum, foreign as it might have been to them.

Passersby watch the dismantling of the Kalachakra Sand Mandala at the IBM Gallery. New York City, 1991.

a lesson in impermanence

With all the hard work and intensity of focus devoted to the construction of the Kalachakra Sand Mandala during its presentation as a cultural offering at the various institutions, Western viewers expected to share in some expression of celebration as the work came to completion. Instead, there was none; the monks were visibly unmoved. This is because the completion of this spectacular mandala is not thought to be the end at all, but only one part of the larger ritual and only one step toward attaining the enlightened state of mind. It is important to note that the same degree of focus and mindfulness exhibited in the construction of the mandala is demonstrated in its dismantling process.

The tradition of the Kalachakra Sand Mandala continues as the sand is poured into the waters of Santa Monica Bay.
Los Angeles, 1989.

Significantly, each time the mandala was swept up, regardless of how methodically or ritualistically this was done, Western onlookers reacted strongly, in many cases with great emotion. Many viewed the act not as a dismantling but rather as a destruction. What transpired was most unexpected, almost inconceivable, to the mind-set of our culture, which places such value on possession. The lesson we came away with—some of us less with the wisdom of understanding, perhaps, than with the shock of disbelief—was the Buddhist view of the impermanence of all phenomena. As the monks said,

251

"Just as foreign as it is to you to think of not preserving the sand mandala, so it is equally unthinkable to us to keep it. The best way of preserving this tradition is to dismantle it and come again to make another one."

The ritual art in its sacred context is unconcerned with "product," as it is always a part of the much larger ritual. In this case, the sand mandala is constructed primarily as the vehicle through which the ritual master confers the empowerment of the tantra to the student. Once this has been accomplished, it has fulfilled its function. Just as the deities were initially invoked into the sand mandala for the purpose of conferring the initiation, so at its conclusion the ritual master thanks them and requests that they return to their sacred home. Thus the deities are removed from the sand mandala, and the sand is swept up and poured into the local waters.

the heart of the matter

At the center of the process is the motivation of altruism. With this intention, the artist meditates on attachment to understand that the things we cling to are impermanent, and none is more insubstantial than the "I" which does the clinging. But the lesson conveyed is actually greater than that of impermanence. The idea of possessing any of the ritual arts, and holding on to either their form or the accomplishment derived from them, defeats the purpose. Rather, the emphasis is placed on perfecting the mind of the practitioner, who generates in himself or herself the body, speech, mind, and wisdom consciousness of the deity. With this as the objective, the idea of permanence in ritual art is as inconceivable as not wanting happiness.

It all depends on our point of view. If we see Buddhist ritual arts in the context of their sacred meaning, then we see them as an integral part of the ritual, which is itself part of the realization of the goal of the enlightened state of mind. But by trying to decipher their meaning through the context of those plastic arts whose intention is to express the artist's inspiration, and which have a beginning (creation) and an end (exhibition, presentation, demonstration, and even sale), we attach a very different meaning. If one is accustomed to preserving a work of art, or seeking material value for it, then of course the idea of creating art as part of a ritual for the sake of the benefit of others would seem foreign indeed. In that case, it would be logical to see the objective of the ritual arts as a lesson in impermanence, and a divergence from the context of possession, which implies attachment.

In fact, the monks object to the phrase "creating the sand mandala," because they claim that they are "constructing" it strictly according to the text. They cannot "create" that which already exists.

At the completion of the ritual, the artist goes a step further by offering any merit he or she has accumulated as a result of this activity to the benefit of others. This act of altruism serves to further the goal of nonattachment, since possessing the fruits of his or her action can also serve as an obstacle to attaining the goal of realizing Buddha mind. More than impermanence, it is a lesson in nonattachment. For the nonpractitioner, it is letting go of the beauty and material value of the object; for the practitioner, it is letting go of the accumulation of merit, joy, and bliss consciousness for the benefit of others.

product vs. process

There is a parallel here to what young children go through as they learn to hold a pencil and write their name or draw. At the end of the day, the child often comes home euphoric from this wonderful learning experience and shows his parents what he has done. Proud of the child, his parents put the work up on the kitchen wall as if to say to the neighbors and to the world, "My child has accomplished this." But this emphasis can easily be misunderstood by the child. What is fundamentally important to his development is not what he brings home, or even the approval of his parents, but what happened inside him during the learning process. The parents' kindhearted attempts to acknowledge the child appears to place value only on the tangible evidence of achievement.

In our product-oriented culture, part of us tends to get suppressed: that of the creative process itself. Children begin to feel they are loved for what they produce rather than for who they are, and who they are in reality is an ongoing process of learning and growing. They come to understand unconsciously that they have to create something that conforms to an external arbitrary standard, and

Young monks practicing a Kalachakra dance at Gomang Tashi Kyil Monastery in eastern Tibet, 1987.

within given time limits, rather than at their own natural pace. Intuitive knowledge gets repressed and often lost. They learn to compete with the other children in the class rather than to explore and appreciate their creative potential and to enjoy the excitement of the learning experience.

The children who come to the museum become transfixed watching the monks and the entire painstaking, unhurried process, and they leave with a sense of having experienced something beyond the day-to-day activity of their circumscribed worlds.

a transformation of values

The answer to the question of why people have been so drawn to the silent observation of a seemingly uneventful spectacle such as the construction of a sand mandala is perhaps that each one of us intuitively recognizes the importance of our own internal process. The sand mandala reminds us of something almost lost in our society: It is not only the intrinsic beauty of the piece that holds us, or the opportunity to assist in the preservation of an ancient living tradition, or even to marvel at the physical dexterity and perfection of mind required of the artist—though these are rare enough in today's mechanized world. What we are reminded of most forcefully, as we come into this space, is the depth and capability of the human mind.

The Dalai Lama agrees with the neuroscientists that we are probably using only one percent of our mental capacity. Tibetan Buddhism, through centuries of the most rigorous study, has learned to harness the power of concentration. Its exotic and mysterious rituals are all based on one all-encompassing "secret"—that the union of wisdom and compassion is the key to the attainment of great bliss. Now we have discovered that, if a handful of monks demonstrate a work of art based on the sincere motivation of compassion for others, even people stepping in from the teeming streets of New York or the freeways of Los Angeles will respond to that which resonates from deep inside us all.

Afterword

THE PRESENTATION of the material in this book is quite revolutionary. Secrets that are centuries old are not only revealed to a mass audience but also translated and interpreted so that their very esoteric concepts and terminology can be widely understood. Traditionally in Tibet, such material was read only by devoted scholars for the purpose of becoming enlightened. But today, the Earth is getting smaller and our cultural mixture is expanding. In this globalization of knowledge and wisdom, Kalachakra, like other Indo-Tibetan tantras, is finding its place. It is my hope that this book will add to its compassionate message, and to the Dalai Lama's statement that Kalachakra is a vehicle for world peace.

I asked the Dalai Lama if he could suggest a simple meditation for those readers who have never been exposed to Buddhist practice before. He said, "Just for a moment, concentrate on the infinite, deep blue sky. When we look at something like a table, our consciousness just stops on it. But when we look at deep space, there is nothing to stop our mind. Then, eventually, the mind automatically withdraws. The mind itself goes deep inside, stopping all thought and restlessness. Meditation is refreshment of the mind. Just breathe, and look at the sky."

I request the forgiveness of Kalachakra, the masters of the lineage, and the reader for any mistakes in this book. May any merit gained from this work be offered for the benefit of all beings.

Opposite: The stupa adjacent to the bodhi tree at the site of the Buddha's enlightenment at Bodhgaya, India. The Dalai Lama says that just as people are blessed by visiting such a place, so too is the power of the place increased by their presence. But he also reminds us that "the actual stupa exists in the heart center in each of us."

Picture Credits

Front Matter

Pages ii–iii, *King Songsten Gampo and two wives.* Photo/copyright: Samaya Foundation. Courtesy Mokotoff Asian Arts Gallery, New York; vi–vii *Dalai Lama applies first sand.* Photo/copyright: John Bigelow Taylor/ Tibet Center.

Introduction

Pages xiv–1, *Monks at prayer, Bodhgaya.* Photo/copyright: John C. Smart; 3, *Lay people at Bodhgaya.* Photo/copyright: Moke Mokotoff. Courtesy Mokotoff Asian Arts Gallery, New York; 4, *Monk prostrating.* Photo/copyright: Moke Mokotoff. Courtesy Mokotoff Asian Arts Gallery, New York; 5, *Monk with chakpu, Kalachakra Mandala (detail).* Photo/copyright: Ernst Haas. Courtesy Ernst Haas Studio, New York; 7, *Bodhgaya crowds.* Photo/copyright: John C. Smart.

The Path of Kalachakra

Pages 8–9, *Kalachakra Mandala, Peking, 1932.* Photo: Basil Crump. Courtesy/copyright: Jacques Marchais Center of Tibetan Art, Staten Island, New York; 12, (left) *Avalokiteshvara.* Photo: Armen. Courtesy/copyright: Newark Museum, anonymous gift, 1981; 12, (right) *Vajrapani.* Photo: Armen. Courtesy/copyright: Newark Museum, gift Doris Weiner, 1969; 13, (both) *Manjusri.* Photo: Stephen Germany. Courtesy/copyright: Newark Museum bequest Estate of Eleanor Olsen, 1982; 15, *Kalachakra.* Photo: Jim Coxe, neg. no. 336754. Courtesy/copyright: Department of Library Services, American Museum of Natural History, New York; 16, (left) *Power of Ten.* Copyright: Kalachakra Program, Deer Park, 1981, David Patt, editor. Courtesy Deer Park, Oregon, Wisconsin; 16, (right) *Mani stone.* Courtesy/copyright: Newark Museum, Carter D. Holton Collection, 1936; 17,

Kalachakra mantra. Copyright: Kalachakra Program, Deer Park, 1981, David Patt, editor. Courtesy Deer Park, Oregon, Wisconsin; 19, *Shakyamuni Buddha, Tibet.* Photo/copyright: John Ford. Courtesy Berthe and John Ford Collection; 22, *Raktayamari Mandala.* Photo/copyright: John Bigelow Taylor. Courtesy Zimmerman Family Collection; 23, *Sand mandala, Tibet, 1935.* Photo: Cutting. Courtesy/copyright: Newark Museum, C. Suydam Cutting Collection.

The Sand Mandala as a Cultural Offering

Pages 26–27, *People viewing mandala at American Museum.* Photo: Gregory Durgin. Copyright: Samaya Foundation/Namgyal Monastery; 29, *Lobsang Samten/Guhyasamaja Mandala.* Photo: Barry Bryant. Copyright: Samaya Foundation/Namgyal Monastery; 30, (left) *Thekpu sketch.* Drawing: Gregory Durgin. Copyright: Samaya Foundation; 30, (center) *Thekpu structural schema.* Drawing: Stanley C. Bryant. Copyright: Samaya Foundation; 30, (right) *Detail of thekpu/brocades.* Photo: Barry Bryant. Copyright: Samaya Foundation/Namgyal Monastery; 31, (left) *Phuntsok Dorje painting thekpu.* Photo: Gregory Durgin. Copyright: Samaya Foundation; 31, (right) *Thekpu.* Photo: Barry Bryant. Copyright: Samaya Foundation; 32, *Monks praying, American Museum.* Photo: Barry Bryant. Copyright: Samaya Foundation/Namgyal Monastery; 33, (left) *People viewing completed mandala.* Photo: Gregory Durgin. Copyright: Samaya Foundation/Namgyal Monastery; 33, (right) *Viewing mandala on monitor.* Photo: Barry Bryant. Copyright: Samaya Foundation/Namgyal Monastery; 34, *Procession to river.* Photo: Warner Bryant Scheyer. Copyright: Samaya Foundation; 35, *Monks praying on beach at Santa Monica Bay.* Photo: Don Farber. Copyright: Samaya Foundation; 37, *Kalachakra Fire Offering Mandala with John Denver.* Photo: Gregory Durgin. Copyright: Samaya Foundation/Namgyal Monastery.

The Life of The Buddha

Pages 38–39, *Maya's Dream.* Courtesy/copyright: Asian Art Museum of San Francisco, Avery Brundage Collection; 40, *Birth of Buddha.* Courtesy/copyright: Newark Museum, gift Heerameneck

Galleries, 1965; 41, *Infant Buddha (detail).* Photo: Joseph Szaszfai. Courtesy/copyright: Yale University Art Gallery, gift Paul F. Walter; 42, *Education of Prince Siddhartha.* Courtesy/copyright: R.M.N.–Musée National des Arts Asiatiques–Guimet, Paris; 43, *The Four Encounters of Prince Siddhartha.* Courtesy/copyright: R.M.N.–Musée National des Arts Asiatiques–Guimet, Paris; 44, *Palace Scenes: Prince Siddhartha and Yasodhara Enthroned.* Courtesy/ copyright: Seattle Art Museum, Eugene Fuller Memorial Collection, 39.34; 45, (top) *Scenes From the Life of the Buddha.* Courtesy/copyright: Los Angeles County Museum of Art, gift Ahmanson Foundation; 45, (bottom) *The Departure and Buddha's Temptation by Mara and His Daughters.* Courtesy/copyright: Metropolitan Museum of Art, Fletcher Fund, 1928; 46, *Ascetic Shakyamuni.* Courtesy/copyright: Mr. and Mrs. James W. Alsdorf, Chicago; 47, *Buddha Attacked by the Evil Forces of Mara.* Courtesy/copyright: Cleveland Museum of Art, purchase from the J.H. Wade Fund, 71.18; 48, *Shakyamuni Buddha.* Photo/copyright: John Bigelow Taylor/Tibet Center. Courtesy Zimmerman Family Collection; 51, *Buddha Teaching.* Photo: Blakeslee-Lane. Courtesy/copyright: Berthe and John Ford Collection; 52, *Stupa at Sarnath.* Photo: Barry Bryant. Copyright: Samaya Foundation; 57, *Paranirvana.* Photo: Otto E. Nelson. Courtesy/copyright: Mary and Jackson Burke Collection.

The Early History of Kalachakra

Pages 58–59, *3-D Palace of Kalachakra.* Photo/copyright: Robin Bath; 60, *King of Shambala.* Courtesy/copyright: Newark Museum, bequest Elizabeth P. Martin, 1976; 61, *Miniature stupa.* Courtesy/copyright: Newark Museum, Crane Collection, 1911; 62, (left) *Cosmic Mt. Meru.* Courtesy/copyright: Zimmerman Family Collection; 62, (right) *Cosmic Mt. Meru schema.* Drawing: Barry Bryant. Copyright: Samaya Foundation; 64, *Vajrapani.* Courtesy/copyright: Newark Museum, purchase 1977 Mr. C. Suydam Cutting Bequest Fund; 66, *Mandala of Shambala, Tibet.* Courtesy/copyright: R.M.N.–Musée National des Arts Asiatiques–Guimet, Paris, 92EN3902 (MA1041); 70, *The Kingdom of Shambala (detail).* Courtesy/copyright: R.M.N.–Musée National des Arts Asiatiques–Guimet, Paris; 73, *Mahasiddha.* Courtesy/copyright: Robert Hatfield Elsworth Private Collection; 74, *Virupa, Naropa, Saraha and Dombi Heruka (detail).* Courtesy/copyright: Museum of Fine Arts, Boston, gift John Goelet, C46-316; 75, *Nalanda University.* Photo/copyright: Sean Jones. Courtesy Tibet Image Bank, London.

Kalachakra Comes to Tibet

Pages 76–77, *Padma Sambhava (detail).* Photo: Logan, neg. no. 333391. Courtesy/copyright: Department of Library Services, American Museum of Natural History; 78, *Tara with Throne and Prabhamandala.* Courtesy/copyright: Newark Museum, Sheldon Collection, 1920; 79, *King Songsten Gampo and two wives.* Photo/copyright: Samaya Foundation. Courtesy Mokotoff Asian Arts Gallery, New York; 80, (left) *Jowo Buddha.* Photo/copyright: Tamara W. Hill, 82:20.2; 80, (right) *Jokhang Temple roof.* Photo/copyright: Valrae Reynolds. Courtesy Newark Museum; 81, *Map of Tibet, 1872.* Photo/copyright: Samaya Foundation. Courtesy Office of Tibet, New York; 82, *Samye Monastery.* Courtesy/copyright: Newark Museum, Sheldon Collection, 1920, 20.271; 83, *Three great religious kings of Tibet.* Photo: Barry Bryant. Copyright: Samaya Foundation, 84, *Milarepa.* Photo: Armen. Courtesy/copyright: Newark Museum, purchase 1975 Anderson Bequest Fund, 75.94; 85, (both) *Kalachakra.* Photo: Armen. Courtesy/copyright: Newark Museum, purchase 1976 Members, Membership Endowment, Bedminster, Inc., Samuel C. Miller and Andrew Sponer Funds, Mary Livingston Griggs and Mary Griggs Burk Foundation; 86, *Buton.* Courtesy/copyright: Asian Art Museum of San Francisco, Avery Brundage Collection; 87, *Tsong Khapa.* Courtesy/copyright: Newark Museum, purchase 1920 Sheldon Collection, 20.270; 88, *Ganden Monastery, 1932.* Photo/copyright: Hugh Richardson. Courtesy British Museum, Tibet Image Bank, London; 89, *Ganden Monastery, 1959.* Photo/copyright: Stone Roots. Courtesy Tibet Image Bank, London.

The Dalai Lamas and Namgyal Monastery

Pages 90–91, *Lhamo Latso.* Photo/copyright: Philippe Goldin; 92, *Shadakshari Avalokiteshvara.* Photo/copyright: Moke Mokotoff. Courtesy Mokotoff Asian Arts Gallery, New York; 94, *Dalai Lama (Third, Gyalwa Sonam Gyatso).* Courtesy/copyright: Robert Hatfield Elsworth Private Collection; 96, *Namgyalma.* Photo: Barry Bryant. Copyright: Samaya Foundation; 97, *Dalai Lama (Fifth, Gyalwa Ngawang Lobsang Gyatso).* Photo/copyright: John Taylor. Courtesy Rose Art Museum, Brandeis University, Waltham, Massachusetts, gift N. and L. Horch to the Riverside Museum Collection; 98, *Panchen Lama (First).* Courtesy/copyright: Newark Museum; 99, *Potala Palace.* Courtesy/copyright: Newark Museum; 100, *Penden Lhamo.* Photo/copyright: Moke Mokotoff. Courtesy Mokotoff Asian Arts Gallery, New York; 101, *Dalai Lama (Seventh, Gyalwa Kalsang Gyatso).* Courtesy/copyright: Newark Museum, purchase 1920 Sheldon Collection; 103, *Dalai Lama (Thirteenth, Gyalwa Thubten Gyatso).* Photo: Th. Paar Studio. Courtesy/copyright: Newark Museum, gift C. Suydam Cutting, 1935; 104, *Kalachakra Temple, St. Petersburg.* Photo: Mikhail Khusidman. Copyright: Samaya Foundation; 105, *Norbulingka.* Photo: Cutting. Courtesy/copyright: Newark Museum; 106, *Dalai Lama (Fourteenth, Tenzin Gyatso), 1940.* Painting: Kanwal Krishna. Photo: Armen. Courtesy/copyright: Newark Museum, gift Mrs. C. Suydam Cutting; 107, *Dalai Lama (Fourteenth, Tenzin Gyatso), 1950.* Photo: AP/Wide World Photos; 108, (top) *Dalai Lama and Mao Tse-tung, 1954.* Photo: AP/Wide World Photos; 108, (bottom) *Ling Rinpoche.* Photo/copyright: Ernst Haas. Courtesy Ernst Haas Studio; 109, *Dalai Lama journey to exile in India.* Photo: AP/Wide World Photos; 111, *Thekchen Choling.* Photo/copyright: Namgyal Monastery; 113, *Kalu Rinpoche.* Photo: Barry Bryant. Copyright: Samaya Foundation; 115, *Dalai Lama with children.* Photo: Gregory Durgin. Copyright: Samaya Foundation; 116, *Namgyal monks debating.* Photo/copyright: Bill Warren; 117, *Dalai Lama, Noble Peace Prize.* Photo: Gregory Durgin. Copyright: Samaya Foundation.

Tibetan Buddhist Philosophy

Pages 118–119, *Teaching mudra.* Photo/copyright: Ernst Haas. Courtesy Ernst Haas Studio; 120, *Avalokiteshvara (1000 arms).* Courtesy/copyright: Zimmerman Family Collection; 122, *Shakyamuni Buddha teaching.* Courtesy/copyright: Mr. and Mrs. James W. Alsdorf, Chicago; 126, *Wheel of Transmigration.* Photo: A. Rota, neg. no. 323100. Courtesy/copyright: Department of Library Services, American Museum of Natural History; 129, *Monks at bodhi tree, 1973.* Photo/copyright: John C. Smart; 130, *Dalai Lama teaching, 1985.* Photo: Barry Bryant. Copyright: Samaya Foundation.

The Kalachakra Initiation

Pages 132–133, *Monks performing a celebratory dance, 1991.* Photo/copyright: John Bigelow Taylor/Tibet Center; 134, *Kalachakra dance, 1981.* Photo/copyright: Marcia Keegan; 135, *Torma.* Photo: Gregory Durgin. Copyright: Samaya Foundation; 136, (left) *Tooth stick.* Photo/copyright: Kim Yeshi. Courtesy Department of Religions and Culture, Central Tibetan Administration of H.H. Dalai Lama, previously published in *Cho-Yang;* 136, (right) *Requesting initiation.* Photo/copyright: Lawrence Lauterborn; 137, (left) *Six energy centers.* Drawing/copyright: Phuntsok Dorje; 137, (right) *Purbas.* Photo: Gregory Durgin. Copyright: Samaya Foundation; 138, (top) *Vase consecration.* Photo/copyright: John Bigelow Taylor/Tibet Center; 138, (center) *Chalk string consecration.* Photo/copyright: John Bigelow Taylor/Tibet Center; 138, (bottom) *Vajra and bell consecration.* Photo/copyright: John Bigelow Taylor/Tibet Center; 139, (top) *Snapping chalk strings.* Photo/copyright: John Bigelow Taylor/Tibet Center; 139, (bottom) *Monks drawing Kalachakra Mandala.* Photo/copyright: Carlos Gonzalez/Thubten Dhargye Ling; 140, (left) *Vajra Vega.* Photo/copyright: Ernst Haas. Courtesy Ernst Haas Studio; 140, (right) *Dalai Lama as vajra master dancing.* Photo/copyright: John Bigelow Taylor/Tibet Center; 141, *Grains of wheat on mandala drawing.* Photo/copyright: John Bigelow Taylor/Tibet Center; 142, (top) *Dalai Lama holding five wisdom strings.* Photo/copyright: John Bigelow Taylor/Tibet Center; 142, (center) *Monks holding wisdom strings.* Photo/copyright: John Bigelow Taylor/Tibet Center; 142, (bottom) *Dalai Lama applies first sand.* Photo/copyright: John Bigelow Taylor/Tibet Center; 143, *Dalai Lama string at heart center.* Photo/copyright: John Bigelow Taylor/Tibet Center; 144, (top) *Monks seated painting mandala.* Photo: Gregory Durgin. Copyright: Samaya Foundation/Namgyal Monastery; 144, (bottom) *Monks standing painting mandala.* Photo/copyright: John Bigelow Taylor/Tibet Center; 145 and 147, *Offering mandala mudra.* Photo: Gregory Durgin. Copyright: Samaya Foundation; 146, *Dalai Lama (Fourteenth, Tenzin Gyatso) as vajra master, 1973.* Photo/copyright: John C. Smart; 149, *Kusha grass.* Photo/copyright: Moke Mokotoff. Courtesy Mokotoff Asian Arts Gallery, New York; 150, *Students wearing costumes of Kalachakra.* Photo/copyright: Lawrence Lauterborn; 151, *Students with blindfolds.* Photo/copyright: John Bigelow Taylor/Tibet Center; 152, *Students with flowers on foreheads.* Photo/copyright: Ernst Haas. Courtesy Ernst Haas Studio; 155, *Kalachakra/Vishvamata (implements).* Drawing/copyright: Sidney Piburn; 158, 160 (both), 161, 162, 163, and 164, *Initiation substances.* Drawings/copyright: Phuntsok Dorje; 165, *Wheel of the Law.* Courtesy/copyright: Newark Museum, Crane Collection, 1911; 166, *Vajrasattva.* Courtesy/copyright: Newark Museum, purchase 1984 Willard W. Kelsey Bequest Fund; 167, *Rice offering mandala.* Photo/copyright: Ernst Haas. Courtesy Ernst Haas Studio; 168, *Students on line at Leh, Ladakh for blessings.* Photo/copyright: Moke Mokotoff. Courtesy Mokotoff Asian Arts Gallery, New York; 169, *Dalai Lama meditates, 1973.* Photo/copyright: Ernst Haas. Courtesy Ernst Haas Studio; 170, (top) *Dalai Lama removes deities from Kalachakra Sand Mandala.* Photo/copyright: John Bigelow Taylor/Tibet Center; 170, (bottom) *Dalai Lama cuts mandala's energy.* Photo/copyright: Don Farber/Thubten Dhargye Ling; 171, (both) *Sweeping up the mandala.* Photo/copyright: John Bigelow Taylor/Tibet Center; 172, *Dalai Lama prays at Hudson River.* Photo/copyright: Lawrence Lauterborn/Tibet Center; 173, (top) *Dalai Lama pours sand into river, Switzerland.* Photo: Barry Bryant. Copyright: Samaya Foundation; 173, (bottom) *Dalai Lama pours sand into Santa Monica Bay.* Photo/copyright: Don Farber/Thubten Dhargye

Ling; 174, *Dalai Lama pours water on mandala base.* Photo/copyright: Lawrence Lauterborn/Tibet Center; 175, (right) *Monks scrub off mandala lines.* Photo/copyright: John Bigelow Taylor/Tibet Center; 175, (left) *Dalai Lama sitting on mandala base.* Photo/copyright: John Bigelow Taylor/Tibet Center.

The Kalachakra Sand Mandala

Pages 176–177, *Monk applying sand/chakpu.* Photo: Barry Bryant. Copyright: Samaya Foundation/Namgyal Monastery; 178, *Deities represented by dots and syllables.* Photo: Barry Bryant. Copyright: Samaya Foundation/Namgyal Monastery; 179, *Elevation 3-D Kalachakra palace.* Auto-cad drawing: Daniel Maciejczyk/Barry Bryant under direction of Christian Lischewski, Associate Professor Pratt School of Architecture. Computer facility courtesy Perkins + Will. Copyright: Samaya Foundation; 180, *Six circles of mandala.* Photo: Gregory Durgin. Copyright: Samaya Foundation/Namgyal Monastery; 182, *Prajnaparamita text.* Newark Museum, Sheldon Collection, 1920. 183, 184, 185, 186, 187, 188, 189, 190, and 191, *Step-by-step drawings of the Kalachakra Mandala.* Auto-cad drawings: Daniel Maciejczyk/Barry Bryant under the direction of Christian Lischewski, Associate Professor Pratt School of Architecture. Computer facility courtesy Perkins + Will; 192, *Mandala gate.* Drawing: Stanley C. Bryant. Copyright: Samaya Foundation; 193, *Monks painting Mandala of Enlightened Mind.* Photo/copyright: John Bigelow Taylor/Tibet Center; 195, *Monk demonstrating chakpu.* Photo/copyright: Natural History Museum of Los Angeles County; 196, (left) *Monk demonstrating shinga.* Photo: Gregory Durgin. Copyright: Samaya Foundation/Namgyal Monastery; 196, (right) *Mandala class.* Photo: Gregory Durgin. Copyright: Samaya Foundation/Namgyal Monastery; 198, *Monk painting central lotus.* Photo/copyright: John Bigelow Taylor/Tibet Center; 199, *Cross section of center of lotus.* Drawing: Barry Bryant. Copyright: Samaya Foundation. 200, 201, 203, 205, 207, 209, 212, 213, 215, 216, 217, 219, 220, 221, 222, 223, 225, 226, and 227 *Details of Kalachakra Sand Mandala.* Photos: Gregory Durgin. Copyright: Samaya Foundation/Namgyal Monastery;

218, *3-D Kalachakra Palace (detail)*. Photo/copyright: Robin Bath; 228 and 229, (top) *Dismantling, monks praying and removing deities*. Photo: Gregory Durgin. Copyright: Samaya Foundation/Namgyal Monastery; 229, *Sweeping up sand*. Photo: Warner Bryant Scheyer. Copyright: Samaya Foundation/Namgyal Monastery; 230, *Procession to river*. Photo: Warner Bryant Scheyer. Copyright: Samaya Foundation/Namgyal Monastery; 231, (both) *Monks praying at Hudson River and pouring sand into river*. Photos: Warner Bryant Scheyer. Copyright: Samaya Foundation/Namgyal Monastery.

World View According to Kalachakra

Pages 232–233, *Astrological calculations*. Photo: Rani Gill. Courtesy/copyright: Department of Religions and Culture, Central Tibetan Administration of H.H. Dalai Lama, previously printed in *Cho-Yang;*. 235, *Personal horoscope*. Photo/copyright: Samaya Foundation; 236 and 237, *Tibetan calendar*. Photo: Stephen Germany. Courtesy/copyright: Newark Museum, gift Carter D. Holton Collection, 1936; 238, *Steer (detail Kalachakra Mandala)*. Photo: Gregory Durgin. Copyright: Samaya Foundation/Namgyal Monastery; 239, *Cosmic Mt. Meru/astrological animals*. Photo: Barry Bryant. Copyright: Samaya Foundation; 240, *Sun and moon*. Photo: Gregory Durgin. Copyright: Samaya Foundation/Namgyal Monastery; 242, *Channels*. Drawings/copyright: Phuntsok Dorje.

Transformation of Consciousness

Pages 245–246, *Child painting with sand*. Photo: Barry Bryant. Copyright: Samaya Foundation/Namgyal Monastery; 246, *Monk contemplating*. Photo/copyright: John Bigelow Taylor/Tibet Center; 247, *Monk making torma*. Photo/copyright: Moke Mokotoff. Courtesy Mokotoff Asian Arts Gallery, New York; 248, *Greg Louganis*. Photo/copyright: Lonnie Majors; 249, *Laura Dean Musicians and Dancers*. Photo/copyright: Johan Elbers. Courtesy Laura Dean; 250, *Dismantling mandala at IBM Gallery*. Photo/copyright: John Bigelow Taylor/Tibet Center; 251, *Pouring sand into Santa Monica Bay*. Photo: Don Farber. Copyright: Samaya Foundation/Namgyal Monastery;

254, *Monks practicing the Kalachakra ritual dance.* Photo/copyright: Robyn Brentano.

Afterword

Page 256, *Bodhgaya stupa.* Photo: Barry Bryant. Copyright: Samaya Foundation.

Book Cover

Kalachakra Sand Mandala. Photo: Samaya Foundation. Copyright: Samaya Foundation/Namgyal Monastery. Back flap. *Dalai Lama and Barry Bryant, Los Angeles, 1989.* Photo/copyright: Natural History Museum of Los Angeles County.

Selected Bibliography

Avedon, John. *In Exile from the Land of Snows.* New York: Alfred A. Knopf, 1984.

Bechert, Heintz and Gombrich, Richard, (eds). *The World of Buddhism.* New York: Facts on File, 1984.

Bernbaum, Edwin Marshall. "The Mythic Journey and Its Symbolism: A Study of the Development of Buddhist Guidebooks to Shambhala in Relation to their Antecedents." Ph.D. dissertation, University of California, Berkeley, 1985.

————. *The Way to Shambhala.* Los Angeles: Jeremy P. Tarcher, Inc., 1980.

Bryant, Barry, and Yignyen, Tenzin. *Process of Initiation: The Indo-Tibetan Rite of Passage into Shambala: The Kalachakra Initiation.* New York: Samaya Foundation and Namgyal Monastery, 1990.

Cho Yog Thubten Jamyang. *Kalachakra Initiation, Los Angeles, 1989,* translated by Sharpa Tulku with David Patt, 2nd rev. ed. by Thubten Dhargye Ling. Oregon, Wisconsin: Deer Park Books, 1989.

Council for Religious and Cultural Affairs. *Cho-Yang.* Issues 1, 2, 3, and "Year of Tibet" edition (1986, 1987, 1991). Dharamsala, India.

Cozort, Daniel. *Highest Yoga Tantra.* Ithaca, NY: Snow Lion Publications, 1986.

Dhargyey, Geshe Ngawang. *A Commentary on the Kalachakra Tantra.* Dharamsala, India: Library of Tibetan Works and Archives, 1985.

Dhondup, K. *Songs of the Sixth Dalai Lama.* Dharamsala, India: Library of Tibetan Works and Archives, 1981.

Gyatso, Tenzin (the Fourteenth Dalai Lama). *The Kalachakra Tantra.* Edited by Jeffrey Hopkins. London: Wisdom Publications, 1985.

————. *Kindness, Clarity and Insight.* Ithaca, NY: Snow Lion Publications, 1984.

————. *My Land and My People.* New York: Potala Publications, 1977.

————. *The Union of Bliss and Emptiness.* Ithaca, NY: Snow Lion Publications, 1988.

————. *Universal Responsibility and the Good Heart.* Dharamsala, India: Library of Tibetan Works and Archives, 1977.

Gyatso, Tenzin (the Fourteenth Dalai Lama), Khapa, Tsong, and Hopkins, Jeffrey. *Tantra in Tibet.* Ithaca, NY: Snow Lion Publications, 1977.

Hopkins, Jeffrey. *The Tantric Distinction, An Introduction to Tibetan Buddhism.* London: Wisdom Publications, 1984.

Kalu Rinpoche. *The Kalachakra Empowerment: Taught by the Venerable Kalu Rinpoche.* Vancouver, B.C.: Kagyu Kunkhyab Chuling, 1986.

Kalupahana, David and Indrani. *The Way of Siddhartha, A Life of the Buddha.* Boston: Shambhala Publications, 1982.

Kyabje Dorje Chang. *The Dharma That Illuminates All Beings Like the Light of the Sun and the Moon.* Albany: State University of New York Press, 1986.

Minke, Gisela. "The Kalachakra Initiation." *Tibet Journal,* vol. 2, nos. 3–4. Dharamsala, India: Library of Tibetan Works and Archives, 1972.

Mitchell, Robert Allen. *The Buddha: His Life Retold.* New York: Paragon House, 1989.

Mullin, Glenn H. *Selected Works of the Dalai Lama I.* Ithaca, NY: Snow Lion Publications, 1985.

————. *Selected Works of the Dalai Lama II.* Ithaca, NY: Snow Lion Publications, 1985.

————. *Selected Works of the Dalai Lama VII.* Ithaca, NY: Snow Lion Publications, 1985.

Newman, John Ronald. *The Outer Wheel of Time.* Ph. D. dissertation, 1987. UMI Dissertation Information Service, 1989.

Niwano, Nikkyo. *Shakyamuni Buddha: A Narrative Biography.* Tokyo: Kosei Publishing Co., 1969.

Olischak, Blanch Christine, with Wangyal, Geshe Thupten. *Mystic Art of Ancient Tibet.* New York: McGraw-Hill Publishing Company, 1973.

Pal, Pratapaditya. *Art of the Himalayas.* New York: Hudson Hill Press; American Federation of Arts, 1991.

———. *Art of Tibet.* Berkeley: Los Angeles County Museum of Art; University of California Press, 1988.

———. *Indian Sculpture Vols. 1–2.* Los Angeles: Los Angeles County Museum of Art; Berkeley: University of California Press, 1986.

———. *Light of Asia: Buddha Sakyamuni in Asian Art.* Los Angeles: Los Angeles County Museum of Art, 1984.

Reigle, David. "The Lost Kalachakra Mula Tantra on the Kings of Shambhala." *Kalachakra Research Publications,* no. 1. Talent, OR: Eastern School Press, 1986.

Reynolds, Valrae. *Tibet: A Lost World.* New York: American Federation of Arts, 1978.

Reynolds, Valrae, and Heller, Amy. *Introduction.* Vol. 1, 2nd ed., *The Newark Museum Tibetan Collection.* Newark, NJ: Newark Museum, 1983.

Reynolds, Valrae, Heller, Amy, and Gyatso, Janet. *Sculpture and Painting.* Vol. 3, 2nd ed., *The Newark Museum Tibetan Collection.* Newark, NJ: Newark Museum, 1986.

Rhie, Marylin, M., and Thurman, Robert A. F. *Wisdom and Compassion, The Sacred Art of Tibet.* San Francisco: Asian Art Museum of San Francisco; New York: Tibet House; New York: Harry N. Abrams, 1991.

Shakabpa, Tsepon W.D. *Tibet: A Political History.* New York: Potala Publications, 1984.

Smith, Huston. *The Religions of Man.* New York: Harper & Row, 1958.

Snellgrove, David. *Indo-Tibetan Buddhism.* Vols 1–2. Boston: Shambhala Publications, 1987.

Snellgrove, David, and Richardson, Hugh. *A Cultural History of Tibet.* Boston: Shambhala Publications, 1986.

Sopa, Geshe Lhundub, Jackson, Roger, and Newman, John. *The Wheel of Time: The Kalachakra in Context.* Oregon, WI: Deer Park Books, 1985.

Tharchin, Geshe Lobsang. *Offering of the Mandala.* Washington, D.C.: Mahayana Sutra and Tantric Center, 1981.

Thondup, Tulku. *Buddhist Civilization in Tibet.* Cambridge, MA: Maha Siddha Nyingmapa Center, 1982.

Trungpa, Chogyam. *Shambhala: The Sacred Path of the Warrior.* Boston: Shambhala Publications, 1988.

———. *The Rain of Wisdom.* Boston: Shambhala Publications, 1980.

Videotapes

The Kalachakra Initiation, Bodhgaya, India, Meridian Trust, 1974.

The Kalachakra Initiation, Bodhgaya, India, Samaya Foundation, 1985.

The Kalachakra Initiation, Los Angeles, California, Thubten Dhargye Ling, 1989.

The Kalachakra Initiation, Madison, Wisconsin, Educational Communications, 1981.

The Kalachakra Initiation, Vancouver, Canada, Samaya Foundation, 1986.